STRESS AND DEFORMATION

STRESS AND DEFORMATION

A Handbook on Tensors in Geology

Gerhard Oertel

University of California, Los Angeles

New York   Oxford

OXFORD UNIVERSITY PRESS

1996

Oxford University Press

Oxford   New York
Athens Auckland Bangkok Bombay
Calcutta Cape Town Dar es Salaam Delhi
Florence Hong Kong Istanbul Karachi
Kuala Lumpur Madras Madrid Melbourne
Mexico City Nairobi Paris Singapore
Taipei Tokyo Toronto

and associated companies in
Berlin   Ibadan

Published by Oxford University Press, Inc.,
198 Madison Avenue, New York, New York 10016

Oxford is a registered trademark of Oxford University Press

Library of Congress Cataloging-in-Publication Data
Oertel, G. F.
Stress and deformation: a handbook on tensors
in geology / Gerhard Oertel.
p. cm. Includes bibliographical references and index.
ISBN 0-19-509503-0
1. Vector analysis. 2. Geology—Mathematics.
3. Strains and stresses. I. Title.
QE33.2.M3037 1996 551'.01'51563—dc20 95-18621

1 3 5 7 9 8 6 4 2

Printed in the United States of America
on acid-free paper

PREFACE

This book began as a set of notes for a graduate course taught by Dr. Ronald L. Shreve, my colleague at the University of California, Los Angeles; he first saw that our geology students needed such a course, and the author was one of his "students" (a significant proportion of the faculty members of what was then the Department of Geology sat in, and worked the problems, when the course was offered for the first few times). Later I took over the teaching of this course and Dr. Shreve's course notes, which consisted of a set of problems, a set of concise answers, and several handouts. I gradually added a few new problems but changed the original set hardly at all. From the form in which I received them, the notes have evolved principally by elaborations of Dr. Shreve's handouts and by the addition of handouts that I distributed after each class session. They contained the worked answers and became more explicit over the years. Students suggested improvements, and elegant solutions found by them replaced or supplemented earlier versions. Especially helpful were Peter A. Craig, Theresa L. Heirshberg, and my colleague Dr. An Yin, who used a first draft of this book as a handout; they turned out to be superb collaborators by discovering numerous mistakes and shortcomings.

J. F. Nye's *Physical Properties of Crystals*, Clarendon Press, Oxford, in its slightly revised 1964 edition, served as the textbook for that course. The attentive reader will recognize many traits of Nye's approach in the present book. This, I hope, will be understood as an indication of his didactic success. Upon Dr. Shreve's and later my recommendation, many of the students used I. S. Sokolnikoff and R. M. Redheffer's *Mathematics of Physics and Modern Engineering*, McGraw-Hill, New York, 1958 or later. Several of the problems have been inspired by that book. My students also found Y. C. Fung's *Foundations of Solid Mechanics*, 1965, and *A First Course in Continuum Mechanics*, 2nd edition, 1977, both published by Prentice-Hall, Englewood Cliffs, lucid and helpful.

By permission of McGraw-Hill Book Company Inc., Problems 1 to 9, 12 to 14, 19 to 23, and 25 of this book were reproduced, some with modifications, from I. S. Sokolnikoff & R. M. Redheffer, *Mathematics of Physics and Modern Engineering*, McGraw-Hill, New York, © 1958. By permission of Oxford University Press, Problems 39, 40, 57, 66, 69 to 71, and 98, and Figures 4.3 and 5.2

of this book were reproduced, some with modifications, from J. F. Nye, *Physical Properties of Crystals*, Clarendon Press, Oxford, © 1957. By permission of Prentice Hall, Inc., Figures 4.1 and 5.5 were redrawn from Y. C. Fung, *Foundations of Solid Mechanics*, © 1965.

Los Angeles
July 1995   G. O.

છ

To the Reader

Geologists at various stages of their careers, but especially as graduate students ready to begin their first own research project, may find that their mathematical training has left them unprepared to deal with the notation and concepts of continuum mechanics and other branches of physics dealing with tensor quantities. Trying to read literature pertinent to their research topic, they may be bewildered by symbols decorated by a multiplicity of letter or number subscripts. When they seek out the appropriate textbooks, their bewilderment may grow because such books are written for specialists and cover much more subject matter than a geologist needs. Even if the text of such a book is lucid enough, students eagerly reading forward to the topic they hope themselves to apply, commonly end up unable to solve their problems. Merely reading about the fundamental operations without practicing them, they tend to forget much of what they eventually need to know. Typically, they find the textbook increasingly confusing and give up before reaching the topic of interest.

In the author's experience, the apparent difficulties of continuum mechanics and of tensor and matrix notation are manageable to any student who comes prepared with some instruction in calculus and the rudiments of differential equations and has not discarded the old textbooks, so that memory may be refreshed by some of the details in these subjects. However, the student must be prepared to spend the time and effort necessary to solve a sequence of problems that lead him gradually from the simple to the complex and from the intuitively obvious to the, sometimes, counterintuitive but true.

The problems presented in this book start with a review of the mathematics of vectors and lead from there to the mathematics of stress and strain, including finite strain. They are arranged so that the solutions to successive problems depend on comprehension of earlier solutions. Although many of the later problems are not intrinsically more difficult than the earlier ones, I recommend only to a well versed reader to skip the seemingly simple problems in the early chapters.

The main purpose of the problems is to train the student in mathematical methods and not to cover any particular field of mathematical physics. I hope, all the same, that the topics on which the problems are based will interest the student in their own right and that they may illuminate one or the other branch of physics pertinent to several divisions of geology, including structural geology, glaciology, crystallography, crystal physics, the geophysics of heat flow, and others. These subjects, however, are not covered systematically.

Chapter 1 leads the reader from the conventional to the subscript notation for vectors and their arithmetic. Chapter 2 deals with scalar and vector fields. Chapter 3 introduces tensors of the second rank describing the properties of anisotropic materials, the so-called matter tensors. Chapters 4 and 5 deal with two field tensors of the second rank, the stress and infinitesimal strain tensors, tensors the effects of which can be combined by simple addition (they are superposable). Chapter 6 provides the methods necessary to calculate the effects of finite strains, which are not superposable. Chapter 7 contains various applications of continuum mechanics to geology and glaciology. Chapter 8, finally, introduces the concept of a strain history and explains a method of factoring a finite strain into finite increments. The most important part of these chapters is the problems, which I encourage the reader to solve initially with as little help as possible from the Answers that follow after the end of Chapter 8.

These answers are meant principally to allow the reader to check whether a problem has been solved correctly. If, however, a problem is found to be too difficult after a serious attempt at solution, the student should allow no more than a glance at the answer, enough to be pointed in the right direction, and then return to independent work. Too much reliance on the prepared solution will defeat the purpose of this book. Thus, look at any of the answers only after finishing a solution or at least after having tried several approaches to one.

A list of symbols precedes Chapter 1. The Summary of Formulæ at the end of the book is meant to spare the reader memorization and to be consulted freely. It is a collection of important identities and working formulæ occurring throughout the book. Appendix A contains a table of pairings of elastic parameters for an isotropic material, and Appendix B lists units of stress.

ↅ

Contents

Symbols and Notation   x

Chapter 1.   Vectors (Problems 1 to 20)   3

Chapter 2.   Fields (Problems 21 to 25)   10

Chapter 3.   Matter Tensors and Coordinate Transformations

                (Problems 26 to 75)   13

Chapter 4.   Stress (Problems 76 to 97)   44

Chapter 5.   Infinitesimal Strain (Problems 98 to 115)   56

Chapter 6.   Finite Strain (Problems 116 to 124)   72

Chapter 7.   Effects of Stress (Problems 125 to 135)   85

Chapter 8.   Strain History and Polar Decomposition (Problem 136)   94

Answers   97

Summary of Formulæ   270

Appendix A.   Elastic Parameters for Isotropic Bodies   286

Appendix B.   Units of Stress   287

References   288

Index   289

☙

Symbols and Notation

$\mathbf{0}$	null vector	$^l\Delta$	linear dilatation
a_i	component, original position vector	Δ_{ij}	component, deviator of stress
A_{ij}	cofactor	δ_{ij}	Kronecker delta (element, identity matrix)
A_{ij}	element, transformation matrix	$\mathfrak{E}_{ij}$	component, Green's tensor
a_{ij}	direction cosine	e_{ij}	component, Almansi's tensor
$\mathbf{A}, \mathbf{B}$	vectors		
$\mathbb{A}, \mathbb{B}$	matrices	e_{ij}	component, displacement gradient
α_{ij}	direction angle		
c	distance, origin to Mohr circle center	ε_{ij}	component, infinitesimal strain
$\mathfrak{C}_{ij}$	component, Lagrangian Cauchy's tensor	ϵ_{ijk}	element, alternating matrix
$\mathfrak{c}_{ij}$	component, Eulerian Cauchy's tensor	$\dot{\varepsilon}_{ij}$	component, strain rate
c_{ij}	element, stiffness matrix	$\mathbb{F}$	asymmetric transformation matrix
c_{ijkl}	component, stiffness tensor	F_i	component, body force
$C_{,i}$	implicit spatial derivative of C	$f(\dots)$	function
D	dip	Φ	angle of rotation
D_{ij}	component, infinitesimal distortion	$\varphi(\dots)$	constraint function
		G_i	component, body torque
$\mathfrak{D}_{ij}$	component, Green's distortion tensor	g_i	component, acceleration due to gravity
$\mathfrak{d}_{ij}$	component, Almansi's distortion tensor	γ	engineering shear strain
		$\dot{\gamma}$	engineering shear strain rate
dS	increment of final length		
ds	increment of initial length	H	Hubble constant
dV	increment of final volume	Hz	hertz (unit of frequency)
dv	increment of initial volume	$\mathbf{i}, \mathbf{j}, \mathbf{k}$	unit vectors parallel to coordinate axes
Δ	dilatation		

$^1I, {}^2I, {}^3I$	invariants, second-rank tensor	ϖ_{ij}	component, infinitesimal rotation tensor
J	joule (unit of energy)	$\dot{\varpi}_{ij}$	component, rotation rate tensor
K	kelvin (unit of temperature)		
K	thermal conductivity	q_i	component, heat flow
K_{ij}	component, dielectric permittivity tensor	$\mathbb{R}$	rotation matrix, polar decomposition
K_{ij}	component, thermal conductivity tensor	R_{ij}	generator of rotations
l_i	component, unit direction-vector	R, a, b	axis-subscripts, modulo 3
λ	elongation	$\dot{\mathbf{R}}, \ddot{\mathbf{R}}$	first and second time derivatives of $\mathbf{R}$
λ	Lagrange multiplier	r	radius of Mohr circle
λ	Lamé's constant	ρ	density
λ	principal tensor magnitude, general	$^{iA}\rho$	particular principal angular density of axes
$^i\lambda$	particular principal elongation	$^{iP}\rho$	particular principal angular density of poles
m	mass	S	strike
m	meter (unit of length)	$S_{\alpha\beta}$	component, two-dimensional tensor
μ	coefficient of "internal friction"	S_{ij}	component, stretch tensor
μ	coefficient of viscosity	S_{ij}	component, symmetric tensor
μ	Lamé's constant, shear modulus	s_{ij}	element, compliance matrix
N	magnitude of normal component	s_{ijkl}	component, compliance tensor
N_i	component, normal force (traction)	s	second (unit of time)
N	newton (unit of force)	σ	magnitude of normal traction
$\boldsymbol{\Omega}$	angular velocity	σ_{ij}	component, stress tensor
ω_i	component, rotation vector	$\sigma_1, \sigma_2, \sigma_3$	principal stresses
P	confining pressure	T	trend
P	plunge	T	magnitude of tangential component
p	pressure		
Pa	pascal (unit of stress)	T	temperature

T_{ij} component, second-rank tensor

T_i component, shear (tangential) force (traction)

$T_{,i}$ component, temperature (thermal) gradient

τ magnitude of shear (tangential) traction

$^{\circ}\tau$ octahedral shear (tangential) stress

τ_{ij} component, shear (tangential) stress

Θ angle of shear

θ angle of coordinate rotation

$\mathbb{U}$ stretch tensor, left-polar decomposition

u_i displacement component

V volume

$\mathbb{V}$ stretch tensor, right-polar decomposition

v_i vector component

v_1, v_2, v_3 cartesian vector components

x_1, x_2, x_3 cartesian coordinates

x_R, x_a, x_b rotation axis, and following axes

x_i general cartesian coordinate, or component, final position vector

$\mathcal{X}, \mathcal{Y}, \mathcal{Z}$ coordinates in Mohr space

$(...)$ ordered array of matrix elements

(a_{ij}) rotation matrix

(M_{ij}) matrix

$[...]$ ordered array of tensor components

$[A_{ij}]$ antisymmetric second-rank tensor

$[S_{ij}]$ symmetric second-rank tensor

$[T_{ij}]$ second-rank tensor

$[UVW]$ crystallographic direction

$[v_i]$ vector

$|M_{ij}|$ determinant (of a matrix or tensor)

$|S|$ absolute value (of a scalar)

$\times$ vector cross-multiplication

$\bullet$ vector dot-multiplication

∇ del (or nabla)

☙

STRESS AND DEFORMATION

1

VECTORS

The reader, even if familiar with vectors, will find it useful to work through this chapter because it introduces notation that will be used throughout this book. We will take vectors to be entities that possess magnitude, orientation, and sense in three-dimensional space. Graphically, we will represent them as arrows with the sense from tail to head, magnitude proportional to the length, and orientation indicated by the angles they form with a given set of reference directions. Two different kinds of symbol will be used to designate vectors algebraically, boldface letters (and the boldface number zero for a vector of zero magnitude), and subscripted letters to be introduced later. The first problems deal with simple vector geometry and its algebraic representation. Multiplying a vector by a scalar affects only its magnitude (length) without changing its direction.

Problem 1. State the necessary and sufficient conditions for the three vectors **A**, **B**, and **C** to form a triangle. (Problems 1 – 9, 12 – 14, 19 – 23, and 25 from Sokolnikoff & Redheffer, 1958.)

Problem 2. Given the sum $\mathbf{S} = \mathbf{A} + \mathbf{B}$ and the difference $\mathbf{D} = \mathbf{A} - \mathbf{B}$, find **A** and **B** in terms of **S** and **D** (a) graphically and (b) algebraically.

Problem 3. (a) State the unit vector **a** with the same direction as a nonzero vector **A**. (b) Let two nonzero vectors **A** and **B** issue from the same point, forming an angle between them; using the result of (a), find a vector that bisects this angle.

Problem 4. Using vector methods, show that a line from one of the vertices of a parallelogram to the midpoint of one of the nonadjacent sides trisects one of the diagonals.

Two vectors are said to form with each other two distinct products: a scalar, the *dot product*, and a vector, the *cross product*. The dot product D of the vectors **A** and **B**, forming with each other an angle θ, is calculated as:

$$D = \mathbf{A} \bullet \mathbf{B} \equiv AB\cos\theta = \mathbf{B} \bullet \mathbf{A} \equiv BA\cos\theta \ . \tag{1.1}$$

Hence the dot product is *commutative* and is A times the projection of **B** on **A** or

B times the projection of **A** on **B**. Two vectors are perpendicular if and only if their dot product is zero. The cross product **C** of **A** and **B** is:

$$C = A \times B \equiv u\,AB\sin\theta \, ,$$
$$B \times A = -C \, ,$$

$$(1.2)$$

where **u** is the unit vector perpendicular to the **AB** plane, positive in the screw-rule direction for the angle θ from **A** to **B**. The cross product is not commutative, and two vectors are parallel or antiparallel (parallel with opposite sense) if and only if their cross product is a null vector.

Problem 5. (a) Which condition makes the otherwise arbitrary vectors **A**, **B**, and **C** coplanar? (b) Give an example.

Problem 6. Can both equations $A \times B = 0$ and $A \cdot B = 0$ be true without either **A** or **B** being a null vector?

Problem 7. Find an example of three unequal vectors such that the cross product of any two of them forms a right angle with the third.

Let three mutually orthogonal unit vectors **i**, **j**, and **k** have their tails at the origin of a right-handed cartesian coordinate system. This allows us to define as the *position vector* for every point in the coordinate system a vector that is the sum of the products of the three unit vectors, each multiplied by an appropriate scalar. Note that all position vectors have their tails at the coordinate origin.

Problem 8. Let the vertices of a triangle be defined by the position vectors **i** + **j** + **k**, 2**j** + **k**, and 2**i** + **j**. Form this triangle by vectors, head against tail, and verify that the vector sum is the null vector **0**.

With the help of the axis-parallel unit vectors the cross product can be redefined:

$$C = A \times B = \begin{vmatrix} i & j & k \\ A_1 & A_2 & A_3 \\ B_1 & B_2 & B_3 \end{vmatrix} \, ,$$

$$(1.3)$$

where the vertical bars enclosing the three-by-three array designate the *determinant* of that array. Explicitly, that determinant is:

$$\mathbf{C} = \left\langle \begin{array}{c} \mathbf{i}\,(A_2\,B_3 - A_3\,B_2) - \mathbf{j}\,(A_1\,B_3 - A_3\,B_1) \\ + \mathbf{k}\,(A_1\,B_2 - A_2\,B_1) \end{array} \right\rangle . \tag{1.4}$$

Angular brackets $\langle\,\rangle$ are chosen in eq. (1.4) because the square brackets are reserved for arrays designating vectors (and other tensors), as in eq. (1.6) below. If we let the traditional cartesian coordinate system $x, y,$ and z coincide with that defined by the three unit vectors of Problem 8, concepts of analytical and vector geometry can be combined.

Problem 9. Which vector is perpendicular to the plane $ax + by + cz + d = 0$?

Problem 10. Three plane mirrors are silvered on both sides and arranged so that each intersects the others at right angles, forming eight concave corners. Show, by vector methods, that a beam of light entering the system from any direction will emerge traveling in exactly the opposite direction. (This is the principle of radar targets and of many highway reflectors.)

Problem 11. Find $\mathbf{A} \cdot \mathbf{B}$, $\mathbf{A} \times \mathbf{B}$, and $\mathbf{B} \times \mathbf{A}$ if $\mathbf{A} = \mathbf{i} + 2\mathbf{j} + 3\mathbf{k}$ and $\mathbf{B} = 5\mathbf{i} + 7\mathbf{j} + 11\mathbf{k}$.

Problem 12. (a) Show that the cross product of any two of the following vectors is parallel to the third: $\mathbf{i} + \mathbf{j} + \mathbf{k}$, $\mathbf{i} - \mathbf{k}$, and $\mathbf{i} - 2\mathbf{j} + \mathbf{k}$. (b) What are the implications for the three vectors?

Vector algebra can be simplified by introducing the subscript notation. Replace the letters $x, y,$ and z for cartesian coordinates by the subscripted variables $x_1, x_2,$ and x_3. To characterize a vector, first decompose it into components parallel to the coordinates:

$$\mathbf{v} \equiv a\mathbf{i} + b\mathbf{j} + c\mathbf{k} , \tag{1.5}$$

and then give the three scalar factors a, b, and c the new designations v_1, v_2, and v_3. Let further a letter subscript, such as i, j, etc., stand collectively for the integer number subscripts 1, 2, or 3. Then the vector of eq. (1.5) can be characterized by the ordered array of scalars $[\ v_1,\ v_2,\ v_3\ ]$. The vector nature of the array as a whole is indicated by the brackets. To simplify the writing, the commas can be eliminated, and the vector in subscript notation is:

$$\left[v_i\right] \equiv \left[\ v_1\ \ v_2\ \ v_3\ \right] . \tag{1.6}$$

Because the brackets indicate a complete array, brackets around one term of an equation require brackets around every other term of the same equation. It is understood that a letter subscript, if repeated in the same term of an expression, implies summation over the numerical range of the subscript by a convention introduced by and named after Albert Einstein. Because the letter repeated in any one term of an equation can be replaced by another letter, it is called a *dummy subscript*. A different letter subscript is used for every separate term that is to be summed. An example follows of what the Einstein summation convention thus implies:

$$A_i B_i \equiv A_j B_j \equiv A_1 B_1 + A_2 B_2 + A_3 B_3 . \tag{1.7}$$

Note that this convention does not hold for numeric subscript and thus does not affect the right-hand side of eq. (1.7).

In subscript notation the sum **S** of the two vectors **A** and **B** can be written as:

$$[S_i] = [A_i + B_i] , \tag{1.8}$$

or more commonly by specifying an arbitrary component of the sum vector as:

$$S_i = A_i + B_i . \tag{1.9}$$

In this usage, the subscript i is called a *free subscript*, and every term of an equation must necessarily possess the same free subscript. In subscript notation, the scalar dot product D of the two vectors of eq. (1.8) is written as:

$$D = A_i B_i , \tag{1.10}$$

where i is a dummy subscript and should be interpreted as in eq. (1.7).

Problem 13. (a) Using subscript notation, determine whether the vectors $^1 v = i + j + k$, $^2 v = i - k$, and $^3 v = i - 2j + k$ are mutually orthogonal. Note: Throughout this book, left superscripts are taken to serve only for identification and not to have any further function. (b) Select scalars x, y, and z so as to make the vectors $^1 w = i + j + 2k$, $^2 w = -i + x k$, and $^3 w = 2i + y j + z k$ mutually orthogonal.

Problem 14. (a) Let **A** be a nonzero vector; does the equation $A \cdot B = A \cdot C$ ensure that $B = C$? If not, provide an example to the contrary. (b) If $A \cdot B = A \cdot C$ for every vector **A**, must **C** equal **B**? (c) Use subscript notation to prove your answer.

To take full advantage of the subscript notation, we must introduce the *alternating matrix* ϵ_{ijk}. Each of its three subscripts is an integer *modulo* 3. By that we understand that the series of integer numbers restarts over and over at 1 every time after the number 3 has been reached, thus 2 > 1 and 3 > 2, but 1 > 3. Sets of three integers modulo 3 have *cyclic permutations*, calculated by adding to them the integers 1 and 2, thus the cyclic permutations of the ascending series 1, 2, 3 are 2, 3, 1 and 3, 1, 2; those of the descending series 3, 2, 1 are 1, 3, 2 and 2, 1, 3. These two sets of series contain all possible complete sets of all three positive integers 1, 2, and 3. The 27 elements of the alternating matrix are determined by the following rules:

$$\epsilon_{ijk} = 1 \text{ if } i, j, k = 1, 2, 3 \text{ or cyclic permutations,}$$
$$\epsilon_{ijk} = -1 \text{ if } i, j, k = 3, 2, 1 \text{ or cyclic permutations,}$$
$$\epsilon_{ijk} = 0 \text{ if neither of these.} \tag{1.11}$$

Problem 15. Find the components of the vector $\mathbf{i}\,n_1 + \mathbf{j}\,n_2 + \mathbf{k}\,n_3$ that is normal to the plane containing the two unit vectors from the origin, $\mathbf{i}\,v_1 + \mathbf{j}\,v_2 + \mathbf{k}\,v_3$ and $\mathbf{i}\,w_1 + \mathbf{j}\,w_2 + \mathbf{k}w_3$. Find an expression for the strike and dip of the plane if $\mathbf{i}$ points north, $\mathbf{j}$ east, and $\mathbf{k}$ down.

Problem 16. Suppose that a force $\mathbf{F}$ acts on the center of one face of a cube of edge $2a$, and forces $-\mathbf{F}$ on the center of the opposite face, $\mathbf{P}$ on that of a third, and $\mathbf{Q}$ on that of its opposite face. (a) Make a diagram and (b) explain the relationships between $\mathbf{P}, \mathbf{Q},$ and $\mathbf{F}$ such that the cube will be in equilibrium (i.e., so as to avoid any acceleration).

To specify the direction of a vector with respect to a coordinate system, it is sufficient and most convenient to find the components, called the *direction cosines* of the unit, or direction, vector pointing in this direction; these components are the direction vector's projections onto the three coordinate axes and hence the cosines of its *direction angles*. This is an alternative to the geological convention of stating the attitude of a line in terms of its trend, the azimuth of its horizontal projection, and its plunge, the angle between the trend and the downward branch of the line. We may want to extend the concept of plunge to vectors if they are referred to geological coordinates. The plunge of a vector with a downward component (x_3 positive) will be regarded as positive and that of a vector with an upward component (x_3 negative) as negative.

Problem 17. In the geological coordinate system $(x_i) = ($ north, east, down $)$, two vectors are given by $[u_i] = [\ 3 \quad 4 \quad 12\ ]$ and $[v_i] = [\ 5 \quad 12 \quad 84\ ]$. Find (a) the trend T and plunge P of u_i and its magnitude and (b) the angle included between u_i and v_i. Angles need be stated only to the nearest degree. (c) Determine the direction cosines of $[u_i]$ and $[v_i]$. (d) What is the angle φ between u_i and x_1?

An alternative way of calculating the trend and plunge of a vector stated in geological coordinates uses the correspondence of the trend T and the angle $(90° - P)$ to the *spherical coordinates* θ and φ at constant $\rho = 1$ in which θ and φ are related to $(x_i) = ($ north, east, down $)$ according to the scheme:

	θ	φ
x_1	0°	90°
x_2	90°	90°
x_3		0°

Cartesian direction cosines l_i have the spherical counterparts:

$$\rho = 1 \, ,$$
$$\theta = \cos^{-1}\left(l_1/\sin\varphi\right) ,$$
$$= \sin^{-1}\left(l_2/\sin\varphi\right) = \tan^{-1}\left(l_2/l_1\right) ,$$
$$\varphi = \cos^{-1} l_3 , \tag{1.12}$$

where the second expression for θ serves to break the sign ambiguity of the first, and the third is useful for greater accuracy at small φ (steep plunge). The trend in these coordinates is $T = \theta$ and the plunge $P = (90° - \varphi)$. Conversely, given θ and $\varphi = 90° - P$ of a line, its direction cosines are:

$$l_1 = \cos\theta \sin\varphi ,$$
$$l_2 = \sin\theta \sin\varphi ,$$
$$l_3 = \cos\varphi . \tag{1.13}$$

Vectors can be differentiated with respect to time; their time derivatives are also vectors.

Problem 18. Let the position of a particle at time t be defined by the vector from the origin: $\mathbf{R} = \mathbf{i}\,a\cos(2\pi t/T) + \mathbf{j}\,a\sin(2\pi t/T)$. (a) Interpret the equation and find the physical significance of the constants a and T. (b) Find the velocity **V** and (c) the acceleration **A** of the particle, and their magnitudes and directions as functions of time.

Problem 19. Show that $\mathbf{R} \times \mathbf{R} = \mathbf{0}$ if **R** is a twice differentiable function of time t as follows:

$$\mathbf{R} = \mathbf{A} + f(t)\,\mathbf{B} \; ,$$

where **A** and **B** are constant.

Problem 20. When a particle of constant mass m is accelerated by the force **F** and has an instantaneous velocity **V**, show that its kinetic energy increases at the rate of:

$$\frac{d}{dt}\left(\tfrac{1}{2}\,m\,\mathbf{V} \bullet \mathbf{V}\right) = \mathbf{F} \bullet \mathbf{V} \; .$$

Hint: Newton's law is: $\mathbf{F} = d(m\mathbf{V})/dt$.

෴

2
FIELDS

A scalar determined at every point in a given domain, analytically or otherwise, constitutes a *scalar field*. Vectors similarly determined constitute a *vector field*. The defining analytical expressions of a three-dimensional field are commonly differentiable with respect to space; hence in a cartesian coordinate system they are amenable to partial differentiation with respect to x_1, x_2, and x_3. In this context it is useful to define several *differential operators*. The operator ∇ is called the *"del"* or the *"nabla"* and is defined as follows:

$$\nabla \equiv \mathbf{i}\,\frac{\partial}{\partial x_1} + \mathbf{j}\,\frac{\partial}{\partial x_2} + \mathbf{k}\,\frac{\partial}{\partial x_3} \ , \tag{2.1}$$

or:

$$(\nabla)_i \equiv \frac{\partial}{\partial x_i} \ . \tag{2.2}$$

It can be seen that the del is a vector. By convention, however, it is not rendered in boldface. Before we define additional differential operators, we extend the subscript notation further and let a subscribed comma indicate partial differentiation. A comma preceding a letter subscript, say i, is taken to imply differentiation with respect to x_i. Thus, if $\varphi(x_i)$ is a scalar function of position and thus defines a scalar field, its *gradient*, another differential operator, is defined by the equation:

$$(\operatorname{grad}\varphi)_i \equiv \varphi_{,i} \equiv \frac{\partial \varphi}{\partial x_i} \ . \tag{2.3}$$

Thus the gradient of a scalar is a vector. A vector field with suitably differentiable functions of position has a *divergence* defined by:

$$\operatorname{div}[V_i] \equiv V_{i,i} \equiv \frac{\partial V_i}{\partial x_i} \ , \tag{2.4}$$

which is short for:

$$\operatorname{div}\mathbf{V} \equiv \nabla\bullet\mathbf{V} \equiv \frac{\partial V_1}{\partial x_1} + \frac{\partial V_2}{\partial x_2} + \frac{\partial V_3}{\partial x_3} \ . \tag{2.5}$$

Because the divergence is the dot product of two vectors it is a scalar.

With the help of the alternating matrix, we define another differential operator, the *curl* of a vector field:

10

$$\text{curl}\left[V_i \right] \equiv \left[\epsilon_{ijk}\, V_{k,\,j} \right] .\qquad (2.6)$$

Note that in this equation i is a free subscript, occurring once and only once in each of the terms; j and k, on the other hand, both occur twice in the same term and thus are dummy subscripts implying summation. To understand the implications of eq. (2.6), the reader should write it out explicitly, term by term, to find that it is short for:

$$\text{curl}\mathbf{V} \equiv \nabla\times\mathbf{V}$$

$$\equiv \begin{vmatrix} \mathbf{i} & \mathbf{j} & \mathbf{k} \\ \dfrac{\partial}{\partial x_1} & \dfrac{\partial}{\partial x_2} & \dfrac{\partial}{\partial x_3} \\ V_1 & V_2 & V_3 \end{vmatrix} . \qquad (2.7)$$

Being a cross product, the curl is a vector.

Problem 21. If $\mathbf{r}$ is a position vector of the form $\mathbf{r} = \mathbf{i}\,x_1 + \mathbf{j}\,x_2 + \mathbf{k}\,x_3$, demonstrate that:

$$\nabla\, r^n = n\, r^{n-2}\, \mathbf{r} .$$

Problem 22. Demonstrate that the following equations hold if $\mathbf{a}$ is a constant vector and $\mathbf{r}$ a variable position vector: (a) $\nabla\,(\mathbf{a}\bullet\mathbf{r}) = \mathbf{a}$, (b) $\nabla\times(\mathbf{a}\times\mathbf{r}) = 2\mathbf{a}$, (c) $\nabla\bullet(\mathbf{a}\times\mathbf{r}) = 0$. Draw general conclusions from your findings.

If vector or scalar fields satisfy certain conditions, they are said to possess one or more of the following characteristics. Vector fields are *solenoidal* if:

$$\nabla\bullet\mathbf{V} = 0 , \qquad (2.8)$$

or:

$$\mathbf{V} = \nabla\times\mathbf{A} , \qquad (2.9)$$

where $\mathbf{V}$ is the vector field and $\mathbf{A}$ is a vector function. *Irrotational* vector fields satisfy either:

$$\nabla\times\mathbf{V} = \mathbf{0} , \qquad (2.10)$$

or:

$$\mathbf{V} = \nabla\varphi . \qquad (2.11)$$

where φ is a scalar function. A *well-behaved* vector or scalar field in a simply connected region (a doughnut-shaped region is doubly connected) satisfies any one of the following four conditions:

$$\mathbf{V} = \nabla \times \mathbf{A} + \nabla \varphi \ ,$$
$$\nabla \times \mathbf{V} = \nabla \times (\nabla \times \mathbf{A}) \ ,$$
$$\nabla \cdot (\nabla \times \mathbf{A}) = 0 \ ,$$
$$\nabla \times \nabla \varphi = \mathbf{0} \ , \tag{2.12}$$

where $\mathbf{V}$ is a vector field, $\mathbf{A}$ a vector function, and φ a scalar field.

Problem 23. A rigid body rotates with a constant angular velocity $\mathbf{\Omega}$ about an axis in its interior that passes through a point O with the position vector $\mathbf{0}$; choose this point as the origin of a local coordinate system. Let the local position vector, relative to O, of an arbitrary point R in the rotating body be $\mathbf{r}$; then the velocity $\mathbf{v}$ of that point relative to the external reference frame is:

$$\mathbf{v} = {}^{\circ}\mathbf{v} + \mathbf{\Omega} \times \mathbf{r} \ ,$$

where ${}^{\circ}\mathbf{v}$ is the velocity of O in the external frame. (a) Show that at R curl $\mathbf{v} = 2\mathbf{\Omega}$ (angular velocity equals one-half the curl of the velocity field). (b) Demonstrate that $\mathbf{v}$ is solenoidal. *Hints:* The velocity ${}^{\circ}\mathbf{v}$ is independent of the local coordinates in which $\mathbf{r}$ is given. Use one of the results of Problem 22. (c) Characterize the velocity field.

Problem 24. The "red shift" of the spectra of distant galaxies is interpreted as indicating that galaxies generally move away from the earth with a velocity $\mathbf{v}$ proportional to their distance $\mathbf{r}$, thus that $\mathbf{v} = H\mathbf{r}$, where H is the so-called Hubble constant. (a) What is the divergence of the velocity field $\mathbf{v}$? (b) Show that observations from another galaxy would indicate the same constant H and thus the same expansion of the universe.

Problem 25. If the position vector $\mathbf{r} = \mathbf{i}\, x_1 + \mathbf{j}\, x_2 + \mathbf{k}\, x_3$, (a) demonstrate that the vector field $\mathbf{v} = r^n \mathbf{r}$ is irrotational. (b) Is it solenoidal?

Fields can be formed by tensors of the second and higher ranks, and we will encounter a few of them in later chapters.

⚬

3
MATTER TENSORS AND COORDINATE TRANSFORMATIONS

Vectors, the subject of the previous two chapters, may be classified as members of a class of mathematical entities called *tensors*, insofar as they can be expressed in the form of ordered arrays, or matrices, and insofar as they further conform to conditions to be explored in the present chapter. Tensors can have various *ranks*, and vectors are tensors of the first rank, which in three-dimensional space have 3^1 or three components. Much of this, and later, chapters deals with tensors of the second rank which in the same space have 3^2 or nine components. Tensors of higher (*n*th) rank do exist and have 3^n components, and so do, at least nominally, tensors of zero rank with a single, or 3^0, component, which makes them *scalars*.

Tensors of the second rank for three dimensions are written as three-by-three matrices with each component marked by two subscripts, which may be either letters or numbers. The first subscript indicates the row, the second the column, of the component as follows:

$$[T_{ij}] \equiv \begin{bmatrix} T_{11} & T_{12} & T_{13} \\ T_{21} & T_{22} & T_{23} \\ T_{31} & T_{32} & T_{33} \end{bmatrix} , \qquad (3.1)$$

where the brackets indicate that T_{ij} is a tensor and not some other matrix. A matrix always is a tensor if it relates two vectors to each other by a matrix product, or *tensor equation*, like the following:

$$\begin{bmatrix} v_1 & v_2 & v_3 \end{bmatrix} = \begin{bmatrix} T_{11} & T_{12} & T_{13} \\ T_{21} & T_{22} & T_{23} \\ T_{31} & T_{32} & T_{33} \end{bmatrix} \begin{bmatrix} w_1 \\ w_2 \\ w_3 \end{bmatrix} , \qquad (3.2)$$

or shorter:

$$[v_i] = [T_{ij} w_j] , \qquad (3.3)$$

or even shorter (as with vectors, a full tensor can be indicated by its subscripted generalized component; i.e., the brackets indicating the complete tensor may be omitted):

$$v_i = T_{ij} w_j , \qquad (3.4)$$

where in the previous two equations the Einstein summation convention implies summation over the dummy subscript j. A more precise condition that discriminates tensors from other matrices will be introduced later. Among other uses, tensors of the second rank serve to describe properties of materials, geological and other, that differ from one direction to another and are called *anisotropic* (sometimes also *heterotropic*), in contradistinction to *isotropic* materials, in which properties are the same in all directions. Tensors of this kind are called *matter tensors* (Nye, 1964).

Problem 26. The relation between the flow of heat q_i and the temperature gradient $T_{,i}$ in a homogeneous medium is given by the tensor equation:

$$q_i = -K_{ij} T_{,j} \, ,$$

in which the second-rank tensor, K_{ij}, is the thermal conductivity tensor. In the coordinate system $(x_i) = ($ north east down $)$ the conductivity of a certain body of rock is given by:

$$\left[K_{ij} \right] = \begin{bmatrix} 50.0 & 0.0 & 0.0 \\ 0.0 & 70.0 & 0.0 \\ 0.0 & 0.0 & 30.0 \end{bmatrix} \text{J m}^{-1}\text{s}^{-1}\text{K}^{-1}.$$

(a) Is the body thermally isotropic or anisotropic? (b) Qualitatively, what is the principal thermal characteristic of this body of rock? This range of variation with direction is typical of many rocks. (c) Suppose, as an example, that the temperature gradient is given by:

$$\left[T_{,i} \right] = \begin{bmatrix} 0.0 \\ 0.0 \\ 1.6 \end{bmatrix} \text{K m}^{-1}.$$

In which direction does the temperature decrease at the maximum rate? (d) Find the flow of heat and its direction in terms of trend and plunge. (e) Suppose that the temperature gradient is given instead by:

$$\left[T_{,i} \right] = \begin{bmatrix} 1.0 \\ 1.0 \\ 1.0 \end{bmatrix} \text{K m}^{-1}.$$

In which direction (trend and plunge) does the temperature decrease at the maximum rate in this case? (f) Find the flow of heat and its direction cosines and trend and plunge; compare the directions of heat flow and temperature gradient. (g) What error would be committed if the geothermal heat flow were based on temperature measurements in a deep vertical well and if it were calculated upon the assumption that the temperature gradient was vertically upward and if the conductivity of the rock were taken to be isotropic at the average of $50\ \mathrm{J\ m^{-1}\ s^{-1}\ K^{-1}}$?

Problem 27. The two vectors p_i and q_i are related according to the tensor equation:

$$p_i = S_{ij}q_j ,$$

in which the tensor S_{ij} is given by:

$$\left[S_{ij}\right] = \begin{bmatrix} 21 & 5\sqrt{3} & -2\sqrt{3} \\ 5\sqrt{3} & 31 & -6 \\ -2\sqrt{3} & -6 & 28 \end{bmatrix} .$$

Suppose that $[q_i] = [\ 1\ \ -\sqrt{3}\ \ 2\sqrt{3}\ ]$. (a) If $(x_i) = (\ \mathrm{north\ east\ down}\ )$, what are the magnitude and direction (trend and plunge) of q_i? (b) Find p_i, given this q_i. (c) What is the magnitude of p_i? (d) Find the angle included between p_i and q_i. Now replace $[q_i]$ by $[q_i'] = [\ 3\ \ 3\sqrt{3}\ \ 6\sqrt{3}\ ]$. (e) What are the magnitude and direction (trend and plunge) of q_i'? (f) What is the magnitude of the resulting p_i'? (g) Find the angle included between p_i' and q_i'.

Second-rank tensors have a distinct *magnitude* for each direction. Where $p_i = S_{ij}q_j$, the scalar magnitude S of $[S_{ij}]$ is the projection of $\mathbf{p}$ onto a line parallel to $\mathbf{q}$ divided by the vector magnitude q, $S = (\mathbf{p} \bullet \mathbf{q})/q$, or its equivalent $S = (p_i q_i)/q$, or substituting $S_{ij}q_j$ for p_i:

$$S = S_{ij}\frac{q_i}{q}\frac{q_j}{q} . \tag{3.5}$$

Because $[q_i/q]$ is a unit vector, the concept of magnitude can be generalized and made independent of the existence of a vector $[q_i]$. Replace $[q_i/q]$ with the unit vector l_i for the direction in which the magnitude of the tensor is sought:

$$S = S_{ij}l_i l_j . \tag{3.6}$$

Being a matrix, a second-rank tensor has an *inverse* which can be calculated by *Cramer's rule*. The inverse T_{ij}^{-1} of the tensor T_{ij} is:

$$T_{ij}^{-1} = A_{ji}/D \ , \tag{3.7}$$

where D is the determinant of the tensor:

$$D = \left| T_{ij} \right| \ , \tag{3.8}$$

and A_{ij} [note the transposition of the subscripts on the right-hand side of eq. 3.7] are the *cofactors*:

$$A_{ij} = \epsilon_{ikl} \, \epsilon_{jmn} \, T_{mk} \, T_{nl}/2 \ . \tag{3.9}$$

In other terms, an individual cofactor A_{ij} is the determinant that remains after deleting the ith row and the jth column of T_{ij}, multiplied by a sign factor of $(-1)^{i+j}$, as in the particular case of :

$$A_{23} = \left| \begin{array}{cc} T_{11} & T_{12} \\ T_{31} & T_{32} \end{array} \right| (-1)^5 . \tag{3.10}$$

Problem 28. Find q_j, given that $p_i = T_{ij} \, q_j$, that $[p_i] = [\ 16 \quad 20 \quad 24\ ]$, and that:

$$\left[T_{ij} \right] = \left[\begin{array}{ccc} 6 & 1 & 8 \\ 7 & 5 & 3 \\ 2 & 9 & 4 \end{array} \right] .$$

(a) Solve for q_j, treating the tensor equation as a system of linear equations.
(b) Solve by matrix inversion, using Cramer's rule.

Up to now, we have considered vectors, and in the last few problems second-rank tensors with reference to a single cartesian coordinate system. It is, however, commonly necessary to refer a tensor, of any rank, to more than a single set of coordinates. If we omit translations, that is, if we let two or more such coordinate systems share their origin, then one cartesian coordinate system is related to another by the set of angles that each of the three coordinate axes of the first system makes with each of those of the second. Figure 3.1 shows the angles that a unit vector along one of the axes of the second system forms with the three axes of the first and its projections onto these axes.

The complete set of these angles can be thought to form a three-by-three array of *direction angles*, which are conventionally ordered so that the angles formed by the same axis of the *old* system are arranged in the three columns of the array, each column in turn being ordered so that the angles formed with each of the axes of

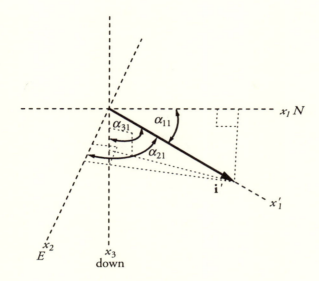

Figure 3.1. Projections of the unit vector **i**' along the axis x'_1 of a second coordinate system, x'_p, onto the axes x_i of the first. Angles α_{i1} are the direction angles of **i**'.

the *new* system form the rows of the array. The array of direction angles α_{ij} is, accordingly:

$$
(\alpha_{ij}) =
\begin{pmatrix}
\alpha_{11} & \alpha_{12} & \alpha_{13} \\
\alpha_{21} & \alpha_{22} & \alpha_{23} \\
\alpha_{31} & \alpha_{32} & \alpha_{33}
\end{pmatrix}
\begin{matrix}
x'_1 \; \uparrow \\
x'_2 \; \text{new} \\
x'_3 \; \downarrow
\end{matrix}
\qquad (3.11)
$$

with header $\xleftarrow{}\ \text{old}\ \xrightarrow{}$ over columns $x_1 \quad x_2 \quad x_3$.

To arrive at the "new" coordinates in terms of the "old," however, we need the projections of the old onto the new axes; hence the cosines of the angles in eq. (3.11). The array, ordered as in eq. (3.11), consists of the three sets of direction cosines of the new coordinate axes, referred to the old system, forming the three rows of the *rotation matrix*. The three columns so formed are also sets of direction cosines, those of the old coordinate axes referred to the new system. The rotation matrix corresponding to the array of direction angles eq. (3.11) is:

$$
(a_{ij}) =
\begin{pmatrix}
a_{11} & a_{12} & a_{13} \\
a_{21} & a_{22} & a_{23} \\
a_{31} & a_{32} & a_{33}
\end{pmatrix}.
\qquad (3.12)
$$

Because each row and each column of this matrix is a unit vector, the following holds:

$$a_{1i} a_{1i} = a_{2i} a_{2i} = a_{3i} a_{3i} = a_{i1} a_{i1}$$
$$= a_{i2} a_{i2} = a_{i3} a_{i3} = 1 \ , \tag{3.13}$$

a property that follows also from the six independent equations of the *orthogonality conditions* that hold for the rotation matrix:

$$a_{ik} a_{jk} = \delta_{ij} \ ,$$
$$a_{ki} a_{kj} = \delta_{ij} \ , \tag{3.14}$$

where δ_{ij}, the *Kronecker delta*, has the following definition:

$$\delta_{ij} = 1 \ , \ \text{if} \ i = j \ ,$$
$$\delta_{ij} = 0 \ , \ \text{if} \ i \neq j \ , \tag{3.15}$$

so that, for example:

$$\delta_{22} = 1 \ , \quad \delta_{13} = 0 \ , \quad \delta_{ii} = 3 \ . \tag{3.16}$$

Note that multiplying the Kronecker delta matrix with another matrix results in that identical matrix, for which reason it is also called the *identity matrix*. Because of the imposition of the six conditions of eq. (3.14) on the rotation matrix (a_{ij}), it is highly redundant, and only three of its nine elements are independent. The conditions on a_{ij} also have the consequence:

$$\left| a_{ij} \right| = \pm 1 \ , \tag{3.17}$$

where the negative sign affects only transformations that change the coordinates from right- to left-handedness, or vice versa. A set of interrelations of the elements of the rotation matrix is:

$$a_{1i} = \pm \epsilon_{ijk} a_{2j} a_{3k} \ ,$$
$$a_{2i} = \pm \epsilon_{ijk} a_{3j} a_{1k} \ ,$$
$$a_{3i} = \pm \epsilon_{ijk} a_{1j} a_{2k} \ ,$$
$$a_{i1} = \pm \epsilon_{ijk} a_{j2} a_{k3} \ ,$$
$$a_{i2} = \pm \epsilon_{ijk} a_{j3} a_{k1} \ ,$$
$$a_{i3} = \pm \epsilon_{ijk} a_{j1} a_{k2} \ , \tag{3.18}$$

or explicitly:

$$a_{11} = \pm(a_{22} a_{33} - a_{23} a_{32}) ,$$
$$a_{12} = \pm(a_{23} a_{31} - a_{21} a_{33}) , \qquad (3.19)$$
$$\vdots$$

where the negative signs apply only to transformations involving a change of handedness, and furthermore:

$$a_{ij}^{-1} = a_{ji} , \qquad (3.20)$$

and:

$$\epsilon_{ijk} \epsilon_{pqk} = \delta_{ip} \delta_{jq} - \delta_{iq} \delta_{jp} . \qquad (3.21)$$

Because of the great number of possible cyclic permutations, it is preferable to state this rule by identifying the four free subscripts i, j, p, and q by the order in which they occur on the left-hand side of eq. (3.21) according to the scheme:

$$\epsilon_{\text{first second dummy}} \, \epsilon_{\text{third fourth dummy}}$$
$$= \delta_{\text{first third}} \, \delta_{\text{second fourth}} - \delta_{\text{first fourth}} \, \delta_{\text{second third}} .$$

The orthogonality of cartesian coordinates has the consequence:

$$\frac{\partial x_i}{\partial x_j} = \delta_{ij} . \qquad (3.22)$$

Problem 29. Why can the elements a_{ij} of a transformation matrix for the rotation of coordinate axes never be greater than $+1$ or less than -1?

Problem 30. (a) Find the matrix (a_{ij}) for the transformation of a set of old axes $(x_i) = (\text{north east down})$ to new $(x_i') = (120°, 0° \quad 30°, -60° \quad 210°, -30°)$. (Directions are stated as trends and plunges, negative if upward, with azimuths in degrees counting from north to east, south, and west.) *Hint:* Use the method of Problem 17. (b) Evaluate the determinant $|a_{ij}|$. (c) Are the (x_i') axes right- or left-handed? (d) Find the matrix (a_{ij}') to transform from new axes (x_i') to newer $(x_i'') = (300°, 0° \quad 210°, 60° \quad 30°, 30°)$; this is called an inversion of axes. (e) Evaluate the determinant $|a_{ij}'|$. (f) Are the (x_i'') axes right- or left-handed? (g) Find the matrix (a_{ij}'') to transform from the old (x_i) to the newer axes (x_i''). (h) Evaluate the determinant $|a_{ij}''|$. (i) Find the transformation matrix $(\overset{*}{a}_{ij})$ for a transformation from new axes (x_i') to the old axes (x_i). (j) Find the determinant of $(\overset{*}{a}_{ij})$. (k) Find the matrix $(\hat{a}_{ij})$ to transform from the new axes (x_i') to reflected axes $(\hat{x}_i) = (120°, 0° \quad 210°, 60° \quad 210°, -30°)$. (l) What is the mirror plane for this reflection? (m) Evaluate the determinant $|\hat{a}_{ij}|$. (n) Are the axes ($\hat{x}_i$) right- or left-handed?

The transformation (a'_{ij}) of Problem 30 is an *inversion*, ($\hat{a}_{ij}$) a *reflection*; these are typical *symmetry operations*.

Problem 31. Find the element a_{23} of the transformation matrix from an old coordinate system with axes (x_i) = (north   east   down) to new axes (x'_i) = (225°, 0°   315°, −30°   315°, 60°).

Problem 32. Suppose that:

$$(a_{ij}) = \begin{pmatrix} -0.721278 & 0.523225 & -0.453865 \\ 0.171491 & -0.499963 & -0.848898 \\ -0.671080 & -0.690125 & 0.270884 \end{pmatrix},$$

and that the old axes were (x_i) = (north   east   down). (a) Find the directions of the new axes (x'_i) obtained from the old by means of the transformation matrix (a_{ij}). (b) Are the new axes orthogonal?

Problem 33. Suppose (a_{ij}) is the transformation matrix for the rotation of a set of right-handed orthogonal coordinates. Express a_{31}, a_{32}, and a_{33} in terms of $a_{11}, a_{12}, a_{13}, a_{21}, a_{22}$, and a_{23}.

The principal application of the rotation matrix is to transform tensors referred to one cartesian coordinate system into the form appropriate for another. Note that the physical entity described by the tensor obviously cannot be affected by the choice of a different reference system. The rule for the transformation from "old" to "new" coordinates of a tensor of the first rank, that is, a *polar vector*, is:

$$v'_i = a_{ij} v_j , \tag{3.23}$$

and that for the transformation of a second-rank tensor is:

$$T'_{ij} = a_{ik} a_{jl} T_{kl} , \tag{3.24}$$

or explicitly:

$$T'_{11} = a_{11} a_{11} T_{11} + a_{11} a_{12} T_{12} + a_{11} a_{13} T_{13}$$
$$+ a_{12} a_{11} T_{21} + a_{12} a_{12} T_{22} + a_{12} a_{13} T_{23}$$
$$+ a_{13} a_{11} T_{31} + a_{13} a_{12} T_{32} + a_{13} a_{13} T_{33} \, ,$$
$$T'_{12} = a_{11} a_{21} T_{11} + a_{11} a_{22} T_{12} + \cdots . \tag{3.25}$$

Problem 34. Transform the tensor:

$$\left[S_{ij} \right] = \begin{bmatrix} 21 & 5\sqrt{3} & -2\sqrt{3} \\ 5\sqrt{3} & 31 & -6 \\ -2\sqrt{3} & -6 & 28 \end{bmatrix} ,$$

to the *new* coordinates defined by the transformation matrix:

$$\left(a_{ij} \right) = \begin{pmatrix} \sqrt{3}/2 & -1/2 & 0 \\ 1/4 & \sqrt{3}/4 & \sqrt{3}/2 \\ -\sqrt{3}/4 & -3/4 & 1/2 \end{pmatrix} .$$

The *new* coordinate axes are called the *principal axes* of the tensor $[S'_{ij}]$, they form its *principal coordinates*. Verify that each of the three terms S'_{ii} (no sum on i) satisfies the cubic equation:

$$\begin{vmatrix} (21 - S'_{ii}) & 5\sqrt{3} & -2\sqrt{3} \\ 5\sqrt{3} & (31 - S'_{ii}) & -6 \\ -2\sqrt{3} & -6 & (28 - S'_{ii}) \end{vmatrix} = 0 \text{ (no sum on } i) .$$

The *secular equation* of Problem 34 for any symmetric tensor of the second rank can be solved to find the principal values (also called *eigenvalues*) of that tensor.

Problem 35. To show formally that a diagonal tensor with equal diagonal components, such as:

$$\left[B_{ij} \right] = \begin{bmatrix} B & 0 & 0 \\ 0 & B & 0 \\ 0 & 0 & B \end{bmatrix} ,$$

is invariant to the transformation of axes (a_{ij}), use the Kronecker delta in an appropriate equation for subscripted variables.

Problem 36.   Let $[\,T_{ij}\,]$ be a tensor.   Using subscript notation, show that T_{ii} is invariant, which is to say that the T_{ii} are the same for all orientations of the coordinate axes.

Problem 37.   Using subscript notation, show that the dot product of two vectors, say p_i and q_i, is invariant to the transformation of axes $(\,a_{ij}\,)$.

Problem 38.   Using subscript notation, show that the square of the magnitude of a vector p_i, and hence the magnitude, is invariant to the transformation of axes $(\,a_{ij}\,)$.

Invariants of a tensor are numbers that are independent of the choice of reference system.   Second-rank tensors have invariants (solutions to Problems 35 and 36), but they require further study.   Tensors of rank zero are scalars and thus obviously invariant to orientation (solution to Problem 37).   Vectors, tensors of the first rank, have only one independent invariant, the magnitude (solution to Problem 38).   One may be tempted to assume that any fully defined, ordered array of three scalars forms a vector and thus a first-rank tensor.

Problem 39.   Let p_i and q_i be vectors; then the three numbers p_1/q_1, p_2/q_2, and p_3/q_3 are defined for any set of axes.   Are they the components of a vector?   (Problems 39, 40, 57, 66, 69 – 71, and 98 from Nye, 1964.)

That an entity is a tensor of some rank or other can be determined by verifying whether its invariants remain intact after the appropriate transformation.

A second-rank tensor is *symmetric* if it is identical with its *transpose*:

$$\left[S_{ij}\right] = \left[S_{ji}\right] = \begin{bmatrix} S_{11} & S_{12} & S_{13} \\ S_{12} & S_{22} & S_{23} \\ S_{13} & S_{23} & S_{33} \end{bmatrix}. \qquad (3.26)$$

Problem 40.   Let $T'_{ij} = a_{im}\,a_{jk}\,T_{mk}$ and $T_{mk} = \pm T_{km}$.   Using subscript notation show that:

$$T'_{ij} = \pm T'_{ji}\ .$$

Symmetry of a tensor is an invariant property but not an "invariant," not being a number.

For the next problem it will be necessary to solve an equation jointly with a *constraint equation*.   As long as both equations are differentiable, this can be done elegantly with the method of *Lagrange multipliers*.   To maximize a function:

$$u = f(x_i) , \tag{3.27}$$

subject to the constraint:

$$\varphi(x_i) = 0 , \tag{3.28}$$

the constraint implying that the x_i in these two equations are *not* independent, multiply the constraint equation by the arbitrary Lagrange multiplier λ. Then solve:

$$\frac{\partial f}{\partial x_i} + \lambda \frac{\partial \varphi}{\partial x_i} = 0 , \tag{3.29}$$

for λ together with:

$$\varphi = 0 . \tag{3.30}$$

We further have to define the *representation quadric* of a symmetric second-rank tensor. Such a tensor may be represented by the quadric centered on the coordinate origin:

$$S_{ij} x_i x_j = \pm 1 , \tag{3.31}$$

with semiaxes $S_i^{-1/2}$. By choosing an appropriate sign, one can avoid imaginary branches of the quadric. Let $[\, r_i \,]$ be the radius vector to x_i; then the magnitude of the tensor is:

$$S = 1/r^2 . \tag{3.32}$$

Spatial differentiation allows us to find the *tangent plane* through an arbitrary point on a surface and an *outward normal* vector through that point. Let the equation of the surface be:

$$u(x_i) = 0 . \tag{3.33}$$

Then at the point $^{\circ}x_i$ the equation of the tangent plane is:

$$(x_i - {^{\circ}x_i}) \, u_{,i}(^{\circ}x_i) = 0 , \tag{3.34}$$

and the normal n_i to the tangent plane is:

$$n_i = u_{,i}(^{\circ}x_i) . \tag{3.35}$$

Note that the expression:

$$u_{,i}(^{\circ}x_i)$$

in eqs. (3.34) and (3.35) does not designate a multiplication, and hence does not imply summation.

Among other useful applications, the representation quadric of a second-rank tensor helps with the visualization of the relative orientations of two vectors that are interrelated by a tensor equation. The relationship is called the *radius-normal property* of the representation quadric. If $p_i = S_{ij} q_j$, then the direction of **p** for a given **q** is found by drawing a radius vector of the representation quadric for S_{ij} parallel to **q** (Fig. 3.2), then drawing the tangent to the quadric surface where it touches the radius vector. The outward normal to that tangent plane then parallels **p**. Figure 3.2 is drawn for an ellipsoidal quadric, but the same rule~holds if the quadric is a paraboloid or hyperboloid.

In the principal directions of a symmetric tensor, the vectors **p** and **q** are parallel. This fact can be used to determine principal directions of a tensor analytically by a method described at the end of this chapter.

Problem 41. Suppose p_i and q_i are related by the tensor equation $p_i = S_{ij} q_j$, where:

$$[S_{ij}] = \begin{bmatrix} S_1 & 0 & 0 \\ 0 & S_2 & 0 \\ 0 & 0 & S_3 \end{bmatrix},$$

and where $S_1 > S_2 > S_3$. (a) Is $[S_{ij}]$ referred to its principal axes? (b) If q_i is a variable-direction unit vector, which direction for q_i will give the greatest magnitude p for p_i? (c) What is the direction of p_i in that case? (d) What is the equation of the representation quadric for S_{ij}? (e) Find the radius vector x_i from the origin to the surface of the quadric, in the direction of a given unit vector u_j. (f) Find the unit vector n_i normal to the quadric at the point where the quadric is touched by x_i. (g) Find p_i given $q_i = u_i$. (h) What is the angle between p_i and n_i?

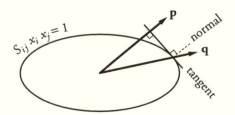

Figure 3.2. The radius-normal property of the representation ellipsoid. Central section of the ellipsoid containing both **p** and **q** and therefore perpendicular to the tangent.

Where the context prevents confusion with a vector, it is common practice to designate principal values of symmetric second-rank tensors by symbols with a single subscript. Use of this convention permits, for example, the simplification of eq. (3.6) for the magnitude of a tensor referred to its principal coordinates:

$$S = S_i l_i^2 , \tag{3.36}$$

where S_i are the principal values of the symmetric second-rank tensor S_{ij} and l_i the direction cosines for the direction in which the magnitude S is sought. Beware, however, of applying the transformation rule for vectors to a singly subscripted variables if, as in eq. (3.36), it does not represent a vector.

Second-rank tensors can be *asymmetric*, and in that case they can be divided into summands, one of which is symmetric [eq. (3.25)] and the other *antisymmetric*. An antisymmetric second-rank tensor A_{ij} has the property:

$$A_{ij} = -A_{ji} , \tag{3.37}$$

which implies:

$$\left[A_{ij} \right] = \begin{bmatrix} 0 & A_{12} & A_{13} \\ -A_{12} & 0 & A_{23} \\ -A_{13} & -A_{23} & 0 \end{bmatrix} . \tag{3.38}$$

Problem 42. Show that any tensor T_{ij} can be expressed as the sum of a symmetrical tensor S_{ij} and an antisymmetrical tensor A_{ij}; that is, show that there exist S_{ij} and A_{ij} such that:

$$p_i = T_{ij} q_j = S_{ij} q_j + A_{ij} q_j ,$$

where $S_{ij} = S_{ji}$ and $A_{ij} = -A_{ji}$. Find expressions for S_{ij} and A_{ij} in terms of T_{ij}.

Problem 43. Find a symmetric tensor S_{ij} and an antisymmetrical tensor A_{ij} such that:

$$T_{ij} = S_{ij} + A_{ij} ,$$

where:

$$\left[T_{ij} \right] = \begin{bmatrix} 1 & 2 & 3 \\ 4 & 5 & 6 \\ 7 & 8 & 9 \end{bmatrix} .$$

Tensors of the second and of higher ranks are used to describe physical properties of anisotropic materials, especially of crystals (Nye, 1964, pp. 20-21).

Physical properties of crystals are subject to *Neumann's principle: The symmetry elements of any physical property of a crystal must include, but may exceed, the symmetry elements of the point group of the crystal.* An example is the optical isotropy of cubic crystals; the symmetry elements of complete isotropy exceed those of even the class with the highest point group in the cubic system, $m\,3\,m$. As an empirical fact, none of the known crystal properties requires an asymmetric second-rank tensor for their description; hence matter tensors of that rank will here be assumed to be symmetric.

Problem 44. (a) Write the transformation matrix $_\theta^2(a_{ij})$ for a rotation of axes through an arbitrary angle θ about the x_2 axis (sense of rotation positive in the direction from x_3 to x_1). (b) Write the other two transformation matrices performing rotations about coordinate axes. (c) Suppose now that $\theta = 180°$; find $_{180°}^2(a_{ij})$. (d) Transform a symmetric tensor $[S_{ij}]$ using the transformation matrix $_{180°}^2(a_{ij})$ to produce the new tensor called $_{180°}[S_{ij}]$. (e) If the tensor $_0[S_{ij}]$ represents some tensor property of a monoclinic crystal, then what does Neumann's principle say about $_{180°}[S_{ij}]$, assuming that the dyad axis of the monoclinic crystal is parallel to x_2? (f) From this, deduce the general form of $_0[S_{ij}]$ for monoclinic crystals.

Problem 45. Using the method of Problem 44, find the general form of a tensor S_{ij} representing some tensor property of a crystal in the trigonal system, in coordinates with x_3 parallel to the threefold axis.

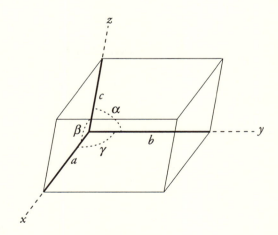

Figure 3.3.   Unit cell of dimensions a, b, and c
measured along crystallographic axes x, y, and z that
form interaxial angles α, β, and γ.

By convention, the generally unequal dimensions of the unit cell of a crystal, designated as a, b, and c, are measured along the generally nonorthogonal crystallographic axes x, y, and z, which form interaxial angles α, β, and γ (Fig. 3.3). The lengths a, b, and c are treated as the respective unit lengths for the axes x, y, and z. Crystallographic directions $[UVW]$ are vectors, referred to this system of crystallographic axes, drawn from an origin at a corner of a unit cell to another corner of the same or of another unit cell. They have integer components U, V, and W, chosen so as not to possess a common denominator (multiplication with a common factor does not change the direction of a vector). In cartesian coordinates with the same origin as the crystallographic coordinates (and with a conventional relative orientation of the two coordinate systems), the vectors $[UVW]$ become $U\mathbf{a} + V\mathbf{b} + W\mathbf{c}$ if $\mathbf{a}$, $\mathbf{b}$, and $\mathbf{c}$ are vectors of lengths a, b, and c in the crystallographic x, y, and z directions, but referred to the cartesian coordinates.

Problem 46. Find an expression for the unit vector $[u_i]$ in the crystallographic direction $[UVW]$ in the cubic system. Follow the usual conventions for the orientation of crystallographic axes, for example, those used by Nye (1964, pp. 276-88). Use $[UVW] = [1\ 1\ 1]$ as an example.

Problem 47. Find an expression for the unit vector $[u_i]$ in the crystallographic direction $[UVW]$ in the monoclinic system. Follow conventions as in Problem 46.

The case of the monoclinic system is still too highly symmetric to allow the derivation of a general rule for finding unit vectors in given crystallographic directions. That rule (stated without derivation) is: *Cartesian components of a unit vector in terms of crystallographic components are:*

$$u_i = (Ua_i + Vb_i + Wc_i)/k \,,\qquad(3.39)$$

where:

$$k = \left\langle \left(Ua_i + Vb_i + Wc_i\right)\left(Ua_i + Vb_i + Wc_i\right)\right\rangle^{1/2}.\qquad(3.40)$$

Note that in evaluating the quantity in brackets it is not permissible to cancel the apparent square represented by the product of the two identical quantities in parentheses against the exponent outside the brackets. The reason is the implied summation of the nine products with repeated subscript that results from the full multiplication of the two parentheses. An expression of this type must be evaluated from the inside out. To evaluate the squared normalization constant k^2 for the monoclinic system in terms of the general rule, we write:

$$k^2 = (U a_i + V b_i + W c_i)(U a_i + V b_i + W c_i)$$

$$= U a_i \, U a_i + \boxed{U a_i \, V b_i} + U a_i \, W c_i$$

$$+ \boxed{V b_i \, U a_i} + V b_i \, V b_i + \boxed{V b_i \, W c_i}$$

$$+ W c_i \, U a_i + \boxed{W c_i \, V b_i} + W c_i \, W c_i \, , \qquad (3.41)$$

where:

$$[a_i] = [\quad a \quad 0 \quad 0 \quad] ,$$
$$[b_i] = [\quad 0 \quad b \quad 0 \quad] ,$$
$$[c_i] = [\; c \cos\beta \quad 0 \quad c \sin\beta \;] . \qquad (3.42)$$

The framed terms in eq. (3.41) vanish, because whenever $a_i \neq 0$, then $b_i = 0$, and whenever $b_i \neq 0$, then $c_i = 0$. Hence:

$$k^2 = U^2 a^2 + V^2 b^2 + W^2 c^2 + 2U \, a \, W \, c \, \cos\beta \, , \qquad (3.43)$$

which agrees with eq. (3.40). In the triclinic system, eq. (3.41) represents 27 explicit products, and none of them vanishes although 9 are duplicated because of the commutative property of multiplication.

Problem 48. The relative permittivity tensor $[K_{ij}]$ (the tensor form of the dielectric constant) of a single crystal has been experimentally determined to be:

$$[K_{ij}] = \begin{bmatrix} 5.6 & 0 & 0 \\ 0 & 5.6 & 0 \\ 0 & 0 & 5.6 \end{bmatrix} ,$$

at an applied frequency of 200 kHz. What can be deduced from this finding concerning the symmetry of the crystal? Explain your reasoning.

Problem 49. (a) Of the four threefold symmetry axes arranged like the body diagonals of a cube that are required for membership in the cubic system, choose one, give it a sense, and find its direction cosines in a conventionally oriented coordinate system. (b) Write out the transformation matrix effecting the symmetry operation consisting of a 120° rotation about the threefold axis specified by the answer to (a). (c) Verify whether the symmetry operations determined in the answer to (b) ensure the second-rank tensor properties required by *Neumann's Principle* for cubic crystals.

Anisotropic physical properties are not restricted to single crystals. Crystalline materials in which the crystallographic axes of constituent grains are not oriented at random but are statistically aligned are anisotropic, and so are layered rock bodies, if considered as a whole and if their inhomogeneities are neglected.

Note that the physical measurement of an anisotropic material property that can be defined by a tensor of the second rank involves the application of an independent vectorial variable to a sample of the material in a specific direction and the measurement of the dependent vectorial variable *in the same direction*, repeated in a sufficient number of *different directions*. This amounts to the determination of a series of tensor magnitudes for the matter tensor to be determined.

Problem 50. The temperature gradient in a certain body of rock is given by:

$$[T_{,i}] = [\quad 1.0 \quad 1.0 \quad 1.0 \quad] \; K \, m^{-1},$$

and the corresponding flow of heat as:

$$[q_i] = -[\quad 50 \quad 70 \quad 30 \quad] \; J \, m^{-2} \, s^{-1}.$$

Find the thermal conductivity [short for magnitude of the thermal conductivity tensor, see eqs. (3.5) and (3.6)] in the direction of $[T_{,i}]$. Note that the same two vectors play a role in Problem 26 and here but that they are not necessarily related to each other by the same tensor $[K_{ij}]$.

Matter tensors can also describe manufactured materials.

Problem 51. In a specimen of insulating material the thermal conductivity K in various directions was found in a set of experiments:

Temperature gradient ($10^4 \; K \, m^{-1}$)	Magnitude of conductivity in the direction of the gradient ($J \, m^{-1} \, s^{-1} \, K^{-1}$)
[1.1 0.0 0.0]	8
[0.0 1.7 0.0]	16
[0.0 0.0 1.0]	8
[0.8 0.8 0.0]	24
[0.0 0.7 0.7]	20
[0.7 0.0 0.7]	12

Find the thermal conductivity $[K_{ij}]$ for this material.

There follow several exercises dealing with general features of second-rank tensors.

Problem 52. Two vector quantities p_i and q_i are related by the tensor property:

$$\left[S_{ij}\right] = \begin{bmatrix} 21 & 5\sqrt{3} & -2\sqrt{3} \\ 5\sqrt{3} & 31 & -6 \\ -2\sqrt{3} & -6 & 28 \end{bmatrix},$$

according to the equation, $p_i = S_{ij}\, q_i$. Find the magnitude of the property S_{ij} in the direction (a) of the vector $[1 \quad -\sqrt{3} \quad 2\sqrt{3}\ ]$ and (b) of the vector $[\,1 \quad \sqrt{3} \quad 2\sqrt{3}\ ]$. *Hint:* Compare with Problem 34.

Problem 53. Suppose a tensor is:

$$\left[T_{ij}\right] = \begin{bmatrix} 6 & 1 & 8 \\ 7 & 5 & 3 \\ 2 & 9 & 4 \end{bmatrix}.$$

(a) Find the magnitude T in the direction of the $x$axis.   (b) Find the magnitude T' in the direction of the vector $[\,1\ 1\ 1\,]$.

Problem 54. Show that the tensor:

$$\left[S_{ij}\right] = \begin{bmatrix} S_1 & 0 & 0 \\ 0 & S_1 & 0 \\ 0 & 0 & S_3 \end{bmatrix},$$

is invariant to rotation of coordinate axes about x_3.

Determinants of matrices (and thus of tensors) have uses beyond those we have already encountered; their usefulness is enhanced by the fact that determinants are unaffected by certain manipulations.

Problem 55. (a) Show that the determinant of a matrix (A_{ij}) is:

$$\left|A_{pq}\right| = \epsilon_{ijk}\, A_{i1}\, A_{j2}\, A_{k3} = \epsilon_{ijk}\, A_{1i}\, A_{2j}\, A_{3k} .$$

(b) Show that the determinant derived from $|A_{pq}|$ by interchanging two columns or two rows is equal to $-|A_{pq}|$, and therefore that, if two columns or two rows of $|A_{pq}|$ are identical, necessarily $|A_{pq}| = 0$. (c) From these results, show that:

$$|A_{ij}| \epsilon_{pqr} = \epsilon_{ijk} A_{pi} A_{qj} A_{rk} .$$

(d) Also show that, if $|A_{ij}|$ and $|B_{ij}|$ are determinants and $C_{ij} = A_{ik} B_{jk}$, then:

$$|C_{ij}| = |A_{pq}||B_{rs}| .$$

Problem 56.   Using one of the results of the previous problem and the orthogonality relations of a rotation matrix for cartesian coordinates a_{ij}, show that $|a_{ij}| = \pm 1$.

The representation quadric [eq. (3.31)] is not the only geometric figure descriptive of a second-rank tensor.

Problem 57.   Let a symmetric second-rank tensor S_{ij} have x_i as its principal coordinates. Prove that the ovaloid surface:

$$\left(S_1 x_1^2 + S_2 x_3^2 + S_3 x_3^2\right)^2 = \left(x_1^2 + x_3^2 + x_3^2\right)^3 ,$$

has in every direction a radius vector r_i with magnitude r equal to the tensor magnitude S in that direction.

This surface represents a second-rank tensor less conveniently than the representation quadric of eq. (3.31) and Figure 3.2.

Let a symmetric tensor relate the vector $[p_i]$ to the vector $[q_i]$:

$$p_i = S_{ij} q_j . \tag{3.44}$$

Principal directions of this tensor are those in which the radius-normal property of the representation quadric ensures that $[p_i]$ parallels $[q_i]$ and thus that:

$$C p_i = q_i , \tag{3.45}$$

where C is a proportionality constant. By definition, the magnitude of our tensor is:

$$S = (p_i q_i)/q . \tag{3.46}$$

Choosing as an instance of $[q_i]$ the unit vector $[l_i]$, eq. (3.44) becomes:

$$p_i = S_{ij} l_j \; , \tag{3.47}$$

and eq. (3.46):

$$S = p_i l_i \; . \tag{3.48}$$

Principal coordinates x_i of the tensor S_{ij} are parallel to its *principal directions*; thus, in each of the three principal directions x_i, both $[p_i]$ and $[l_i]$ have only one component, and this component equals their magnitudes, p and 1. Accordingly, in principal coordinates the vector magnitudes $^i p$ in the three principal directions must be identical with the three principal tensor magnitudes $^i S$ of $[S_{ij}]$ and thus with the three solutions $^i \lambda$ of the secular equation and also with the proportionality constants C in eq. (3.45). Therefore, in each of the principal directions:

$$^i p = {}^i S \equiv {}^i \lambda \; . \tag{3.49}$$

This finding cannot depend on the choice of coordinates, and it can therefore be generalized for principal directions of a symmetric tensor referred to an arbitrary coordinate system; for this purpose, we substitute for the components p_i of a vector with the magnitude p in eq. (3.47) those of a vector of magnitude λ, or λl_i:

$$S_{ij} l_j = \lambda l_i \; , \tag{3.50}$$

where l_i are the direction cosines of one of the principal directions of $[S_{ij}]$. Equation (3.50) represents three homogeneous linear equations in the variables l_i. They have a solution only if the following condition is met:

$$\left| S_{ij} - \delta_{ij} \right| = 0 \; , \tag{3.51}$$

which is the secular equation we have seen before.

Problem 58. Find (a) the coefficients of the secular equation and (b) the three principal values of the tensor:

$$[S_{ij}] = \begin{bmatrix} 21 & -3\sqrt{6} & -5 \\ -3\sqrt{6} & 14 & 3\sqrt{6} \\ -5 & 3\sqrt{6} & 21 \end{bmatrix} .$$

(c) Find the direction cosines of the shortest principal axis of the representation quadric of $[S_{ij}]$ and (d) those of the other two axes. (e) Write the secular equation for the same tensor in principal coordinates.

General rule: Invariance to rotation of the coefficients of the secular equation is a necessary property of a second-rank tensor. They are, therefore, commonly called the first, second, and third invariant of a second-rank tensor T_{ij} and are defined as follows:

$$^1I \equiv T_{ii} ,$$

$$^2I \equiv \begin{vmatrix} T_{22} & T_{23} \\ T_{32} & T_{33} \end{vmatrix} + \begin{vmatrix} T_{11} & T_{13} \\ T_{31} & T_{33} \end{vmatrix} + \begin{vmatrix} T_{11} & T_{12} \\ T_{21} & T_{22} \end{vmatrix} ,$$

$$^3I \equiv \begin{vmatrix} T_{ij} \end{vmatrix} . \tag{3.52}$$

In general, it is less convenient to solve for the principal directions of a symmetric tensor than in Problem 58. A universally applicable method is needed. Unit vectors l_i in the principal or eigendirections of a symmetric tensor being subject to eq. (3.50), in which λ is any one of the principal values $^i\lambda$ of the tensor, the individual equations for the three eigendirections can be restated as:

$$S_{ij} \,^1X_j = \,^1\lambda \,^1X_i ,$$

$$S_{ij} \,^2X_j = \,^2\lambda \,^2X_i ,$$

$$S_{ij} \,^3X_j = \,^3\lambda \,^3X_i , \tag{3.53}$$

where the ^iX_j are vectors of arbitrary magnitude parallel to the principal directions of the tensor; it is generally impossible to solve for these vectors directly because of their uncertain magnitude. Instead, one solves for the ratios of two of the components of each ^iX_j. Define:

$$^i\widehat{X}_1 \equiv \,^iX_1 / \,^iX_3 ,$$

$$^i\widehat{X}_2 \equiv \,^iX_2 / \,^iX_3 , \tag{3.54}$$

and arbitrarily set:

$$^i\widehat{X}_3 = \pm 1 . \tag{3.55}$$

Find the direction cosines (components of a unit vector) for the ith eigendirection by normalization:

$$^il_j = \,^i\widehat{X}_j / \left(\,^i\widehat{X}_1^2 + \,^i\widehat{X}_2^2 + 1 \right)^{1/2}. \tag{3.56}$$

Note that a number of arbitrarily accurate numerical methods exist for the simultaneous determination of eigenvalues and eigendirections.

The orthogonality conditions of the rotation matrix [eq. (3.14)] and the $\varepsilon\delta$ rule [eq. (3.21)] commonly introduce the Kronecker delta, or identity matrix, into tensor equations. The Kronecker delta has a useful attribute, the *substitution property*:

$$\delta_{ij} T_{jk} = T_{ik} \; . \tag{3.57}$$

Where a subscripted variable shares a dummy subscript with the Kronecker delta, the second subscript of the Kronecker delta can be substituted for the dummy subscript of the variable, while the Kronecker delta itself is eliminated. This property can easily be verified by executing eq. (3.57) explicitly. Note that numerical subscripts cannot be dummy subscripts and do not, therefore, convey the substitution property to the Kronecker delta; the Kronecker delta with numerical subscripts simply has the value of 0 or of 1.

Problem 59. Eliminate the Kronecker deltas from the following expressions by means of the substitution property and characterize them as either scalars (in their simplest form they have no free subscripts), vectors (one subscript), or second-rank tensors (two subscripts): (a) $\delta_{ij} T_{kj}$; (b) $\delta_{pq} \delta_{rq} S_{pr}$; (c) $\delta_{ij} \delta_{mn} A_{jm} B_{np} C_{pq}$; (d) $\delta_{pq} \delta_{qr} \delta_{rs} T_{st}$; (e) $\delta_{ij} \delta_{ji}$.

Problem 60. Transform the following expressions by rotation, simplify the results where possible using the orthogonality relations of rotation matrices and the substitution property of the Kronecker delta, and characterize the expressions as scalars, vectors, or tensors: (a) $A B_{mn}$; (b) $B_{tu} C_{vw}$; (c) $p_k q_k$; (d) $p_m A_{mk}$; (e) $A_{mp} B_{np}$.

Problem 61. Eliminate the Kronecker deltas from the following expressions: (a) $\delta_{pq} \delta_{rs} T_{ps}$; (b) $\delta_{23} T_{13}$; (c) $(\delta_{ii})^2$; (d) δ_{ii}^2.

The most reliable criterion that distinguishes tensors of any rank from other subscripted variables (matrices) is the conservation of the tensor invariants when it is rotated according to the rule appropriate for its rank. Equations (3.23) and (3.24) are the rules for tensors of the first and second rank, and tensors of higher ranks are rotated analogously. Thus, for a tensor of the third rank with three subscripts, the rule is:

$$T'_{ijk} = a_{il} a_{jm} a_{kn} T_{lmn} \; , \tag{3.58}$$

for one of the fourth rank:

$$T'_{ijkl} = a_{im} a_{jn} a_{ko} a_{lp} T_{mnop} \; , \tag{3.59}$$

and so on.

Problem 62. Using the Kronecker delta and the orthogonality relations, show that if A_{ij} and B_{ij} are tensors, then C_{ij}:

$$C_{ik} = A_{ij} B_{jk} ,$$

is also a tensor.

Problem 63. Using the Kronecker delta and the orthogonality relations, show that if A_{ij}, B_{ij}, and C_{ij} are tensors, then D_{ij}:

$$D_{im} = A_{ij} B_{jk} C_{km} ,$$

is also a tensor.

Problem 64. Using the Kronecker delta and the orthogonality relations, show that if:

$$p'_i = a_{ij} p_j ,$$

then the following is true:

$$p_j = a_{ij} p'_i .$$

Problem 65. Transform the vector:

$$\left[p_i \right] = \left[\ \sqrt{3}\,(\sqrt{2} - 1) \quad (\sqrt{2} - 1) \quad 1 \ \right] ,$$

by means of the transformation matrix:

$$\left(a_{ij} \right) = \begin{pmatrix} 1/2 & \sqrt{3}/2 & 0 \\ -\sqrt{6}/4 & \sqrt{2}/4 & \sqrt{2}/2 \\ \sqrt{6}/4 & \sqrt{2}/4 & \sqrt{2}/2 \end{pmatrix} .$$

What geometric relationship does this vector bear to the transformation of axes represented by the matrix (a_{ij})?

Antisymmetric tensors $V_{ji} = -V_{ij}$ (none describes material properties) have diagonal components that are necessarily zero, and an antisymmetric tensor can be expressed in terms of an *axial vector*. In contradistinction to the ordinary, or *polar*, vectors, axial vectors transform according to the rule:

$$r'_i = |a_{km}| a_{ij} r_j . \qquad (3.60)$$

As a consequence of this rule, axial vectors do not change sign upon inversion, whereas polar vectors do. Inversion is the transformation by means of the negative identity matrix [eq. (A.30.4)]. All vectors that describe a rotation are axial.

Problem 66. Given two arbitrary vectors $[p_i]$ and $[q_i]$ and the relationship:

$$V_{ij} = -p_i q_j + p_j q_i \, ,$$

show (a) that $[V_{ij}]$ is antisymmetric and (b) that it can be stated in terms of an axial vector $[r_i]$. (c) Demonstrate that $[V_{ij}]$ is a tensor by showing that its secular equation is invariant to rotation. (d) Does $[V_{ij}]$ have real principal values?

General rule: No real solutions exist for the secular equations of antisymmetric tensors.

The antisymmetric tensor $[V_{ij}]$ can be normalized by dividing the axial vector $[r_i]$ by its magnitude. The resulting antisymmetric tensor is called the *generator of rotations* $[R_{ij}]$:

$$R_{ij} = \frac{V_{ij}}{\left(p_k p_k q_l q_l - p_m q_m p_n q_n \right)^{1/2}} \, . \tag{3.61}$$

This makes $[r_i]$ a unit vector and hence the generator of rotations is:

$$\left[R_{ij} \right] = \begin{bmatrix} 0 & -l_3 & l_2 \\ l_3 & 0 & -l_1 \\ -l_2 & l_1 & 0 \end{bmatrix} . \tag{3.62}$$

It represents a rotation of $+90°$ *relative to a fixed reference frame*. The rotation is about a line with the direction cosines l_i and *clockwise*, seen in the direction of $[l_i]$.

The rotation matrix (a_{ij}) and the generator of rotations $[R_{ij}]$ are necessarily interrelated. Since it transforms the reference frame, the comparison rotation matrix must rotate *the reference frame itself counterclockwise about* $[l_i]$. This is the effect of a matrix with the elements:

$$a_{ij} = \delta_{ij} \cos\theta + \epsilon_{ijk} l_k \sin\theta + l_i l_j \left(1 - \cos\theta \right) , \tag{3.63}$$

for a rotation through the angle θ. In terms of the generator of rotations the same rotation is:

$$a_{ij} = \delta_{ij} - R_{ij}\sin\theta + R_{ik}R_{kj}\left(1 - \cos\theta\right) . \qquad (3.64)$$

Conversely, $\sin\theta$ and $\cos\theta$, and also l_i and thus R_{ij} by eq. (3.62), can be calculated from a known rotation matrix (a_{ij}):

$$\sin\theta = \tfrac{1}{2}\left\langle (a_{23} - a_{32})^2 + (a_{31} - a_{13})^2 + (a_{12} - a_{21})^2 \right\rangle^{1/2}, \qquad (3.65)$$

$$\cos\theta = (a_{11} + a_{22} + a_{33} - 1)/2 , \qquad (3.66)$$

$$l_1 = (a_{23} - a_{32})/(2\sin\theta) ,$$

$$l_2 = (a_{31} - a_{13})/(2\sin\theta) ,$$

$$l_3 = (a_{12} - a_{21})/(2\sin\theta) . \qquad (3.67)$$

Problem 67. Show that an antisymmetric tensor has no principal directions; that is, show that, if p_i and q_i are real, nonzero, finite vectors, if further A_{ij} is a tensor such that $A_{ij} = -A_{ji}$, and if $p_i = A_{ij}q_j$, then the vectors p_i and q_i cannot be parallel.

Problem 68. By applying all pertinent tests, show that S_{11} is a principal component and x_1 a principal axis in the tensor:

$$\left[S_{ij}\right] = \begin{bmatrix} S_{11} & 0 & 0 \\ 0 & S_{22} & S_{23} \\ 0 & S_{23} & S_{33} \end{bmatrix} .$$

General rule: A nonzero diagonal tensor component accompanied by zeroes in the remainder of the same column and of the same row is a principal component of that tensor.

Rotation of a symmetric second-rank tensor about one of its principal axes has a geometrical analog, called the *Mohr circle* after its discoverer (1882, 1914). Let the coordinate axes coincide with the principal tensor axes, and let the tensor thus be:

$$\left[S_{ij}\right] = \begin{bmatrix} S_1 & 0 & 0 \\ 0 & S_2 & 0 \\ 0 & 0 & S_3 \end{bmatrix} . \qquad (3.68)$$

The matrix that rotates this tensor about the x_3 axis, expressed as a function of the positive angle of coordinate rotation θ, is:

$$\left(a_{ij}\right) = \begin{pmatrix} \cos\theta & \sin\theta & 0 \\ -\sin\theta & \cos\theta & 0 \\ 0 & 0 & 1 \end{pmatrix}. \tag{3.69}$$

The rotation of a second-rank tensor by this matrix, $S'_{ij} = a_{ik}\, a_{jl}\, S_{kl}$, apart from the trivial $S'_{13} = S'_{23} = S'_{31} = S'_{32} = 0$ and $S'_{33} = S_3$, then becomes:

$$[S'_{\alpha\beta}] = \begin{bmatrix} \begin{array}{l} S_1\cos^2\theta \\ + S_2\sin^2\theta \end{array} & \begin{array}{l} S_2\sin\theta\cos\theta \\ - S_1\sin\theta\cos\theta \end{array} \\[2ex] \begin{array}{l} S_2\sin\theta\cos\theta \\ - S_1\sin\theta\cos\theta \end{array} & \begin{array}{l} S_1\sin^2\theta \\ + S_2\cos^2\theta \end{array} \end{bmatrix}. \tag{3.70}$$

(*Greek letter subscripts conventionally indicate tensors in two dimensions, thus* $\alpha, \beta = 1, 2$.) Equation (3.69) can be rewritten by means of the double angle identity (see *Summary of Formulæ, Trigonometry*):

$$[S'_{\alpha\beta}] = \begin{bmatrix} \begin{array}{l} \tfrac{1}{2}\left(S_1 + S_2\right) \\ - \tfrac{1}{2}\left(S_2 - S_1\right)\cos 2\theta \end{array} & \tfrac{1}{2}\left(S_2 - S_1\right)\sin 2\theta \\[2ex] \tfrac{1}{2}\left(S_2 - S_1\right)\sin 2\theta & \begin{array}{l} \tfrac{1}{2}\left(S_1 + S_2\right) \\ + \tfrac{1}{2}\left(S_2 - S_1\right)\cos 2\theta \end{array} \end{bmatrix}. \tag{3.71}$$

Equation (3.70) lends itself to a plot of the nondiagonal components, $\mathcal{Y} = S'_{\alpha\beta}$, $\alpha \neq \beta$, versus the diagonal components $\mathcal{X} = S'_{\alpha\beta}$, $\alpha = \beta$, in cartesian *Mohr space*. With variation of 2θ (corresponding to a variation of the angle θ in real space), $\mathcal{Y}$ describes a circle with the invariant radius $r = (S_2 - S_1)/2$, half the difference between the principal values of $S_{\alpha\beta}$, about the point $(S_1 + S_2)/2$ on the $\mathcal{X}$ axis, the mean principal value of $S_{\alpha\beta}$ (Fig. 3.4).

Both the radius r and the center of the circle can be seen to be invariants of the two-dimensional second-rank tensor $S_{\alpha\beta}$. As θ increases from zero to 45°, $\mathcal{Y}$ increases from zero to r, and as θ increases further to 90°, $\mathcal{Y}$ decreases back to zero. With the same 90° change in θ, S_{22} decreases from S_2 to the value of S_1; hence principal directions have been exchanged. Further increase of θ results in negative values for $\mathcal{Y}$, and at $\theta = 180°$ the original state has been restored (because the tensor is symmetric). Rotations in real space through angles between 180° and 360° are mapped into Mohr space by using negative angles 2θ.

To avoid ambiguities in the correlation of Mohr and real space, it is necessary to use a set of conventions. Let coordinate axes x_R, x_a, and x_b be the

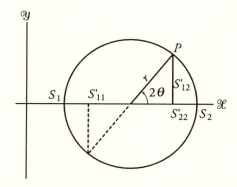

Figure 3.4. Mohr circle construction. In Mohr space, diagonal components are plotted along the $\mathcal{X}$ axis; nondiagonal components along the $\mathcal{Y}$ axis. A circle with the radius r is the locus for these components at various angles of clockwise rotation θ about the x_3 axis in real space. The double angle 2θ is positive counterclockwise, and the point P has coordinates S'_{22} and S'_{12} (solid vertical line) for a particular angle 2θ. The dashed lines serve to localize S'_{11}.

rotation axis and the axes with subscripts following R in ascending order modulo 3 (if $R = 1$, then $a = 2$ and $b = 3$, or in the present case, if $R = 3$, then $a = 1$ and $b = 2$), and let S'_{ij} be the result of a rotation about x_R of S_{ij}, then erect in Mohr space a positive S'_{ab} in the $\mathcal{Y}$ direction from the point $\mathcal{X} = S'_{bb}$ on the abscissa (a negative S'_{ab} is dropped from the same point). In Mohr space, the ordinates for S'_{ab} at $\mathcal{X} = S'_{ab}$ have the opposite sign of that in the tensor itself; they are therefore drawn dashed throughout this book.

Use of this convention enforces a definite association of subscript assignments and sense of rotation between Mohr and real space. A positive Mohr angle 2θ (counterclockwise in the drawing plane because the third Mohr axis $\mathcal{Z}$ points toward the viewer) corresponds to a positive (clockwise seen in the positive direction of the rotation axis) rotation of the two moving coordinate axes in real space. This means that the axis x'_a moves from its original orientation *toward* the original orientation of x_b (Fig. 3.5). (Some authors use conventions for Mohr space that depend on the order of the tensor's principal values. This is inconvenient, because within a coherent tensor field principal values commonly change their sequential position along one and the same trajectory.)

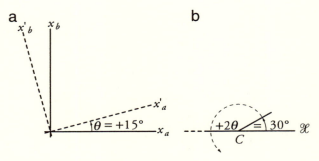

Figure 3.5. Rotation in real space (a) and Mohr space (b). A positive real rotation through an arbitrary 15°, and of the corresponding 30° in Mohr space, appears counter-clockwise in both because the rotation axes x_R in real space and $\mathcal{X}$ in Mohr space point toward the viewer.

Mohr circle constructions could be used to obtain, rather laboriously, reasonably accurate graphic solutions; their real usefulness, however, is the possibility to sketch a construction and then find the appropriate trigonometric relationships between the desired solutions for the unknowns of a problem and the data. Inspection of Figure 3.4 yields, among others:

$$S_1 = \tfrac{1}{2}\left(S'_{11} + S'_{22}\right) - r ,$$
$$S_2 = \tfrac{1}{2}\left(S'_{11} + S'_{22}\right) + r , \tag{3.72}$$

where:

$$r = \left\langle \tfrac{1}{4}\left(S'_{11} - S'_{22}\right)^2 + S'^2_{12}\right\rangle^{1/2}, \tag{3.73}$$

and:

$$\tan 2\theta = 2S'_{12}/\left(S'_{22} - S'_{11}\right) . \tag{3.74}$$

Problem 69. (a) Show with the help of a Mohr circle construction that the determinant:

$$\left| S_{ij}\right| = \begin{vmatrix} S_{11} & S_{12} \\ S_{12} & S_{22} \end{vmatrix}$$

is invariant to rotation about x_3. (b) Is there a geometric interpretation (in Mohr space) for your finding?

The rotation matrices for rotations through θ of coordinate axes about the x_1 and x_2 axes are:

$$(a_{ij}) = \begin{pmatrix} 1 & 0 & 0 \\ 0 & \cos\theta & \sin\theta \\ 0 & -\sin\theta & \cos\theta \end{pmatrix} \quad (3.75)$$

and:

$$(a_{ij}) = \begin{pmatrix} \cos\theta & 0 & -\sin\theta \\ 0 & 1 & 0 \\ \sin\theta & 0 & \cos\theta \end{pmatrix}. \quad (3.76)$$

For tensors stated in descending order of principal values $S_1 > S_2 > S_3$ (the most common convention), the following holds because of equations (3.69), (3.72), and (3.73): after a positive coordinate rotation of less than $180°$ about x_1, S_{23} is *negative*; about x_2, S_{31} is *positive*; and about x_3, S_{12} is *negative*. Changing either the convention to ascending, $S_1 < S_2 < S_3$, or the sense of coordinate rotation to negative, changes the signs of the nondiagonal tensor components. Note that a clockwise physical rotation with respect to fixed coordinates is the equivalent of a counterclockwise (negative) rotation of the coordinates with respect to a physical object. Note also that for this rule only the ordering of the two principal values in the plane perpendicular to the rotation axis is relevant.

Problem 70. Transform the following tensors to their principal coordinates, using the Mohr circle construction (a):

$$[S_{ij}] = \begin{bmatrix} 11.06 & 3.08 & 0 \\ 3.08 & 18.94 & 0 \\ 0 & 0 & 43 \end{bmatrix},$$

(b):

$$[U_{ij}] = \begin{bmatrix} -6 & -3\sqrt{3} & 0 \\ -3\sqrt{3} & 0 & 0 \\ 0 & 0 & 10 \end{bmatrix},$$

(c):

$$[V_{ij}] = \begin{bmatrix} 2 & 2 & 0 \\ 2 & 2 & 0 \\ 0 & 0 & 9 \end{bmatrix},$$

(d):

$$[W_{ij}] = \begin{bmatrix} 8 & 0 & -4 \\ 0 & 12 & 0 \\ -4 & 0 & 2 \end{bmatrix}.$$

Problem 71. There are two rotations of less than $90°$ about x_3 that make the S_{22} component of the following tensor zero:

$$[S_{ij}] = \begin{bmatrix} -1 & 3 & 8 \\ 3 & 10 & 6 \\ 8 & 6 & 2 \end{bmatrix}.$$

Find these rotations by the Mohr circle method.

Problem 72. (a) Using the Mohr circle construction, find the tensor S_{ij} after a coordinate rotation of 45° about x_1 (rotation of x_2 from its original orientation toward x_3 is considered positive), where:

$$[S_{ij}] = \begin{bmatrix} -1 & 0 & 0 \\ 0 & 7 & -3 \\ 0 & -3 & -1 \end{bmatrix}.$$

(b) By the same construction, transform S_{ij} to its principal coordinates. What is the orientation of the principal axes relative to the original coordinates (before the 45° rotation)?

Problem 73. (a) Using Mohr circle construction, find the tensors U'_{ij} and V'_{ij} after a coordinate rotation of 30° about x_2 and x_3, respectively (rotation of x_3 from its original orientation toward x_1 is considered positive, and so is rotation of x_1 from its original orientation toward x_2):

$$[U_{ij}] = \begin{bmatrix} -8 & 0 & 5\sqrt{3} \\ 0 & 5\sqrt{3} & 0 \\ 5\sqrt{3} & 0 & 2 \end{bmatrix},$$

$$[V_{ij}] = \begin{bmatrix} -8 & -2\sqrt{3} & 0 \\ -2\sqrt{3} & -4 & 0 \\ 0 & 0 & 2\sqrt{3} \end{bmatrix}.$$

(b) By the same construction, transform U_{ij} and V_{ij} to principal coordinates and find the necessary rotations.

Problem 74. Using Mohr circle construction, transform the following tensors to the principal coordinates with axes nearest those given and find the necessary rotations.

$$[W_{ij}] = \begin{bmatrix} 12.52 & 0 & -7.12 \\ 0 & 7 & 0 \\ -7.12 & 0 & 1.48 \end{bmatrix},$$

$$[X_{ij}] = \begin{bmatrix} -7.87 & 0 & -4.16 \\ 0 & -4.16 & 0 \\ -4.16 & 0 & -0.45 \end{bmatrix}.$$

Problem 75. Rotate the tensor:

$$[S_{ij}] = \begin{bmatrix} -3 & 0 & 0 \\ 0 & 4 & -6 \\ 0 & -6 & -10 \end{bmatrix},$$

about x_1 by the smallest angle that makes $S_{22} = 6$. Find the direction and angle of rotation and the rotated tensor $[S'_{ij}]$.

The Mohr circle approach is restricted to two-dimensional problems. It is, however, valid for rotations of three-dimensional second-rank tensors about one of the coordinate axes even if this axis is not a principal axis of the tensor. Such a rotation does modify those nondiagonal components of the tensor that share a subscript with the rotation axis. No information about these changes is obtained from the Mohr circle construction. "Principal values" and "principal directions" found in that case are not those of the tensor as a whole but of a two-dimensional section through it. Rotation of the stage of a microscope with crossed polarizers to "extinction" orientation in a thin section through some grain of a birefringent mineral is an analogous operation.

૭૦

4
STRESS

Stress is a tensor quantity that describes the mechanical force density (force per unit area) on the complete surface of a domain inside a material body. A stress exists wherever one part of a body exerts a force on neighboring parts. Its orientation is not tied to any particular directions that are intrinsic to the material like, say, crystallographic axes. It is thus distinct from the matter tensors that were discussed in the preceding chapter, all of which have definitive orientations within a crystal or other anisotropic material; it is called a *field tensor* (and so is strain). The definition of stress depends on the concept of a *continuum*. Let $f(x_i)$ be a single-valued function defined for every point x_i in a region. This function is said to be continuous at the point x_i if the following holds for all paths of approach of x_i to $^{\circ}x_i$:

$$f(x_i) \rightarrow f(^{\circ}x_i) \quad \text{as} \quad x_i \rightarrow {}^{\circ}x_i \ . \tag{4.1}$$

Equivalently, for any number ϵ, no matter how small, there exists a neighborhood of nonzero radius around the point x_i in which:

$$\langle f(x_i) - f(^{\circ}x_i) \rangle^2 < \epsilon \ , \tag{4.2}$$

for all points x_i in that neighborhood. A continuum is an idealized material whose physical attributes are continuous functions of position. Thus neighboring points remain neighbors, and a continuum cannot have gaps or jumps (*discontinuities*) in its properties. Surfaces bounding gaps or defining discontinuities must be specially treated in *continuum mechanics*. Examples are surfaces between two fluids of differing density or viscosity, or between solids with different thermal conductivity or elastic properties. Real materials are never continua; they are discontinuous at the atomic scale, and often at larger scales as well. The notion of a continuum is, therefore, only a macroscopic approximation, but it allows useful mathematical approaches to the treatment of real phenomena.

The definition of a *body force* may serve as an example:

$$\frac{\Delta f_i}{\Delta V} \rightarrow F_i \quad \text{as} \quad \Delta V \rightarrow 0 \ , \tag{4.3}$$

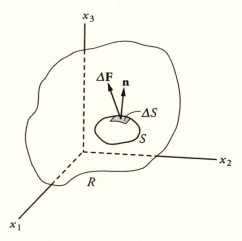

Figure 4.1. Euler's stress principle. The continuous region R contains a domain surrounded by the closed surface S. An element ΔS on that surface has the outward normal $\mathbf{n}$, and upon it the surrounding material exerts the force $\Delta \mathbf{F}$. (After Fung, 1965.)

where Δf_i is the total gravitational (or magnetic, or other) force on the material in the volume ΔV. This definition implies the assumption that the material is a continuum. In a real material, the volume ΔV, if chosen at the scale of, say, 1 m^{-20}, will at most points enclose nothing but empty space, but at a few points extremely dense subatomic matter. Evidently, some averaging is required; that is, ΔV must not be allowed to shrink to zero but only to a volume, small compared to macroscopic dimensions, yet large compared to atoms. In treating polycrystals, turbulent fluids, suspensions, or emulsions, or fluid-saturated rocks as continua, ΔV must be large compared to the volumes of constituent crystals, microcracks, eddies, suspended grains, or pores.

With these provisos in mind, the idea (if not the reality) of stress can be defined by way of *Euler's principle*. Consider a material region R to be a continuum and, inside this region, a domain bounded by the closed surface S (Fig. 4.1). Consider further a small portion of that surface, ΔS, with the outward normal $\mathbf{n}$. Designate as $\Delta \mathbf{F}$ the force exerted by the portions of R neighboring the positive side of $\mathbf{n}$ upon the material inside S adjacent to ΔS neighboring the negative side of $\mathbf{n}$. This force is a function of both the size and the orientation of ΔS. Now *assume* that as ΔS goes to zero, the ratio $\Delta \mathbf{F}/\Delta S$ tends to a definite limit $d\mathbf{F}/dS$, the *force density*, and that any *moment of force* (or *torque*) acting about any point on the surface ΔS vanish in the limit (the second condition may be violated if a rare state exists that is caused by a *body couple* or, according to Nye, 1964, a *body torque*; in this book a state that encompasses effects of a body torque shall be called a *stress in the wider sense*). The limiting vector, called a *traction* or *stress vector*, can then be stated as:

$$^{n}T_i = \frac{dF_i}{dS} \; ,$$

(4.4)

where the superscript n refers to the specific outward normal **n** of ΔS. It can be seen that traction has the dimensions of force per unit area, or $[m \; l^{-1} \; t^{-2}]$. *Stress* at some point completely defines the tractions for all orientations of **n** at that point.

For simplicity, the domain on the surface of which the forces act is commonly taken to be a cube with edges parallel to the axes of the reference coordinate system (Fig. 4.2). The illustration may either be regarded as a magnification to finite size of a point-sized cube or, with less difficulty for the imagination, as the representation of an actually finite cube inside a region subject to a *homogeneous* stress (a stress that is independent of position).

Figure 4.2 shows a set of force densities, or tractions, acting on the surfaces of a cubical domain inside a region subject to a homogeneous stress (in the narrow sense). The tractions that act on the concealed faces are antiparallel to those shown (they appear on Fig. 4.3), and the domain is therefore not subject to an

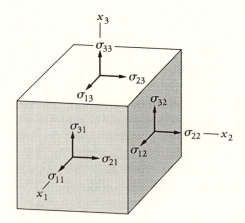

Figure 4.2. Components of force density (force per unit area) acting on faces of a homogeneously stressed cube. The nine components σ_{ij} constitute the stress tensor.

acceleration.   Since both the outward normals of the backward faces and the
tractions on them have the opposite signs of those in front, tractions and stress
components have the same sign back and front. The stress tensor (in the narrow
sense) is symmetric by definition (because moments of force are taken to vanish in
the limit):

$$\sigma_{ij} = \sigma_{ji} \, . \tag{4.5}$$

In the context of geology it is commonly permissible to neglect all fields,
except the gravitational field caused by the Earth's mass, and also the dynamic
transient states of stress occurring during acceleration, say, during an earthquake.
The remaining stresses are said to be in static equilibrium, a condition that can
be expressed precisely as:

$$\sigma_{ij,j} + \rho \, g_i = 0 \, , \tag{4.6}$$

where ρ is the local density and g the acceleration due to gravity. The three
equations (4.6) are the static *equations of equilibrium,* and ρg_i is called a *body force.*
Although other body forces do exist, the one due to gravity is most significant in
geology.

It is common practice to use the terms *normal* and *tangential* stress for the
diagonal and the nondiagonal components of the stress tensor. (Some authors use
two different symbols for the same tensor, σ_{ij}, $i=j$, for the normal, and τ_{ij}, $i \neq j$,

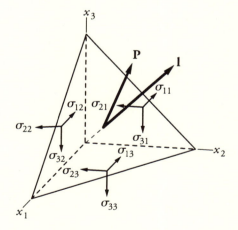

Figure 4.3. Tractions on the faces of a
tetrahedron. The front face has the outward normal **1** of unit length, and the
traction on it is **P**. (After Nye, 1964.)

for the tangential components.) A traction, too, can be decomposed into its normal and tangential components, and so can the force be integrated over a defined area.

How one can find the traction P_i upon an arbitrarily oriented surface element with the unitary outward normal l_i at a point where the stress is σ_{ij} can most easily be demonstrated for a body at static equilibrium subject to a homogeneous stress. Let there be a tetrahedron-shaped element of this body bounded by three surfaces, each parallel to two coordinate axes, and a fourth with an outward normal unit vector l_i as in Fig. 4.3. Since the stress is homogeneous and the size differences of the four bounding surfaces are irrelevant because tractions are normalized to unit area, we can vector-add the contributions to P_i as follows:

$$\begin{aligned}
P_1 &= \sigma_{11}\, l_1 + \sigma_{12}\, l_2 + \sigma_{13}\, l_3 \ , \\
P_2 &= \sigma_{21}\, l_1 + \sigma_{22}\, l_2 + \sigma_{23}\, l_3 \ , \\
P_3 &= \sigma_{31}\, l_1 + \sigma_{32}\, l_2 + \sigma_{33}\, l_3 \ .
\end{aligned} \tag{4.7}$$

In subscript notation, this simplifies to *Cauchy's equation*:

$$P_i = \sigma_{ij}\, l_j \ . \tag{4.8}$$

The existence of this formula, which relates one vector to another linearly, also proves that σ_{ij} is a tensor and that stress is a tensor quantity. When the stress is not homogeneous, or body forces are acting, or when the body is not in static equilibrium, Cauchy's formula still holds because the contributions of these departures from the simple case become negligible as the size of the tetrahedron of Fig. 4.3 becomes vanishingly small.

Problem 76. Let a body be homogeneously stressed, and let the stress be:

$$\left[\sigma_{ij}\right] = \begin{bmatrix} 1 & 3 & 2 \\ 3 & -9 & -6 \\ 2 & -6 & 16 \end{bmatrix} \text{ kPa} \ .$$

Find (a) the normal force N_i and (b) the tangential force T_i, in newtons, that are exerted across 1 m^2 of the plane perpendicular to the vector $[v_i] = [1,\ 1,\ 1]$ by the material on the $+v_i$ side upon that on the $-v_i$ side. (One pascal, Pa, is the traction exerted by a force of one newton, N, acting on one m^2 of surface.)

Several special forms of the stress tensor have specific names. Referred to their principal axes, they are the following:

(i) *Uniaxial stress*, such as the stress in a vertical rod or wire loaded by a weight suspended on its lower end:

$$[\sigma_{ij}] = \begin{bmatrix} \sigma & 0 & 0 \\ 0 & 0 & 0 \\ 0 & 0 & 0 \end{bmatrix}. \tag{4.9}$$

(ii) *Biaxial stress*, such as the stress in a thin plate loaded at the edges:

$$[\sigma_{ij}] = \begin{bmatrix} \sigma_1 & 0 & 0 \\ 0 & \sigma_2 & 0 \\ 0 & 0 & 0 \end{bmatrix}. \tag{4.10}$$

(iii) *Triaxial stress* is an alternative name for a general state of stress.

(iv) *Hydrostatic pressure*, such as the stress in a fluid at rest:

$$[\sigma_{ij}] = \begin{bmatrix} -p & 0 & 0 \\ 0 & -p & 0 \\ 0 & 0 & -p \end{bmatrix}, \tag{4.11}$$

where p is the pressure. Note that pressure is always stated positive for compression, whereas in physics and geophysics, and in this book, extensile stress and traction are conventionally positive. (In the fields of engineering and geology, the opposite convention is common, so that a compressive stress is taken to be positive.)

(v) *Pure shear stress*, a special form of biaxial stress:

$$[\sigma_{ij}] = \begin{bmatrix} -\sigma & 0 & 0 \\ 0 & \sigma & 0 \\ 0 & 0 & 0 \end{bmatrix}, \tag{4.12}$$

which, upon a 45° rotation about the x_3 axis, the *axis of shear*, becomes:

$$[\sigma_{ij}] = \begin{bmatrix} 0 & \sigma & 0 \\ \sigma & 0 & 0 \\ 0 & 0 & 0 \end{bmatrix}, \tag{4.13}$$

hence its name.

Problem 77. (a) Determine the kind of stress that is represented by the tensor:

$$[\sigma_{ij}] = \begin{bmatrix} 4 & -2\sqrt{3} & 6 \\ -2\sqrt{3} & 3 & -3\sqrt{3} \\ 6 & -3\sqrt{3} & 9 \end{bmatrix} \text{ MPa .}$$

Hints: A determinant is multiplied by a given factor by multiplying one row or one column by that factor, and a determinant is unaltered if a multiple of one row (or column) is added to another row (or column). (b) Find the tensor's orientation.

Any stress may be divided into two components, a *mean stress* and a *deviator of stress*. The mean stress, a hydrostatic pressure, is the arithmetic mean of the three principal stresses. Because the trace of the stress tensor is an invariant, the mean stress is $\sigma_{ii}/3$ independently of coordinate orientation. To calculate the deviator of stress, Δ_{ij}, for some general stress, one subtracts from it the mean stress.

Problem 78. Show that, if the stress in a particular coordinate system is σ_{ij}, then the stress deviator Δ_{ij} in the same coordinates is defined by the equation:

$$\Delta_{ij} = \sigma_{ij} - \delta_{ij}\sigma_{kk}/3 .$$

Problem 79. Write the secular equation in terms of the principal stresses λ for the following kinds of stress: (a) uniaxial, (b) biaxial, (c) triaxial, (d) hydrostatic pressure p, and (e) pure shear stress.

If the moment of forces acting on a surface element does not, in the limit, vanish as the element area goes to zero, a body torque does exist, and the state is one of stress in the wider sense. This is a rare phenomenon occurring in certain single crystals of low symmetry when they become polarized or magnetized by long-range forces in an electromagnetic field. The stress due to a body torque is usually superposed on a stress in the narrow sense. By itself the stress τ_{ij} due to a body torque is antisymmetric; that is $\tau_{ji} = -\tau_{ij}$. The body torque can be characterized by the axial vector G_i, which is related to τ_{ij} by:

$$G_i = \epsilon_{ijk}\tau_{jk} . \tag{4.14}$$

Problem 80. (a) Find the body torque G_i per unit volume that is being exerted on a medium in which the stress (in the wider sense) is homogeneous and given by $\sigma_{ij} = A_{ij}$, where A_{ij} is an antisymmetric tensor. Designate as τ_{ij} all nondiagonal components of σ_{ij}, a common practice. (b) Demonstrate that G_i is either a polar or an axial vector? (c) Find the hydrostatic pressure in the medium.

Problem 81. The homogeneous stress (in the wider sense) in a particular medium is:

$$[\sigma_{ij}] = \begin{bmatrix} 10 & 30 & 20 \\ 30 & -90 & -60 \\ 30 & 40 & 50 \end{bmatrix} \text{ kPa} .$$

(a) Find the body torque per m^3 exerted on and (b) the hydrostatic pressure in that medium.

Henceforth, the rare and geologically insignificant phenomenon of body torque will be dropped from consideration. The term stress will therefore be understood only in its narrow sense and be taken to be symmetric. Furthermore, in the context of this book, dynamic states of stress will be neglected, even though their study is essential for the study of earthquakes, and static equilibrium is assumed to hold. In a *stress field*, the stress is defined at every point in some domain.

Problem 82. The stress field in a medium as a function of position is given by:

$$[\sigma_{ij}] = \begin{bmatrix} x_1 + 1 & x_1 + x_2 & x_1 + x_3 \\ x_2 + x_1 & x_2 + 2 & x_2 + x_3 \\ x_3 + x_1 & x_3 + x_2 & x_3 + 3 \end{bmatrix} \text{ Pa} .$$

(a) Find the body force F_i per m^3 in that medium, and (b) calculate it at the point:

$$[x_i] = \begin{bmatrix} 1 & 2 & 4 \end{bmatrix} \text{ m} .$$

Problem 83. Assume that the stress in a medium has the property:

$$\sigma_{ij, j} = 0 .$$

What does this imply about the stress?

Problem 84. The customary "invariants" of a symmetric tensor, say, the stress tensor σ_{ij}, are:

$$^1I = \sigma_{ii} \,,$$

$$^2I = \begin{vmatrix} \sigma_{22} & \sigma_{23} \\ \sigma_{23} & \sigma_{33} \end{vmatrix} + \begin{vmatrix} \sigma_{33} & \sigma_{31} \\ \sigma_{31} & \sigma_{11} \end{vmatrix} + \begin{vmatrix} \sigma_{11} & \sigma_{12} \\ \sigma_{12} & \sigma_{22} \end{vmatrix} \,,$$

$$^3I = \begin{vmatrix} \sigma_{ij} \end{vmatrix} \,.$$

Given the secular equation (in determinant form) for $[\sigma_{ij}]$ in terms of principal values λ, restate it in terms of these three invariants by repeated application of the following rule for determinants: *If each element of a row (or column) of a determinant is the sum of two terms, then this determinant equals the sum of two determinants, each having a row (or column) consisting of the separate summands and remaining rows (or columns) that are identical with those of the original determinant* (see *Determinants* in *Summary of Formulæ*).

Problem 85. Show by direct transformation that the determinant of a tensor is invariant to an arbitrary rotation of the coordinate axes. Use the following rule for determinants: If a new determinant C is formed from two determinants A and B so that the element in the ith row and jth column of C is obtained by multiplying each element in the ith row of A by the corresponding element in the jth row of B and adding the products, then $C = AB$.

Problem 86. Given the three standard invariants of a tensor in principal coordinates:

$$^1I = \sigma_1 + \sigma_2 + \sigma_3 = 4 \,,$$

$$^2I = \sigma_2\sigma_3 + \sigma_3\sigma_1 + \sigma_1\sigma_2 = -11 \,,$$

$$^3I = \sigma_1\sigma_2\sigma_3 = -30 \,,$$

find the principal values σ_1, σ_2, and σ_3 of the tensor.

Problem 87. The stress in a certain body is given by:

$$[\sigma_{ij}] = \begin{bmatrix} -5 & 2 & -3 \\ 2 & 2 & 1 \\ -3 & 1 & -1 \end{bmatrix} \text{ kPa} \,.$$

Find the normal traction N_i and the tangential traction T_i on the plane perpendicular to the vector:

$$[v_i] = \begin{bmatrix} 1 & 3 & 5 \end{bmatrix} \text{ m },$$

by the material on the $+v_i$ side upon that on the $-v_i$ side.

Problem 88. The static stress in a cube, $-1 \leq x_1 \leq 1, -1 \leq x_2 \leq 1, -1 \leq x_3 \leq 1$, is given as:

$$[\sigma_{ij}] = \begin{bmatrix} -3x_1^2 x_2 & x_1^2 x_3 & 6x_1 x_2 x_3 \\ x_1^2 x_3 & 6x_1 x_2 x_3 & -4x_1 x_3^2 \\ 6x_1 x_2 x_3 & -4x_1 x_3^2 & -3x_2 x_3^2 \end{bmatrix},$$

in arbitrary units. (a) Is a body force acting on the material in the cube? Find expressions for the components (b) of the normal traction N_i and (c) of the shear traction T_i acting on the face of the cube defined by $x_3 = -1$, assuming that the cube is in equilibrium.

Problem 89. Determine the type and magnitude (but not the orientation) of the stress represented by the tensor:

$$[\sigma_{ij}] = \begin{bmatrix} -4 & -2\sqrt{3} & 6 \\ -2\sqrt{3} & 1 & -\sqrt{3} \\ 6 & -\sqrt{3} & 3 \end{bmatrix} \text{ kPa }.$$

Problem 90. (a) Show that a stress is a pure shear stress if its first and third invariants are zero. (b) Find the magnitude σ of such a pure shear stress in terms of the second invariant ^2I.

Problem 91. The stress in an octahedron with corners at $(x_i) = (1, 0, 0)$, $(0, 1, 0), (-1, 0, 0), (0, -1, 0), (0, 0, -1)$ is:

$$[\sigma_{ij}] = \begin{bmatrix} (x_2 + 1) & x_3 & x_1 \\ x_3 & (x_1 + 1) & x_2 \\ x_1 & x_2 & (x_3 - 1) \end{bmatrix} \text{ Pa }.$$

The octahedron is stationary. Find (a) the body force and (b) the body torque, if any, acting on it. (c) Find the normal and tangential tractions acting on the face whose outward-pointing normal is:

$$[l_i] = \begin{bmatrix} -1/\sqrt{3} & -1/\sqrt{3} & -1/\sqrt{3} \end{bmatrix}.$$

Problem 92. Along a wire consisting of a single metal crystal acts a tensile stress σ. Resolve the shear stress τ on a glide plane and in the glide direction and show that it is:

$$\tau = \sigma \cos\varphi \cos\lambda \ ,$$

where φ is the angle between the wire axis and the normal to the glide plane and λ the angle between the wire axis and the glide direction.

Problem 93. Let the stress tensor for a certain medium be:

$$\left[\sigma_{ij}\right] = \begin{bmatrix} 9 & -5 & 5 \\ -5 & 4 & 3 \\ 5 & 3 & 8 \end{bmatrix} \text{ kgwt cm}^{-2}.$$

One kilogramweight (kgwt) is the force exerted on a mass of one kilogram by a gravitational field, usually that of the Earth. Find the normal force N_i and the tangential force T_i exerted across 2 cm^2 of the plane perpendicular to the vector:

$$\left[V_i\right] = \begin{bmatrix} 1 & 2 & 2 \end{bmatrix} \text{ cm} \ ,$$

by the material on the $-V_i$ side upon that on the $+V_i$ side. *Hint:* This problem is intentionally posed so as to violate several conventions.

Problem 94. Rotate the stress tensor:

$$\left[\sigma_{ij}\right] = \begin{bmatrix} 1 & 0 & -7 \\ 0 & 0 & 0 \\ -7 & 0 & -7 \end{bmatrix} \text{ kPa} \ ,$$

by the smallest angle θ (stated in degrees) that will make $\sigma'_{31} = -8$ kPa. Find (a) the rotated tensor $[\sigma'_{ij}]$, (b) the angle of rotation, (c) the sense of rotation, and (d) the rotation matrix (in decimal form).

Problem 95. Let a stress have principal components $\sigma_1 = 8$, $\sigma_2 = \sigma_3 = 4$ in arbitrary units. (a) Draw the general Mohr diagram for this stress (the two orthogonal principal planes through the unique axis of this axially symmetric stress are identical). (b) Find the relationship that exists between the angle θ, formed by the unique axis of the stress with the pole of the plane on which a traction acts, and the values of the normal component σ and the tangential component τ of that traction. (c) Let a unit sphere be contoured, at an interval of one unit (half-unit intervals may be helpful where contours are too far apart), for the value of the normal stress σ and for that of the tangential stress τ, plotted where reference-plane poles intersect the unit sphere. Sketch the all-positive octant of this unit sphere, the octant for which $l_1 \geq 0,\ l_2 \geq 0,\ l_3 \geq 0$.

Problem 96. Find the magnitude $^0\tau$ of the shear traction across the faces of a regular octahedron oriented with its fourfold axis parallel to the principal axes of stress in a homogeneously stressed medium. This magnitude is called the *octahedral shear stress*. Show that this magnitude can be stated in terms of the conventional stress invariants ^1I and ^2I as:

$$9\,^0\tau^2 = 2\left(^1I\right)^2 - 6\left(^2I\right) .$$

Problem 97. Show (a) that the second invariant ^2I of the stress deviator Δ_{ij} (defined in Problem 78) simplifies to:

$$^2I = \Delta_{ij}\Delta_{ij}/2 ,$$

and (b) that the third invariant ^3I can be stated as:

$$^3I = \Delta_{ij}\Delta_{jk}\Delta_{ki}/3 .$$

Hint: The first invariant of a deviator vanishes.

Whereas the result (a) is a useful simplification, the result (b) implies 27 nonzero terms in general coordinates and is thus less convenient than the standard expression for $^3I = |\Delta_{ij}|$, evaluated as $\epsilon_{ijk}\Delta_{1i}\Delta_{2j}\Delta_{3k}$, which has only six nonzero terms. Equation (A.97.7) combined with (A.97.9) demonstrates the somewhat surprising general fact that, if $a + b + c = 0$, then $a^3 + b^3 + c^3 = 3\,a\,b\,c$.

☙

5
INFINITESIMAL STRAIN

An elastic material responds to a stress by a change of volume and shape, or *strain*, which stays constant as long as the stress is maintained. Materials for which strains are completely reversible and proportional to the stresses that cause them are called *ideally elastic* and are said to follow *Hooke's law*. Many actual materials are nearly ideally elastic as long as the stress-induced strains are small. Strain in such materials is the usual means of observing stress, which itself is an abstraction and not directly observable.

Strain, as treated in continuum mechanics, is also an abstraction, but one that more closely approaches observable reality. The element of abstraction comes from treating the deformed body as a continuum, with the implication that neighboring material points in an undeformed body remain arbitrarily close neighbors after deformation.

Let one point on a stretchable material line (imagine a rubber band) be held in place at the origin of a one-dimensional coordinate system (Fig. 5.1a) and stretched, throughout but not necessarily uniformly (the rubber band may vary in thickness), by pulling on its free end with the position Δx (Fig. 5.1b). Let the end, as a consequence of the stretching, be moved by the *displacement* Δu. Then any original length element Δx of the line will be changed to a new length $\Delta x + \Delta u$, say, the particular segment starting at the points P before and P' after the

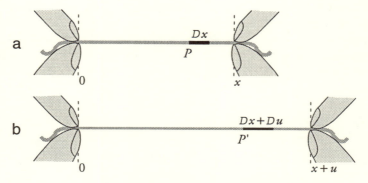

Figure 5.1. One-dimensional strain. A stretchable string, held on the left at the origin (a), is extended at the right by a displacement u to the new length $x + u$. (b) The extension changes the original length Δx at P to $\Delta x + \Delta u$ at P'.

deformation. Because we treat the line as a continuum, we can postulate that a *displacement gradient e* exists at the material point $P = P'$:

$$\frac{\Delta u}{\Delta x} \rightarrow \frac{du}{dx} \equiv e \text{, as } \Delta x \rightarrow 0 \ . \tag{5.1}$$

This dimensionless quantity expresses the rate of change of displacement with distance, and it can be seen that the absolute distance x from the origin of the point P, and hence the choice of origin of our coordinate does not affect this measure. Inversely:

$$\Delta u = e \Delta x \equiv \frac{du}{dx} \Delta x \ . \tag{5.2}$$

When one applies similar arguments to two dimensions, or to a stretchable sheet, complications arise, unless the resulting dimensionless components of the displacement gradient $e_{\alpha\beta}$ (Greek subscripts for two dimensions) are kept infinitesimally small (or for practical purposes, small compared with 1). They are defined analogously to the one-dimensional e:

$$\frac{\Delta u_\alpha}{\Delta x_\beta} \rightarrow \frac{\partial u_\alpha}{\partial x_\beta} \equiv u_{\alpha,\beta} \equiv e_{\alpha\beta} \text{ as } \Delta x_\beta \rightarrow 0 \ , \tag{5.3}$$

which inverts to:

$$\Delta u_\alpha = \frac{\partial u_\alpha}{\partial x_\beta} \Delta x_\beta \equiv e_{\alpha\beta} \Delta x_\beta \ . \tag{5.4}$$

The geometric significance of these equations, sketched with exaggerated length and displacement elements, is shown in Fig. 5.2. Note that neither the arbitrarily large *translation* of the point P to P' nor its position relative to the coordinate origin has any influence on the displacement gradient. Inspection of Fig. 5.2 shows that:

$$\tan^1\theta = \frac{\partial u_2}{\partial x_1} \Delta x_1 \bigg/ \left(\Delta x_1 + \frac{\partial u_1}{\partial x_1} \Delta x_1 \right) , \tag{5.5}$$

which simplifies to:

$$\tan^1\theta = \frac{\partial u_2}{\partial x_1 + \partial u_1} \ . \tag{5.6}$$

Considering that, although not so shown in Fig. 5.2, Δu_1 is actually much smaller than Δx_1, that the same holds for ∂u_1 and ∂x_1, and that for small angles the tangent of an angle equals the angle itself (in radians), we find:

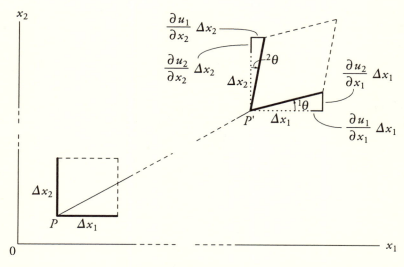

Figure 5.2. Geometry of a small displacement gradient $\partial u_\alpha / \partial x_\beta \equiv e_{\alpha\beta}$. The original point P is displaced to P', and the originally orthogonal line elements Δx_α (heavy lines) are stretched and form angles $^\alpha\theta$ with their original orientations (dotted lines). (After Nye, 1964.)

$$^1\theta \approx \tan^1\theta = \frac{\partial u_2}{\partial x_1} \equiv e_{21} \; , \qquad (5.7)$$

and by the same method:

$$^2\theta \approx \tan^2\theta = \frac{\partial u_1}{\partial x_2} \equiv e_{12} \; . \qquad (5.8)$$

Note that in Fig. 5.2 the displacement gradient components e_{12} and e_{21} are shown unequal. This implies that the transformation involves a rotation of the area element in addition to its changes in size and shape; the rotation is counterclockwise. To find the strain alone, the rotational aspect of the displacement gradient must be removed. For a displacement gradient with infinitesimal (or at least very small) components, this is a simple task. The strain, $\varepsilon_{\alpha\beta}$, is the symmetric and the rotation tensor, $\varpi_{\alpha\beta}$ (the symbol ϖ is read as "script pi"), the antisymmetric part of the generally asymmetric displacement gradient, $e_{\alpha\beta}$. The effect of an antisymmetric two-dimensional displacement gradient, $e_{\alpha\beta} = \varpi_{\alpha\beta}$, $e_{\beta\alpha} = -\varepsilon_{\alpha\beta}$, a physical rigid-body rotation, is shown in Fig. 5.3. (A physical rotation must be distinguished from a rotation of reference coordinates by means of a rotation matrix; the latter may describe a physical rotation, in the sense opposite to the coordinate rotation, if one assumes that the "rotated" coordinates actually remain fixed in space.) In this case

$\partial u_1/\partial x_1 = \partial u_2/\partial x_2 = 0$, the diagonal components of the displacement gradient, are zero, and (apart from the exaggeration of the angle) the transformed line elements forming the hypotenuse in each of the two right triangles enclosing the angle θ are not sensibly longer than the long orthogonal sides Δx_α, because the short orthogonal side is infinitesimally (or at least very) short.

To separate strain from rotation for a given small displacement gradient, we separate its symmetric and antisymmetric parts:

$$e_{\alpha\beta} = \varepsilon_{\alpha\beta} + \varpi_{\alpha\beta} \; , \tag{5.9}$$

where:

$$\varepsilon_{\alpha\beta} = \varepsilon_{\beta\alpha} = \left(e_{\alpha\beta} + e_{\beta\alpha}\right)/2 \; ,$$
$$\varpi_{\alpha\beta} = -\varpi_{\beta\alpha} = \left(e_{\alpha\beta} - e_{\beta\alpha}\right)/2 \; . \tag{5.10}$$

Although it is more difficult to illustrate, infinitesimal strain and rotation in three dimensions follow the same rules as in two, and we can substitute the subscripts i for α and j for β in eqs. (5.9) and (5.10). The three-dimensional *rotation tensor* ϖ_{ij} for small rotations is related to the axial *rotation vector* [see eq. (4.33)] ω_i as follows:

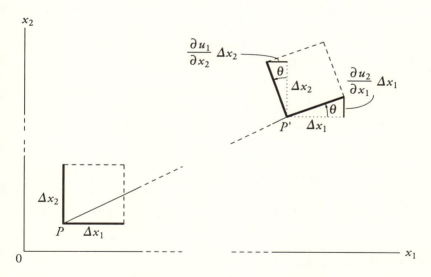

Figure 5.3. Small two-dimensional rigid-body rotation and translation. The original point P is displaced to P', and the original line elements Δx_α (heavy lines) remain orthogonal but form the angle θ with their original orientations (dotted lines).

$$\omega_i = -\epsilon_{ijk}\,\varpi_{jk}/2 \;, \tag{5.11}$$

or explicitly:

$$[\omega_i] = \begin{bmatrix} \varpi_{32} & \varpi_{13} & \varpi_{21} \end{bmatrix} \;, \tag{5.12}$$

where the magnitude ω measures in radians a physical rotation about an axis parallel to ω_i, clockwise seen in the vector's positive direction. In contrast to the rotation tensor, however, the rotation vector is suitable for arbitrarily large rotations. Another method of mathematically describing large rotations will be mentioned in the chapter on finite strain.

Problem 98. A small deformation of a certain crystal is defined by the displacement gradient:

$$[e_{ij}] = \begin{bmatrix} 8 & -1 & -1 \\ 1 & 6 & 0 \\ -5 & 0 & 2 \end{bmatrix} \times 10^{-6}.$$

(a) Determine the strain $[\varepsilon_{ij}]$ and the rotation $[\varpi_{ij}]$. Find (b) the principal magnitudes ε_1, ε_2, and ε_3 and their directions, and (c) the axis, angle, and vector for the rotation $[\varpi_{ij}]$.

The displacement gradient in this problem is, within the specified domain, independent of position and thus homogeneous.

Problem 99. Show that the displacement gradient $[e_{ij}]$ for a body undergoing a small rotation φ but no strain (a so-called rigid-body rotation) is an antisymmetric tensor, whether or not the axis of rotation passes through the origin.

Problem 100. If the axis of a small rotation of a rigid body passes through the origin of the reference coordinate system, show that the displacement u_i at the point x_i is related to the displacement gradient tensor $[e_{ij}]$ by the equation:

$$e_{ij} = u_{i,j} \;.$$

Problem 101. Show that the displacement gradient e_{ij} for a small rigid-body rotation is homogeneous.

Problem 102. Let a body rotate by a small angle φ about an axis parallel to the unit vector l_i, which passes through the point x_i. Find the corresponding displacement gradient tensor e_{ij}.

If, in contrast to those in the previous problems, the local displacement gradient components within a given region are functions of position, the displacement gradient field specified by nine functions is inhomogeneous, and so is the resulting strain field specified by six functions.   Not all sets of six functions, however, can be components of a strain tensor. Mathematically, this is because the six components are the derivatives of only three components of displacement. The conditions that must be satisfied to ensure the validity of a strain tensor are the *compatibility equations*. The physical necessity for them can be understood by realizing that the strain tensor specifies the strain in the neighborhood of the material point, or *particle* identified by the position vector x_i and that all local deformed neighborhoods must fit together to form a continuous region free of gaps, overlaps, or abrupt displacements across faultlike surfaces.

To derive the compatibility conditions, we investigate what ensures that the displacements u_i calculated from the strains along one curve of material points between two specified particles, say, ^oP and P', are the same as displacements u'_i calculated along any other curve between the same particles.   The defining equation for strain (5.10), extended to three dimensions, can be written as:

$$u_{i,j} + u_{j,i} = 2\varepsilon_{ij} , \qquad (5.13)$$

where we consider the displacements, their space derivatives, and thus the strain components to be prescribed functions of the coordinates x_i. Inasmuch as eq. (5.13) is a system of six partial differential equations for the determination of only three components of displacement u_p certain conditions must be fulfilled for the components ε_{ij} to be valid.

Let ^cu_i be the displacement at some point $^cP(^cx_i)$ on a simple, continuous curve C joining the points $^oP(^ox_i)$ and $P'(x'_i)$ within a simply connected region in a deformed body; and let s be the arc length along C measured from some arbitrary point ^aP on C (Fig. 5.4). If at the point $^cP, s = {}^cs$, then at the arc distance $^cs + ds$ on the curve the displacement is $^cu_i + (\partial u_i/\partial s) ds$.   The displacement changes by $(\partial u_i/\partial s) ds$ for each increment of length along the curve, ds. Thus the total change of displacement Δu_i from ^oP to P' is:

$$\Delta u_i = \int_{o_s}^{s'} \frac{\partial u_i}{\partial s} ds ; \qquad (5.14)$$

and, if the displacement at ^oP is ^ou_i, then the displacement at P' is given by:

$$u'_i = {}^ou_i + \int_{o_s}^{s'} \frac{\partial u_i}{\partial s} ds ; \qquad (5.15)$$

or else by:

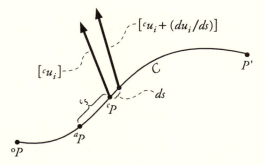

Figure 5.4. Change of displacement at the point ^cP as it slides along the curve C from ^oP to P'. The arc length s along the curve is measured from an arbitrary point ^aP.

$$u'_i = {}^ou_i + \int_{^oP}^{P'} u_{i,j}\, d x_j$$
$$= {}^ou_i + \int_{^oP}^{P'} \varepsilon_{ij}\, d x_j + \int_{^oP}^{P'} \varpi_{ij}\, d x_j \ . \tag{5.16}$$

Substitution into the last integral of $d(x'_j - x_j)$ for dx_j and integration by parts yields:

$$\int_{^oP}^{P'} \varpi_{ij}\, d x_j = \left(x'_j - {}^ox_j\right){}^o\varpi_{ij} + \int_{^oP}^{P'} \left(x'_j - x_j\right) \varpi_{ij,k}\, d x_k \ , \tag{5.17}$$

and thus:

$$u'_i = {}^ou_i + \left(x'_j - {}^ox_j\right){}^o\varpi_{ij} + \int_{^oP}^{P'} \left\langle \varepsilon_{ik} + \left(x'_j - x_j\right) \varpi_{ij,k} \right\rangle d x_k \ . \tag{5.18}$$

Further, $\varpi_{ij,k}$ can be expressed in terms of the strain components and their derivatives:

$$\varpi_{ij,k} \equiv \frac{\partial \left(u_{i,j} - u_{j,i}\right)/2}{\partial x_k}$$
$$= \left(u_{i,jk} - u_{j,ik}\right)/2 + \left(u_{k,ij} - u_{k,ij}\right)/2$$
$$= \frac{\partial \left(u_{i,k} + u_{k,i}\right)/2}{\partial x_j} - \frac{\partial \left(u_{j,k} + u_{k,j}\right)/2}{\partial x_i}$$
$$= \varepsilon_{ik,j} - \varepsilon_{jk,i} \ . \tag{5.19}$$

Note that the last parenthetical expression in the first line is zero but has been inserted to allow the rearrangement shown in the second line (the order of differentiation is arbitrary). Substitution of eq. (5.19) into (5.18) gives the displacement at the point P' in terms of the displacement $^{\circ}u_i$ and rotation $^{\circ}\varpi_{ij}$ at the point $^{\circ}P$ and the components of strain ε_{ij}:

$$u'_i = {}^{\circ}u_i + \left(x'_j - {}^{\circ}x_j\right){}^{\circ}\varpi_{ij}$$
$$+ \int_{{}^{\circ}P}^{P'} \left\langle \varepsilon_{ik} + \left(x'_j - x_j\right)\left(\varepsilon_{ik,j} - \varepsilon_{jk,i}\right)\right\rangle dx_k .$$ (5.20)

Thus, although the displacements cannot be uniquely determined unless $^{\circ}u_i$ and $^{\circ}\varpi_{ij}$ are known, the mutual fit of neighboring small regions is not affected by the integration constants of eq. (5.20), which by themselves imply only a rigid-body translation and rotation of the region. Thus all we have to investigate is whether the u_i are single-valued. A necessary and sufficient condition that u_i be single-valued is that it be independent of the path of integration; that is, the integrand must be an *exact differential.* Let the integrand be denoted by:

$$F_{ik} dx_k = \left\langle \varepsilon_{ik} + \left(x'_j - x_j\right)\left(\varepsilon_{ik,j} - \varepsilon_{jk,i}\right)\right\rangle dx_k ,$$ (5.21)

then, to be an exact differential it is necessary and sufficient that:

$$F_{il,k} - F_{ik,l} = 0 ,$$ (5.22)

or explicitly:

$$\varepsilon_{il,k} - \delta_{jk}\left(\varepsilon_{li,j} - \varepsilon_{jl,i}\right) - \varepsilon_{ik,l} + \delta_{jl}\left(\varepsilon_{ki,j} - \varepsilon_{jk,i}\right)$$
$$+ \left(x'_j - x_j\right)\left(\varepsilon_{li,kj} - \varepsilon_{jl,ik} + \varepsilon_{jk,il} - \varepsilon_{ki,jl}\right) = 0 .$$ (5.23)

Use of the substitution property of the Kronecker delta and of the fact that ε_{ij} is symmetric shows that the first line of eq. (5.23) is zero; therefore, inasmuch as this equation must be true for all x_i it follows that:

$$\varepsilon_{li,jk} - \varepsilon_{jl,ik} + \varepsilon_{jk,il} - \varepsilon_{ki,jl} = 0 ,$$ (5.24)

or, by rearrangement to its usual form:

$$\varepsilon_{ij,kl} + \varepsilon_{kl,ij} = \varepsilon_{ik,jl} + \varepsilon_{jl,ik} .$$ (5.25)

Many of these 81 *compatibility equations* are identical or tautological. Only six of them are *essential,* and they are:

$$\varepsilon_{11,23} + \varepsilon_{23,11} = \varepsilon_{31,12} + \varepsilon_{12,31} \; ,$$

$$\varepsilon_{22,31} + \varepsilon_{31,22} = \varepsilon_{12,23} + \varepsilon_{23,12} \; ,$$

$$\varepsilon_{33,12} + \varepsilon_{12,33} = \varepsilon_{23,31} + \varepsilon_{31,23} \; ,$$

$$2\varepsilon_{12,12} = \varepsilon_{11,22} + \varepsilon_{22,11} \; ,$$

$$2\varepsilon_{23,23} = \varepsilon_{22,33} + \varepsilon_{33,22} \; ,$$

$$2\varepsilon_{31,31} = \varepsilon_{33,11} + \varepsilon_{11,33} \; . \tag{5.26}$$

If we accepted as a description of the strain field $\varepsilon_{ij} = (u_{i,j} + u_{j,i})/2$ in some region an arbitrary set of functions, one that violates the compatibility equations (5.26), we should expect consequences like those shown diagrammatically in Fig. 5.5. Let Fig. 5.5a represent an original triangle somewhere in the region. Integrating the inappropriate functions starting at B and proceeding along the edges of the triangle toward both A and D, these endpoints may either gape as in Fig 5.5b, or the "deformed" edges of the triangle may overlap as in Fig. 5.5c.

Before we begin to apply the compatibility equations to specific cases, we briefly return to a corollary of the statement eq. (5.16) about the displacement at an arbitrary point in a displacement gradient field defined in cartesian coordinates. Where this field is homogeneous, that is, where the displacement derivatives $e_{ij} \equiv u_{i,j}$ are independent of position, the integrands of eq. (5.16) are constants. If we further take the integration constant $^{\circ}u_i$ of that equation to represent the displacement at the coordinate origin, eq. (5.16) simplifies to:

$$u_i = {^{\circ}u_i} + u_{i,j}\, x_j \equiv e_{ij}\, x_j \; . \tag{5.27}$$

If, furthermore, the components of e_{ij} are infinitesimal and thus allow separation into the symmetric strain and antisymmetric rotation tensors (the latter, if homogeneous, representing rigid-body rotations, as we know from the results of Problems 99 to 102), we obtain:

$$u_i = {^{\circ}u_i} + \varpi_{ij}\, x_j + \varepsilon_{ij}\, x_j \; . \tag{5.28}$$

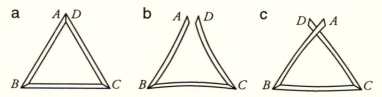

Figure 5.5. An original triangle ($D = A$), B, C (a) is transformed into configurations (b) and (c) according to functions that violate the compatibility conditions. (After Fung, 1965.)

Problem 103. Find the displacement of material points in a body for which the rigid-body translation and the rotation are zero and in which the strain is described by:

$$\left[\varepsilon_{ij}\right] = \begin{bmatrix} 1 & 2 & 1 \\ 2 & 3 & 1 \\ 1 & 1 & 3 \end{bmatrix} \times 10^{-4}.$$

In contrast to homogeneous strain fields, like that of the preceding problem, which trivially obey the continuity equations, inhomogeneous strain fields, whether given in analytical form or calculated, should be checked for continuity.

Strain does, in general, describe both a change in shape, called *distortion*, and one in volume, called *dilatation*. The dilatation Δ is:

$$\Delta = \varepsilon_{ii} , \tag{5.29}$$

a dimensionless scalar that results from summing the diagonal components of the infinitesimal strain tensor (this sum is commonly called the *trace* of the tensor). It is useful to define also the *linear dilatation* $'\Delta$ or mean normal strain:

$$'\Delta = \Delta/3 \equiv \varepsilon_{ii}/3 , \tag{5.30}$$

the relative change in length, in any direction, caused by the dilatation. Subtracting the linear dilatation from each of the normal (diagonal) components of the strain, we obtain the infinitesimal distortion D_{ij}:

$$D_{ij} = \varepsilon_{ij} - \delta_{ij}\varepsilon_{kk}/3 . \tag{5.31}$$

The distortion and linear dilatation are mathematically analogous to the deviator of stress and (negative of) pressure, and isotropic, ideally elastic materials respond to deviatoric stress and pressure by distortion and negative dilatation.

Strain can be classified by type in the same way as stress. Distortion and dilatation are strain types that are analogous to the deviator of stress and to the

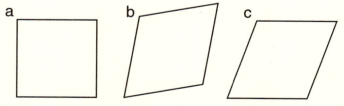

Figure 5.6. Simple shear "strain," a combination of pure strain (from the original shape a to b) and rotation (from b to c).

hydrostatic pressure (except for the sign). A strain can be *uniaxial, biaxial,* or *triaxial.* Another designation for a biaxial strain is *plane strain,* and it is a *pure shear strain* if its dilatation is zero, with the effect of making the two principal values equal but opposite in sign. Somewhat misleadingly, a certain kind of displacement gradient is called *simple shear strain.* It combines pure shear strain with rotation about an axis parallel to the principal axis of zero strain in the special way illustrated in Fig. 5.6.

Problem 104. (a) Find the displacement gradient tensor e_{ij} and the strain tensor ε_{ij} given the displacement vector:

$$\left[u_i \right] = \left[\begin{array}{ccc} x_2^2 x_3 & 2x_1 x_2 x_3 & x_1 x_2^2 \end{array} \right] \times 10^{-5}.$$

(b) Find the strain tensor, the principal strains, and the type of strain at $[x_i] = [2 \ -3 \ 0]$. (c) Find the type of strain everywhere on the plane $x_3 = 0$. (d) Find the dilatation everywhere on the plane $x_1 = 0$. (e) Find the strain tensor, the principal strains, and the type of strain at $[x_i] = [-1 \ 0 \ 3]$. (f) Find the type of strain everywhere on the plane $x_2 = 0$. (g) Find the strain on the x_1 axis. (h) Does the strain tensor ε_{ij} satisfy the equations of compatibility?

Problem 105. (a) Find the dilatation Δ at the point:

$$[x_i] = \left[\begin{array}{ccc} \sqrt{3} & \sqrt{3} & \sqrt{3} \end{array} \right],$$

with arbitrary length units for x_i, in a body in which the displacement field is given by:

$$[u_i] = \left[\begin{array}{ccc} -x_1^2 + 2x_2^2 + 3x_3^2 & 3x_1^2 - x_2^2 + 2x_3^2 & 2x_1^2 + 3x_2^2 - x_3^2 \end{array} \right] \times 10^{-6}.$$

(b) What is the linear dilatation $'\Delta$ at the same point in the x_1 direction?

Problem 106. Find the rotation tensor ϖ_{ij} and the strain tensor ε_{ij} for a body in which the displacement is given by:

$$u_i = \frac{p a^3 x_i}{4 \mu (x_k x_k)^{3/2}},$$

where p, a, and μ are positive constants. Note that spatial differentiation of the position vector x_i yields the Kronecker delta: $\partial x_i / \partial x_j = \delta_{ij}$. *Hint:* Insight into the significance of an equation can in some instances be improved by factoring common elements of several terms.

Problem 107. Prove that, in a homogeneously strained body, rotation is zero everywhere if it is zero at any point.

Problem 108. (a) Prove that, if $b_{ij} x_i x_j = 0$ for all x_i, then $b_{ij} = -b_{ji}$.
(b) Hence, prove that, if b_{ij} is a symmetrical tensor and if $b_{ij} x_i x_j = 0$ for all x_i, then all components of b_{ij} are zero.

From eq. (5.20) we know that one can recover the infinitesimal displacements, apart from an integration constant, from a valid strain tensor. Validity can be checked beforehand by verifying whether the compatibility conditions are satisfied. The integration must proceed along a path without gaps. In practice, this is most conveniently achieved by using a three-dimensional path parallel to the orthogonal coordinate axes. In the following equations, when solving for a particular displacement component u_i, we choose among the three possibilities of letting the subscript on the variable x_i ascend from 1 to 3, modulo 3, that path on which, for each u_i, we last integrate from $^{\circ}x_i$ to x'_i; other orthogonal integration paths can be used and may be more advantageous in specific cases:

$$u_1 = {}^{\circ}u_1 + {}^{\circ}\varpi_{11}(x'_1 - {}^{\circ}x_1) + {}^{\circ}\varpi_{12}(x'_2 - {}^{\circ}x_2) + {}^{\circ}\varpi_{13}(x'_3 - {}^{\circ}x_3)$$

$$+ \int_{\substack{^{\circ}x_2 \\ x_3 = {}^{\circ}x_3 \\ x_1 = {}^{\circ}x_1}}^{x'_2} \left\langle \varepsilon_{12} + (x'_2 - x_2)(\varepsilon_{12,2} - \varepsilon_{22,1}) \right.$$

$$\left. + (x'_3 - {}^{\circ}x_3)(\varepsilon_{12,3} - \varepsilon_{32,1}) \right\rangle dx_2$$

$$+ \int_{\substack{^{\circ}x_3 \\ x_1 = {}^{\circ}x_1 \\ x_2 = x'_2}}^{x'_3} \left\langle \varepsilon_{13} + (x'_3 - x_3)(\varepsilon_{13,3} - \varepsilon_{33,1}) \right\rangle dx_3$$

$$+ \int_{\substack{^{\circ}x_1 \\ x_2 = x'_2 \\ x_3 = x'_3}}^{x'_1} \varepsilon_{11} dx_1 , \tag{5.32}$$

$$u_2 = {}^o u_2 + {}^o\varpi_{21}\left(x'_1 - {}^o x_1\right) + {}^o\varpi_{22}\left(x'_2 - {}^o x_2\right) + {}^o\varpi_{23}\left(x'_3 - {}^o x_3\right)$$

$$+ \int_{\substack{{}^o x_3 \\ x_1 = {}^o x_1 \\ x_2 = {}^o x_2}}^{x'_3} \left\langle \varepsilon_{23} + \left(x'_3 - x_3\right)\left(\varepsilon_{23,\,3} - \varepsilon_{33,\,2}\right) \right.$$

$$\left. + \left(x'_1 - {}^o x_1\right)\left(\varepsilon_{23,\,1} - \varepsilon_{13,\,2}\right) \right\rangle d x_3$$

$$+ \int_{\substack{{}^o x_1 \\ x_2 = {}^o x_2 \\ x_3 = x'_3}}^{x'_1} \left\langle \varepsilon_{21} + \left(x'_1 - x_1\right)\left(\varepsilon_{21,\,1} - \varepsilon_{11,\,2}\right) \right\rangle d x_1$$

$$+ \int_{\substack{{}^o x_2 \\ x_3 = x'_3 \\ x_1 = x'_1}}^{x'_2} \varepsilon_{22}\, d x_2 \;, \tag{5.33}$$

$$u_3 = {}^o u_3 + {}^o\varpi_{31}\left(x'_1 - {}^o x_1\right) + {}^o\varpi_{32}\left(x'_2 - {}^o x_2\right) + {}^o\varpi_{33}\left(x'_3 - {}^o x_3\right)$$

$$+ \int_{\substack{{}^o x_1 \\ x_2 = {}^o x_2 \\ x_3 = {}^o x_3}}^{x'_1} \left\langle \varepsilon_{31} + \left(x'_1 - x_1\right)\left(\varepsilon_{31,\,1} - \varepsilon_{11,\,3}\right) \right.$$

$$\left. + \left(x'_2 - {}^o x_2\right)\left(\varepsilon_{31,\,2} - \varepsilon_{21,\,3}\right) \right\rangle d x_1$$

$$+ \int_{\substack{{}^o x_2 \\ x_3 = {}^o x_3 \\ x_1 = x'_1}}^{x'_2} \left\langle \varepsilon_{32} + \left(x'_2 - x_2\right)\left(\varepsilon_{32,\,2} - \varepsilon_{22,\,3}\right) \right\rangle d x_2$$

$$+ \int_{\substack{{}^o x_3 \\ x_1 = x'_1 \\ x_2 = x'_2}}^{x'_3} \varepsilon_{33}\, d x_3 \;. \tag{5.34}$$

Problem 109. (a) Does the following strain tensor satisfy the equations of compatibility?

$$[\varepsilon_{ij}] = \begin{bmatrix} 6x_1 x_2 & 3x_1^2 & 0 \\ 3x_1^2 & 3x_2^2 & 0 \\ 0 & 0 & 2x_3 \end{bmatrix} \times 10^{-6}.$$

(b) Find the displacement vector u_i for this strain if it acts in a body for which the translation and rotation of the material at the origin are zero.
(c) Find the components of the rotation ϖ_{ij}.

Problem 110. The strain in a certain body is given by:

$$[\varepsilon_{ij}] = \begin{bmatrix} 0 & \alpha \sin(2\pi x_1/\lambda) & 0 \\ \alpha \sin(2\pi x_1/\lambda) & 0 & 0 \\ 0 & 0 & 0 \end{bmatrix},$$

where $\alpha^2 \ll 1$. (a) Does ε_{ij} satisfy the equations of compatibility? Given further that both translation and rotation at the origin are zero, find (b the displacement u_i and (c) the rotation ϖ_{ij} everywhere. (d) Given, on the other hand, that at the origin the displacement is:

$$[\overset{\circ*}{u}_i] = [\gamma_1 \quad \gamma_2 \quad 0],$$

and a clockwise rotation through a small angle φ about an axis:

$$[l_i] = [0 \quad 0 \quad 1],$$

find the displacement $\overset{*}{u}_i$ and the rotation $\overset{*}{\varpi}_{ij}$ everywhere.

If the strain in a material region changes with time, or if it is preferable to consider a reference region through which different material moves in time, it is more convenient to deal with velocities of particles than with their positions and displacements. Let v_i be the velocity of a particle at the point x_i and at a given time. Consider also the velocity, at the same time, of a neighboring particle at $x_i + dx_i$; then the difference between the two velocities is:

$$dv_i = \frac{\partial v_i}{\partial x_j} dx_j. \tag{5.35}$$

In this equation the vectors dv_i and dx_i are related by the *velocity-gradient tensor* $\partial v_i/\partial x_j$; its symmetric part is the *strain rate tensor:*

$$\dot{\varepsilon}_{ij} = \frac{1}{2}\left(\frac{\partial v_i}{\partial x_j} + \frac{\partial v_j}{\partial x_i}\right),\tag{5.36}$$

and its antisymmetric part the *rotation rate tensor:*

$$\dot{\omega}_{ij} = \frac{1}{2}\left(\frac{\partial v_i}{\partial x_j} - \frac{\partial v_j}{\partial x_i}\right).\tag{5.37}$$

These two tensors have exactly the same properties as the infinitesimal strain and rotation tensors, except that their components are not restricted to infinitesimal values; they may be expressed in terms of the sometimes large strains that would result if the instantaneous strain rate were to be maintained for the whole interval of the chosen time unit, whether it be a second or a year. The strain occurring over the instant to which a time interval is reduced by the differentiation remains infinitesimally small (unless the strain rate is infinitely large).

Problem 111. The strain rate tensor for the slow viscous flow of a sheet of Newtonian liquid of depth h down an indefinite plane bed inclined at an angle θ is given by:

$$[\dot{\varepsilon}_{ij}] = \frac{\rho g x_3 \sin\theta}{2\mu}\begin{bmatrix} 0 & 0 & -1 \\ 0 & 0 & 0 \\ -1 & 0 & 0 \end{bmatrix},$$

where ρ and μ are, respectively, the density and viscosity of the liquid, and g is the acceleration due to gravity. For convenience, the coordinate origin is chosen at the upper surface of the liquid, the x_1 coordinate is parallel to that surface and directed down its slope, and the x_3 coordinate is perpendicular to the surface and points downward into the liquid sheet. At the point:

$$[^{\circ}x_i] = \begin{bmatrix} a & b & h \end{bmatrix},$$

the velocity of flow is zero, and the rotation rate is:

$$[^{\circ}\dot{\omega}_i] = \frac{\rho g h \sin\theta}{2\mu}\begin{bmatrix} 0 & 0 & -1 \\ 0 & 0 & 0 \\ 1 & 0 & 0 \end{bmatrix}.$$

Find the velocity v_i and the rotation rate $\dot{\omega}_{ij}$ everywhere in the liquid. *Hint:* Equation (5.19) is useful.

Problem 112. (a) Verify that the equation:

$$\dot{\varepsilon}_{ij} = A\,\sigma_{ij} + B\,\sigma_{kk}\,\delta_{ij} + C\,\delta_{ij}\;,$$

is invariant to a coordinate rotation. (b) Assume that it is the most general form of a linear *flow law* connecting strain rate and stress in an isotropic fluid without memory and that all strain rate components vanish when all stress components are zero. Find the form of the law when the stress is specified in terms of stress deviator and pressure. (c) Find the form giving distortion rate and rate of dilatation in terms of stress deviator and pressure. (d) Show that this form is really two separate equations, and state them separately. (e) Discuss whether the equation involving pressure is physically reasonable. (f) Find the form of the law for an *incompressible* fluid (that is, one that does not change volume during deformation).

∽

6
FINITE STRAIN

Before dealing with finite strain, we demonstrate an important property of the determinant of a three by three matrix the rows of which are vectors.

Problem 113. Show that for three vectors, u_i, v_i, and w_i, the determinant:

$$|V_{ij}| = \begin{vmatrix} u_1 & u_2 & u_3 \\ v_1 & v_2 & v_3 \\ w_1 & w_2 & w_3 \end{vmatrix} ,$$

is invariant to a rotation of axes.

The result can be used to calculate the volume of a general parallelepiped (a solid bounded by parallelograms). The following problems also introduce the concept of dual position vectors, an original set, a_i, and a final set, x_i, where one set is a function of the other. This concept may alternatively be taken to produce two distinct cartesian coordinate systems, a concept used throughout this chapter. The description of a transformation from coordinates a_i to x_i, is designated as *Lagrangian* if it is referred to the system of original coordinates a_i and as *Eulerian* if it is referred to the system of final coordinates x_i. The origins and orientations of the two coordinate systems are independent of each other. For certain purposes, however, it is useful to let the two coordinate systems coincide in origin and orientation, being distinct only by the fact that they indicate positions before and after some event (such as a deformation).

Problem 114. (a) Show that the volume V of a parallelepiped the edges of which are defined by three noncoplanar vectors ^kA_i is given by the dot product of one of the vectors with the cross product of the other two, that is, by:

$$V = \left| {}^kA \right| .$$

Hint: The results of Problem 55 are useful for this task. (b) Using the result of (a), show that the dilatation $\Delta = \Delta V / V$ of the material carried from the point a_i to the point $x_i(a_1, a_2, a_3)$ is given by:

$$\Delta = \left| \partial x_i / \partial a_i \right| - 1 ,$$

in terms of the initial coordinates a_i of material points, and (c) show that the dilatation of material carried to the point x_i from the point $a_i(x_1, x_2, x_3)$ is:

$$\Delta = \left| \partial a_i / \partial x_i \right|^{-1} - 1 \ ,$$

in terms of the final coordinates x_i of material points.

Problem 115. Using the results of Problem 114, show that for an infinitesimal strain ε_{ij} the dilatation is:

$$\Delta = \varepsilon_{ii} \ .$$

Let a material point before and after a deformation have coordinates (position vectors) a_i and x_i, respectively, in coinciding cartesian coordinate systems, and let the deformation of a material region be described either as a function of the original positions:

$$x_i = x_i(a_1, a_2, a_3) \ , \tag{6.1}$$

for which the *Jacobian*, $|\partial x_i/\partial a_j|$ [as in eq. (A.114.12)], must be *nonzero* and *finite* for continuity and *positive* to ensure that the volume does not contract beyond zero:

$$\left| \partial x_i / \partial a_j \right| > 0 \ , \tag{6.2}$$

or conversely as a function of the final positions:

$$a_i = a_i(x_1, x_2, x_3) \ , \tag{6.3}$$

where:

$$\left| \partial a_i / \partial x_j \right| > 0 \ . \tag{6.4}$$

Equations (6.1) and (6.3) express not only the strain (changes in size and shape) of the described region and all its subregions but also any changes they may have undergone in position and orientation. To find the strain alone involves abstraction of only the necessary information.

Consider an arbitrarily small triangle defined by three material points at its apices. The local strain, but not the change in position and orientation, is completely determined by the length changes of its sides; and the strain in a region is similarly determined by the length changes of the sides of all possible, arbitrarily small, material triangles; and these in turn can be found if the changes in distance between all possible pairs of arbitrarily close material points are known. Thus the analysis of strain is concerned with the change of distance between pairs of points.

The square of the distance ds from the point a_i to the neighboring point $a_i + da_i$ in the undeformed material is:

$$ds^2 = da_k\, da_k \,, \tag{6.5}$$

and the square of the corresponding distance dS in the deformed material:

$$dS^2 = dx_k\, dx_k \,. \tag{6.6}$$

Hence:

$$ds^2 = \frac{da_k}{dx_i}\frac{da_k}{dx_j}\, da_i\, da_j \,, \tag{6.7}$$

and:

$$dS^2 = \frac{dx_k}{da_i}\frac{dx_k}{da_j}\, dx_i\, dx_j \,. \tag{6.8}$$

Thus the changes in the squares of the lengths are:

$$dS^2 - ds^2 = \left(\frac{dx_k}{da_i}\frac{dx_k}{da_j} - \delta_{ij}\right) da_i\, da_j \,, \tag{6.9}$$

or:

$$dS^2 - ds^2 = \left(\delta_{ij} - \frac{da_k}{dx_i}\frac{da_k}{dx_j}\right) dx_i\, dx_j \,. \tag{6.10}$$

The actual change of length can be calculated from these equations. Note that the change in distance is a scalar (all subscripts are dummy subscripts), that da_i and dx_i are vectors, and that accordingly the quantities in parentheses are tensors of the second rank. From them we derive two new tensors capable of describing strain, even if it be large; they are *Green's tensor* $\mathfrak{E}_{ij}$ defined as:

$$\mathfrak{E}_{ij} = \frac{1}{2}\left(\frac{dx_k}{da_i}\frac{dx_k}{da_j} - \delta_{ij}\right), \tag{6.11}$$

and *Almansi's tensor* e_{ij} defined as:

$$e_{ij} = \frac{1}{2}\left(\delta_{ij} - \frac{da_k}{dx_i}\frac{da_k}{dx_j}\right). \tag{6.12}$$

Green's tensor, which measures the change of the *initial* distance between the neighboring points a_i and $a_i + da_i$, and eq. (6.9) from which it is derived involve differentiation with respect to the original material coordinates a_j; this makes Green's tensor one of the class of strain tensors that are called *Lagrangian*. Almansi's tensor, on the other hand, which measures the change of the *final* distance between the neighboring points x_i and $x_i + dx_i$, and eq. (6.10) from which it is derived involve differentiation with respect to the deformed material coordinates x_j; this makes Almansi's tensor one of the class of strain tensors called *Eulerian*.

Note that expressions occurring in eqs. (6.11) and (6.12) are commonly referred to as *Cauchy's tensors*. The *Lagrangian Cauchy's tensor* is:

$$\mathfrak{C}_{ij} \equiv \frac{\partial x_k}{\partial a_i} \frac{\partial x_k}{\partial a_j} \; ,$$

and the *Eulerian Cauchy's tensor*:

$$c_{ij} \equiv \frac{\partial a_k}{\partial x_i} \frac{\partial a_k}{\partial x_j} \; .$$

Problem 116. Using the results of Problem 114, show that the dilatation for finite strain is given by:

$$\Delta = \left| \delta_{ij} + 2 \, \mathfrak{C}_{ij} \right|^{1/2} - 1 = \left| \delta_{ij} - 2 \, e_{ij} \right|^{-1/2} - 1 \; ,$$

in which the square roots are taken to be positive .

A collection of additional equations for the calculation of finite dilatation can be found in the *Summary of Formulæ* .

Whereas the definitions of Green's and Almansi's tensor's in eqs. (6.11) and (6.12) are independent of the choice of reference system for material coordinates a_i and x_i, their definition in terms of displacements $u_i = x_i - a_i$ does require reference to a single set of coordinate axes:

$$\mathfrak{C}_{ij} = \frac{1}{2} \left(\frac{d u_i}{d a_j} + \frac{d u_j}{d a_i} + \frac{d u_k}{d a_i} \frac{d u_k}{d a_j} \right) , \qquad (6.13)$$

and:

$$e_{ij} = \frac{1}{2} \left(\frac{d u_i}{d x_j} + \frac{d u_j}{d x_i} - \frac{d u_k}{d x_i} \frac{d u_k}{d x_j} \right) . \qquad (6.14)$$

It can be seen that Almansi's tensor e_{ij} of eq. (6.14) reduces to the infinitesimal strain tensor ε_{ij} if the partial derivatives $\partial u_i/\partial x_j$ are infinitesimal:

$$\varepsilon_{ij} = \frac{1}{2}\left(\frac{d u_i}{d x_j} + \frac{d u_j}{d x_i}\right) . \tag{6.15}$$

If, further, the displacements u_i themselves are also infinitesimal, Green's tensor $\mathfrak{E}_{ij}$, too, reduces to the tensor for infinitesimal strain ε_{ij} of eq. (6.15), known to us from Chapter 4. Finite displacements have the effect of mislocating infinitesimal strains derived from Green's tensor in the reference coordinates x_i. The factor of $1/2$ in eqs. (6.11) – (6.14) has the effect of making these two finite strain tensors the equivalents of the conventional infinitesimal-strain tensor.

The strain rate tensor can also be derived from Green's tensor. Let x_i be the coordinates of a particle at some initial time $^{\circ}t$, and $x_i + \Delta u_i$ the final coordinates of the same particle at some time t, shortly after $^{\circ}t$. Let, further, the displacements be a function of time, $u_i = u_i(t)$. Then, for a general, nonlinear displacement history evaluated at x_i, the displacement can be described as:

$$u_i\bigg|_{t=^{\circ}t+\Delta t} = u_i\bigg|_{t=^{\circ}t} + \frac{\partial u_i}{\partial t}\bigg|_{t=^{\circ}t}\Delta t + \frac{\partial^2 u_i}{2\partial t^2}\bigg|_{t=^{\circ}t}(\Delta t)^2 + \cdots , \tag{6.16}$$

and:

$$\Delta u_i\bigg|_{t=^{\circ}t} = u_i\bigg|_{t=^{\circ}t+\Delta t} - u_i\bigg|_{t=^{\circ}t}$$

$$= v_i\bigg|_{t=^{\circ}t}\Delta t + \frac{\dot{v}_i}{2}\bigg|_{t=^{\circ}t}(\Delta t)^2 + \cdots , \tag{6.17}$$

where $v_i = \partial u_i/\partial t$ is the velocity of the particle at the point x_i in space and at the time $^{\circ}t$. From this, the strain rate can be obtained by going to the limit as the time interval Δt goes to zero:

$$\frac{\Delta \mathfrak{E}_{ij}}{\Delta t} \equiv \frac{1}{2\Delta t}\left(\frac{\partial \Delta u_i}{\partial x_j} + \frac{\partial \Delta u_j}{\partial x_i} + \frac{\partial \Delta u_k}{\partial x_i}\frac{\partial \Delta u_k}{\partial x_j}\right)$$

$$\equiv \frac{1}{2\Delta t}\left(\frac{\partial v_i}{\partial x_j} + \frac{\partial v_j}{\partial x_i} + \text{terms in } (\Delta t)^2 \text{and higher powers}\right)$$

$$\rightarrow \dot{\varepsilon}_{ij} \equiv \frac{1}{2}\left(\frac{\partial v_i}{\partial x_j} + \frac{\partial v_j}{\partial x_i}\right), \text{ as } \Delta t \rightarrow 0 . \tag{6.18}$$

Another approach is to differentiate Green's or Almansi's tensor with respect to time and to take the limit as $a_i \to x_i$:

$$\frac{d\mathfrak{E}_{ij}}{dt} \equiv \frac{1}{2}\left(\frac{\partial v_i}{\partial a_j} + \frac{\partial v_j}{\partial a_i} + \frac{\partial v_k}{\partial a_i}\frac{\partial u_k}{\partial a_j} + \frac{\partial u_k}{\partial a_i}\frac{\partial v_k}{\partial a_j}\right)$$

$$\to \dot{\varepsilon}_{ij} \equiv \frac{1}{2}\left(\frac{\partial v_i}{\partial x_j} + \frac{\partial v_j}{\partial x_i}\right), \quad \text{as} \quad a_i \to x_i, \tag{6.19}$$

inasmuch as:

$$\frac{\partial u_k}{\partial a_j} = \frac{\partial x_k}{\partial a_j} - \delta_{kj} \to 0, \quad \text{as} \quad a_i \to x_i. \tag{6.20}$$

We wish to interpret the local *changes of length* of material lines, and the *changes of angles* between such lines, as a consequence of a strain measured by Green's and Almansi's tensor. To do so, consider the two vectors da_i and $d\bar{a}_i$ from point a_i to the neighboring points $a_i + da_i$ and $a_i + d\bar{a}_i$ in the undeformed material. The dot product of these two vectors is:

$$ds\, d\bar{s} \cos\theta = da_i\, d\bar{a}_i, \tag{6.21}$$

where ds and $d\bar{s}$ are the lengths of the two vectors and θ is the angle between them. The corresponding dot product in the deformed material is:

$$dS\, d\bar{S} \cos\Theta = dx_i\, d\bar{x}_i. \tag{6.22}$$

Thus, using eqs. (6.9) and (6.11):

$$dS\, dS \cos\Theta - ds\, d\bar{s}\cos\theta = 2\mathfrak{E}_{ij}\, da_i\, d\bar{a}_j, \tag{6.23}$$

and eqs. (6.10) and (6.12):

$$dS\, dS \cos\Theta - ds\, d\bar{s}\cos\theta = 2e_{ij}\, dx_i\, d\bar{x}_j. \tag{6.24}$$

As a consequence, a right angle $\theta = 90°$ enclosed by original lines with directions l_i and l_i generally changes after deformation into the different angle Θ, enclosed by the deformed lines with directions L_i and L_j. The new Θ is:

$$\Theta = \cos^{-1}\left(2e_{ij}\, L_i\, \bar{L}_j\right). \tag{6.25}$$

A right angle $\Theta = 90°$ enclosed by lines in the deformed body with directions L_i and L_j generally corresponds to a different original angle θ, enclosed by lines with directions l_i and l_j:

$$\theta = \cos^{-1}\left(-2\mathfrak{E}_{ij} l_i \bar{l}_j\right) . \tag{6.26}$$

Let $d\bar{a}_i = da_i$ [note that in this case, or if $d\bar{x}_i = dx_i$, eqs. (6.21) – (6.24) reduce to eqs. (6.5), (6.6), (6.11), and (6.12)], and define the *elongation* (also called *length ratio* or *stretch*) λ as follows:

$$\lambda \equiv dS/ds . \tag{6.27}$$

(Note that Jaeger 1962, Jaeger & Cook 1979, Ramsay 1967, and others use the symbol λ for the *quadratic elongation*, hence for λ^2 in our notation.) Then, from eqs. (6.9) and (6.23):

$$\lambda^2 - 1 = 2\mathfrak{E}_{ij} l_i l_j , \tag{6.28}$$

and thus:

$$\lambda = \left(1 + 2\mathfrak{E}_{ij} l_i l_j\right)^{1/2}, \tag{6.29}$$

where l_i is the unit vector in the direction of da_j; that is, $l_i = da_i/ds$. Similarly, with $d\bar{x}_i = dx_i$, from eqs. (6.10) and (6.24):

$$1 - \lambda^2 = 1/\left(2e_{ij} L_i L_j\right) , \tag{6.30}$$

and thus:

$$\lambda = \left(1 - 2e_{ij} L_i L_j\right)^{-1/2}, \tag{6.31}$$

where L_i is the unit vector in the direction of dx_i; that is, $L_i = dx_i/dS$.

Let Green's tensor $\mathfrak{E}_{ij}$ be referred to principal coordinates and let $^k da_i$ be vectors along the three coordinate axes. Then, taking them two at a time, $\cos\theta = 0$ and thus:

$$\mathfrak{E}_{ij} \, {}^p da_i \, {}^q da_j = 0 , \tag{6.32}$$

hence $\cos\Theta = 0$. Thus the principal directions of $\mathfrak{E}_{ij}$ are mutually perpendicular material directions in the undeformed state that remain perpendicular (though not necessarily with unchanged orientations) in the deformed state. Similarly, the principal directions of e_{ij} are mutually perpendicular material directions in the deformed state that were also perpendicular before being deformed.

Reference of Green's and Almansi's tensors to principal coordinates allows a simple calculation of the principal elongations $^i\lambda$. From the principal values $^i\mathfrak{E}$ of Green's tensor, they are:

$$^i\lambda = \left(1 + 2\,{}^i\mathfrak{G}\right)^{1/2}, \tag{6.33}$$

and from those of Almansi's tensor, ^{i}e:

$$^i\lambda = \left(1 - 2\,{}^i\mathrm{e}\right)^{-1/2}. \tag{6.34}$$

Green's and Almansi's tensors are most suitable for the analytic expression of finite strain in a strain field. They are less useful in dealing with quantitative strain data from individual samples or sample sets. Strain must necessarily be treated as homogeneous within a sample, but usually also throughout the domain for which the sample is taken as representative.

If a finite strain is *homogeneous* throughout a given domain, it reduces to a linear coordinate transformation. Let α_{ij}, γ_{ij}, and A_{ij} be constants related to each other by:

$$\delta_{ij} + \alpha_{ij} = A_{ij}\,, \tag{6.35}$$

and:

$$\delta_{ij} - \gamma_{ij} = A_{ij}^{-1}\,, \tag{6.36}$$

and let A_{ij} transform original coordinates a_i into final coordinates x_i as follows:

$$x_i = A_{ij}\,a_j + \beta_i = \left(\delta_{ij} + \alpha_{ij}\right) a_j + \beta_i\,, \tag{6.37}$$

and:

$$a_i = A_{ij}^{-1} x_j - \beta_i = \left(\delta_{ij} - \gamma_{ij}\right) x_j - \beta_i\,. \tag{6.38}$$

Then, if β_i is taken to be a constant translation, A_{ij} implies because of its constant components a homogeneous strain along with a possible rotation. If and only if the translation β_i is zero does the matrix A_{ij} alone transform a_i into x_i. Whether or not $\beta_i = 0$, Green's and Almansi's tensors are:

$$\mathfrak{G}_{ij} = \left(A_{ki} A_{kj} - \delta_{ij}\right)\big/2 = \left(\alpha_{ij} + \alpha_{ji} + \alpha_{ki}\alpha_{kj}\right)\big/2\,, \tag{6.39}$$

and:

$$\mathrm{e}_{ij} = \left(\delta_{ij} - A_{ki}^{-1} A_{kj}^{-1}\right)\big/2 = \left(\gamma_{ij} + \gamma_{ji} - \gamma_{ki}\gamma_{kj}\right)\big/2\,. \tag{6.40}$$

Differentiating eq. (6.37) with respect to a_j, we find $\partial x_i / \partial a_j = A_{ij}$. From eq. (A.112.13), the homogeneous dilatation implied by the transformation matrix A_{ij} thus is:

$$\Delta \equiv \frac{V - v}{v} = \left| A_{ij} \right| - 1\,. \tag{6.41}$$

a result obviously also compatible with the two solutions of Problem 116.

Problem 117. A body is deformed according to the equations:

$$x_i(a_1, a_2, a_3) = A_{ij} a_j \, ,$$

where $x_i(a_1, a_2, a_3)$ is the final position of the material point initially at a_i and the transformation matrix is:

$$\left(A_{ij}\right) = \begin{pmatrix} 1 & 1 & -4 \\ -3 & -2 & 2 \\ 5 & 3 & 1 \end{pmatrix} .$$

Find (a) Green's strain tensor $\mathfrak{E}_{ij}$, (b) Almansi's strain tensor e_{ij}, and (c) the dilatation Δ.

The rule for linear coordinate transformation encountered in Problem 117 belongs to the general category of transformation matrices that includes also the rotation matrices and the symmetry matrices of Problems 30 and 31; they can be expressed in *matrix notation*. As part of Problem 30 we performed some consecutive transformations but did not formally investigate how to calculate the compound transformation that results from two (or several) consecutive individual transformations. Let a transformation effected by matrix $\mathbb{A}$ be followed by one effected by matrix $\mathbb{B}$, then the compound transformation, the one that transforms directly from the pre-$\mathbb{A}$ state to the post-$\mathbb{B}$ state, is effected by a third matrix $\mathbb{C}$, which is the matrix product of the first two:

$$\mathbb{C} = \mathbb{B}\mathbb{A} \neq \mathbb{A}\mathbb{B} \, . \tag{6.42}$$

$\mathbb{C}$ is the result of *premultiplying* $\mathbb{A}$ with $\mathbb{B}$, a noncommutative product performed by multiplying each consecutive element in the ith row of $\mathbb{B}$ with the corresponding element in the jth column of $\mathbb{A}$ and adding these three products to obtain the element in the ith row and jth column of $\mathbb{C}$. The compound transformation $\mathbb{C}$ can also be said to be the result of *postmultiplying* $\mathbb{B}$ by $\mathbb{A}$, the matrix performing the earlier transformation. The counterpart in subscript notation of the matrix multiplication eq. (6.42) is:

$$C_{ij} = B_{ik} A_{kj} \equiv A_{kj} B_{ik} \, , \tag{6.43}$$

which can be seen to produce the same result as the rule previously stated in words. To ensure correspondence, the factor matrices must be in the proper pre- or postmultiplication order while the free subscripts appear in the same order on both sides of the equation and the dummy subscripts common to the factor matrices are placed nearest each other. Note, however, that in subscript notation the two factors need not be written in this order, as shown to the right of the equivalence sign of eq. (6.43).

In converting expressions or equations from subscripted to matrix notation, a further rule must be obeyed. Because matrix elements are identified by position only, it is necessary to order the free subscripts in an equation in the same sequence throughout; for this purpose, tensors or transformation matrices may have to be replaced by their transposes.

Problem 118. State the following equation in matrix notation:

$$T_{ij} = a_{ik}\, a_{jl}\, U_{mk}\, V_{ml} - W_{ji} \ .$$

Note that a rotation matrix, usually shown as (a_{ij}) in subscript notation, is conventionally given the symbol $\mathbb{R}$ in matrix notation.

Problem 119. Given the equations (a):

$$\mathbb{D} = \mathbb{A}^{\mathsf{T}}\, \mathbb{B}^{\mathsf{T}}\, \mathbb{B}\, \mathbb{A} + \mathbb{C} \ ,$$

and (b):

$$\mathbb{E} = \mathbb{B}\, \mathbb{A}\, \mathbb{A}^{\mathsf{T}}\, \mathbb{B}^{\mathsf{T}} + \mathbb{C} \ ,$$

state them in subscript notation and execute the transpositions by changing the subscript order.

Part of the information conveyed by the two given equations of Problem 119 resides in the order of noncommutative factors, and $\mathbb{D}$ and $\mathbb{E}$ differ only as a consequence of differences in that order. If all the matrices in these equations are taken to be transformation matrices, then the transformation on the left-hand side is the equivalent of a sequence of transformations starting with the rightmost on the right-hand side and proceeding systematically to the left. In correct answers this information will be faithfully portrayed by the choice of subscripts.

Multiplication in subscript notation being commutative, however, and the order of factors therefore being arbitrary, sequences of transformations become less obvious.

If a transformation matrix A_{ij} expresses a finite homogeneous deformation, then the necessarily symmetric product $A_{ki}\, A_{kj}$ is a tensor with magnitudes the positive square roots of which are the elongations:

$$\lambda = \left(A_{ki}\, A_{kj}\, l_i\, l_j\right)^{1/2} = \left(A_{ki}^{-1}\, A_{kj}^{-1}\, L_i\, L_j\right)^{-1/2}. \tag{6.44}$$

If A_{ij} represents only strain and no rotation, then it is symmetric, $^SA_{ij} = {}^SA_{ji}$, and turns into the *stretch tensor*, the magnitudes of which are the elongations:

$$\lambda = {}^SA \equiv {}^SA_{ij}\, l_i\, l_j = \left({}^SA_{ij}^{-1} L_i\, L_j\right)^{-1}.\qquad(6.45)$$

This symmetric transformation matrix (and stretch tensor) has principal directions and values, which in principal coordinates are the three principal elongations:

$$^i\lambda = {}^SA_i = (1 + 2\mathfrak{E}_i)^{1/2} = (1 - 2\mathfrak{e}_i)^{-1/2}.\qquad(6.46)$$

The dilatation in terms of principal elongations is:

$$\Delta = {}^1\lambda\, {}^2\lambda\, {}^3\lambda - 1 = {}^SA_1\, {}^SA_2\, {}^SA_3 - 1\ .\qquad(6.47)$$

Since it is a transformation matrix, the stretch tensor can be matrix multiplied with other transformation matrices, such as a rotation matrix, to yield a new transformation matrix that describes the compound transformation and is, in general, asymmetric. Lack of symmetry of a transformation matrix implies that it describes not only a strain but also a rigid-body rotation. By first extracting Green's or Almansi's tensor from a general transformation matrix, we can determine the part of that transformation that represents strain alone. Using the stretch tensor for the strain so found, we can attempt to factor the compound transformation into a stretch tensor and a rotation matrix.

Problem 120. A body is deformed according to the equations $x_i = A_{ij}\, a_j$, where $x_i(a_1, a_2, a_3)$ is the final position of a material point initially at a_i and where the transformation matrix has the elements:

$$\left(A_{ij}\right) = \begin{pmatrix} 3.103\ 716\ 5 & 1.998\ 225\ 3 & -0.611\ 587\ 9 \\ -0.324\ 590\ 4 & -0.611\ 587\ 9 & 4.529\ 967\ 7 \\ 5.026\ 090\ 4 & 3.103\ 716\ 5 & -0.324\ 590\ 4 \end{pmatrix}.$$

(a) Find Green's strain tensor $\mathfrak{E}_{ij}$. (b) Determine its principal values and orientations, specified as the rotation *from* principal *to* original coordinate axes, as expressed in terms of a rotation matrix and the corresponding set of rotation angles. (c) Find the three principal elongations $^i\lambda$ and inverse elongations $^i\lambda^{-1}$ implicit in Green's tensor and the dilatation Δ. (d) Find the transformation matrix, or stretch tensor, S_{ij} that is due to strain only, and its inverse. (e) Perform checks of the numerical work. (f) Find the matrix

a_{ij} for the rotation implicit in A_{ij}, assuming that it followed the strain.
(g) Describe that rotation in terms of its axis, angle, and sense.

Problem 121. A body is deformed according to the equations $x_i = B_{ij} a_j$, where $x_i(a_1, a_2, a_3)$ is the final position of a material point initially at a_i and where the transformation matrix has the elements:

$$(B_{ij}) = \begin{pmatrix} -0.324\ 590\ 4 & 5.026\ 090\ 4 & 3.103\ 716\ 5 \\ -0.611\ 587\ 9 & 3.103\ 716\ 5 & 1.998\ 225\ 3 \\ 4.529\ 967\ 7 & -0.324\ 590\ 4 & -0.611\ 587\ 9 \end{pmatrix} .$$

(a) Find Almansi's strain tensor e_{ij} and the dilatation Δ. (b) Find the principal elongations produced by the strain e_{ij} and the transformation matrix S_{ij} that is due to strain only; determine the direction cosines and angles of the principal strains, specified as the rotation from principal to original coordinate axes. (c) find the rotation matrix c_{ij} of a rotation that may have preceded the strain, and describe this rotation with respect to a fixed coordinate system.

Success, in the two previous problems, in resolving one particular asymmetric transformation matrix into a sequence of a strain followed by a rotation, and another into a sequence of a rotation followed by a strain, could mislead us to assume that every asymmetric transformation matrix can be resolved either way. Not so!

Problem 122. (a) Solve Problem 120 again, but under the new assumption that the strain followed rather than preceded the rotation. (b) Similarly, solve Problem 121, but under the assumption that the strain preceded rather than followed the rotation. (c) Resolve which of the two possible assumptions about the temporal order of strain and rotation produces physically interpretable results for the transformation matrix A_{ij} of Problem 117.

We see that the appropriate order of strain and rotation can be found by trial and error. A more systematic procedure will be described in the last chapter. Investigation of some geometric consequences of finite strain may be helpful for later applications.

Problem 123. Show (a) that planes of material points before deformation remain planes after a homogeneous finite strain, and (b) using these results, that straight lines also remain straight lines. (c) Show that original spheres of material points become ellipsoids after a homogeneous strain, and (d) using your results, that ellipsoids become ellipsoids.

Problem 124. (a) Show that, for homogeneous finite strain in which after deformation a sphere given by $c a_i a_i = 1$ becomes an ellipsoid $S_{ij} x_i x_j = 1$ ($S_{ij} = S_{ji}$), the principal axes of Almansi's strain tensor e_{ij} are parallel to those of the ellipsoid. Note that the transformation taking x_i to a_i is of the form $a_i = (\gamma_{ij} + \delta_{ij}) x_j$, neglecting translations, which are immaterial to principal directions. (b) Given your results, what can you say about the principal axes of Green's strain tensor $\mathfrak{E}_{ij}$?

To find the transformation matrix and stretch tensor for distortion only, without volume change, for a homogeneous strain, a general stretch tensor must be normalized. To do so, one divides each stretch tensor component by the linear dilatation, which is the cube root of the volume dilatation:

$$^{D}A_{ij} = {}^{S}A_{ij} \left| {}^{S}A_{pq} \right|^{-1/3}. \tag{6.48}$$

Additional formulæ for finite distortion can be found in the *Summary of Formulæ*.

‹›

7
EFFECTS OF STRESS

The simplest relationship between stress and strain is Hooke's law, describing the linear elastic response of solids to stress. Elastic strain (almost in all cases small) is proportional to the applied stress, with one proportionality factor expressing the relationship between normal, and another that between tangential stress and strain. An ideally elastic strain is completely reversed upon removal of the stress that has caused it. Most materials obey Hooke's law somewhat imperfectly, and that only up to a critical *yield stress* beyond which they begin to flow and to acquire, in addition to the elastic strain, a permanent strain that does not revert upon stress release. Hooke's law in this form is applied to materials that are elastically isotropic, or can be assumed to be approximately so. Crystals, however, never are elastically isotropic, nor are crystalline materials consisting of constituent grains with a distribution of crystallographic orientations that departs from being uniform.

The response of a crystal to a stress (at a level below the yield stress) consists of a strain determined by a matter tensor of the fourth rank, the *compliance tensor* s_{ijkl}:

$$\varepsilon_{ij} = s_{ijkl}\sigma_{kl}, \tag{7.1}$$

the 81 components of which are constants. Any tensor that describes the linear relationship between two tensors of the second rank is necessarily of the fourth rank, and like other tensors of the fourth rank, the compliance tensor can be referred to a new set of reference coordinates by means of a rotation matrix a_{ij}:

$$s'_{ijkl} = a_{im}a_{jn}a_{ko}a_{lp}s_{mnop}. \tag{7.2}$$

The components of the compliance tensor are highly redundant, first because both the stress and the strain tensors are symmetric, and second because the tensor itself is symmetric. The number of independent components for crystals of the lowest, triclinic (both classes) symmetry is 21, and with increasing crystal symmetry the redundancies become more numerous; only three independent compliances are needed to describe the elastic properties of a cubic crystal.

The relationship between stress and strain in elastic solids can be expressed equally well in terms of the fourth-rank *stiffness tensor* c_{ijkl}:

$$\sigma_{ij} = c_{ijkl}\varepsilon_{kl}, \tag{7.3}$$

in which redundancies again reduce the number of independent components to a maximum of 21 for triclinic crystals and to fewer for more highly symmetric crystals and isotropic materials. For isotropic materials, two suffice, such as *Lamé's constants* λ and μ (other pairs of elastic constants are in use, but all can be expressed in terms of λ and μ; see Appendix A).

Equations relating stress to strain and vice versa can be made somewhat more tractable by transforming the stress and strain tensors each into a one-by-six matrix and the stiffness or compliance tensors into a six-by-six matrix. The rules for this transformation are:

i.) Replace pairs of tensor subscripts by single subscripts as follows:

tensor subscript pairs	11	22	33	23 or 32	31 or 13	12 or 21
matrix subscripts	1	2	3	4	5	6

ii.) Multiply tensor strain (but not stress) components with unequal subscripts (to be converted to matrix subscripts 4, 5, and 6) by a factor of 2.

iii.) Multiply tensor compliance (but not stiffness) components by factors of $1/2$ and $1/4$ according to the scheme:

$$s_{mn} \leftarrow s_{ijkl}/2 \text{, if either } m \text{ or } n \text{ is 4, 5, or 6,}$$

$$s_{mn} \leftarrow s_{ijkl}/4 \text{, if both } m \text{ and } n \text{ are 4, 5, or 6.}$$

Thus, for example:

$$s_{1111} \rightarrow s_{11} \text{ ,} \quad s_{2213}/2 \rightarrow s_{25} \text{ ,} \quad s_{3113}/4 \rightarrow s_{55} \text{ ,}$$

$$c_{3322} \rightarrow c_{32} \text{ ,} \quad c_{2311} \rightarrow c_{41} \text{ ,} \quad c_{1212} \rightarrow c_{66} \text{ ,}$$

$$\sigma_{11} \rightarrow \sigma_1 \text{ ,} \quad \sigma_{23} \rightarrow \sigma_4 \text{ ,}$$

$$\varepsilon_{22} \rightarrow \varepsilon_2 \text{ ,} \quad 2\varepsilon_{13} \rightarrow \varepsilon_5 \text{ .}$$

Equations (7.1) and (7.3) then obtain the matrix form:

$$\varepsilon_i = s_{ij}\,\sigma_j \text{ ,} \tag{7.4}$$

and:

$$\sigma_i = c_{ij}\,\varepsilon_j \text{ .} \tag{7.5}$$

For details, see Nye (1964, p. 131). Even so simplified, problems of elastic anisotropy require considerable numerical calculations. They are, therefore, best solved by computer.

An analytically expressed relationship between a strain rate in a material and the stress acting upon it is called a *flow law*. Flow laws may be based on theory; commonly, however, they are found experimentally by exerting on a piece of the material a series of different controlled stress levels and observing the strain rates. The series of paired magnitudes so obtained can then be fitted to an appropriate function. Alternatively, one may impose a series of different controlled strain rates on the experimental material and measure the stress.

An especially simple flow law, the linear one for a Newtonian viscous liquid, was the subject of Problem 111, but most solids and many liquids obey nonlinear flow laws.

Problem 125. Nye (1953) has suggested that the flow law for ice is well represented by the equation:

$$\dot{\varepsilon} = (\sigma/B)^n,$$

where B and n ($n \approx 3$) are constants and where $\dot{\varepsilon}$ is defined by the equations:

$$2\dot{\varepsilon}^2 = \dot{\varepsilon}_{ij}\,\dot{\varepsilon}_{ij}\ ,$$

and:

$$\dot{\varepsilon}_{ii} = 0\ ,$$

and σ by:

$$2\sigma^2 = \Delta_{ij}\,\Delta_{ij}\ ,$$

(a) Make rough graphs of $\dot{\varepsilon}$ versus σ for both ice and a Newtonian liquid ($n = 1$). From inspection of that graph, do ice and the Newtonian liquid become "stronger" or "weaker" with increasing shear stress? (b) Glen (1952) performed uniaxial compression experiments on ice at zero confining pressure. (" *Confining pressure*," a term used by experimentalists, indicates the usually compressive stress acting uniformly on the mantling surface but not the ends of a cylindrical test specimen and is conventionally stated with the positive sign for compressive stress. It is thus distinct from "pressure," the negative of the mean normal stress.) Glen found the rate of shortening $\dot{L}$ to be related to the axial compressive stress S according to the equation:

$$\dot{L} = G L S^q,$$

where L is the axial length of the specimen and G and q are experimental constants. Assuming that ice is incompressible, write out the strain-rate and stress-deviator tensors, and (c) find B and n in terms of G and q. (d) In another set of experiments involving simple ("card deck") shear, Steinemann (quoted by Glen, 1958) found the engineering shear strain rate $\dot{\gamma}$ (the *engineering shear strain* γ and *engineering shear rate* $\dot{\gamma}$ are defined such that they have twice the magnitude of the corresponding tensor components; i.e., $\gamma_{ij} \equiv 2\varepsilon_{ij}$) to be related to the shear stress τ according to the equation:

$$\dot{\gamma} = K\tau^{s},$$

where K and s are experimental constants. Assuming again that ice is incompressible, write out the strain-rate and stress-deviator tensors, and (e) find B and n in terms of K and s. (f) Find also the mutual relationships between the constants in the two different experiments by Glen.

A flow regime may be influenced not only by a flow law but also by specified conditions at the boundaries of the domain under consideration. Both the flow law and the boundary conditions need not be as simple as in Problem 111. The melting or evaporation loss of ice at the surface of a glacier is called *ablation.*

Problem 126. The strain-rate tensor in the ablation region of an ice sheet that flows like a perfectly plastic body is given by:

$$[\dot{\varepsilon}_{ij}] = (A/h)\begin{bmatrix} -1 & -x_2\big/\big(h^2 - x_2^2\big)^{1/2} & 0 \\ -x_2\big/\big(h^2 - x_2^2\big)^{1/2} & 1 & 0 \\ 0 & 0 & 0 \end{bmatrix},$$

in which A is the constant rate of ablation and h the constant depth of the ice. The coordinate origin is, as usual in glaciology, at the surface of the ice; x_1 and x_3 are directed along the surface, and x_2 downward perpendicular to that surface. At the origin the flow velocity is:

$$[^{\circ}u_{i}] = \begin{bmatrix} U & -A & 0 \end{bmatrix},$$

and the rotation is zero. (a) Find the velocity u_i of the ice at a point x_i. *Hint:* determine u_3 first, then u_2, and u_1 last. Note also the following integrals applicable for $t \equiv (h^2 - x^2)^{1/2}$:

$$\int dx/t^3 = x/h^2 t \quad \text{and} \quad \int x^3 \, dx / t^3 = t + h^2/t \ .$$

(b) Find the rate of rotation at that point. (c) Explain the physics of glacier flow at the surface, (d) at the bed, and (e) in between.

When, in some particular material, the tensile normal traction on a surface reaches a critical level, the material's cohesion is overcome, and the material breaks. The critical stress for tensile fracture is commonly called a material's *tensile strength.* Susceptibility to fracture does not exclude flow, as is evident from the existence of *crevasses* in glaciers (by definition, a mass of ice becomes a glacier only if it flows).

Problem 127. The stress near the surface of a deep, narrow valley glacier is given approximately by:

$$
[\sigma_{ij}] = \begin{bmatrix} L & -\rho g\, x_2 \sin\alpha & 0 \\ -\rho g\, x_2 \sin\alpha & 0 & 0 \\ 0 & 0 & -\rho g\, x_3 \cos\alpha \end{bmatrix},
$$

where L is the longitudinal stress, g the acceleration due to gravity, and ρ and α are the density and the surface slope of the ice. The coordinate origin is located at the surface of the glacier midway between the walls, which are planes $x_2 = \pm w$, and the coordinate x_3 is measured downward into the ice perpendicularly to the glacier surface at the plane $x_3 = 0$. (a) Show that, regardless of whether L is tensile (positive) or compressive, one and only one of the principal stresses in the ice will be tensile. (b) Crevasses form perpendicular to the direction of this tensile principal stress. Show that they will be perpendicular to the ice surface. For the three cases (c) $L < 0$, (d) $L = 0$, and (e) $L > 0$, characterize the resulting stresses with Mohr diagrams and sketch the crevasse patterns on the glacier surface.

Failure can occur under other than tensile traction conditions. Many geological materials yield to compressive stress according to the *Coulomb* (1776) *criterion*; that is, if the compressive stress is gradually increased, failure occurs when the shear stress τ across some plane in the material first reaches the value $\tau = {}^{\circ}\tau + \mu\sigma$, where σ is the compressive stress (conventionally taken positive in this context) across the plane, and μ and ${}^{\circ}\tau$ are positive constants termed, respectively, the *coefficient of internal friction* and the *cohesion*. Failure occurs by faulting parallel to this plane.

Problem 128. Suppose that a body of rock that yields according the Coulomb criterion is subjected to a homogeneous stress consisting of the compressive stress S along x_1 and a confining pressure P along x_2 and x_3, and that failure occurs when $S = {}^{f}S$. (a) Write out the stress tensor. (b) Draw Mohr diagrams for $S < {}^{f}S$ and for $S = {}^{f}S$. (c) Find ${}^{f}S$ in terms of P, μ, and ${}^{\circ}\tau$. (d) Find the angle θ between x_1 and the normal to the fault plane. (e) Explain in words and by means of diagrams that $\theta \geq 45°$ for all $\mu \geq 0$.

Various portions of an inhomogeneous material may react differently to a given imposed stress. Certain isolated, large crystals act like rigid bodies in a stress regime that lets the surrounding matrix flow. Commonly the crystals continue to grow in a flowing matrix (Rosenfeld, 1970; Christensen et al., 1989). This can happen with crystals in either metamorphic or igneous rocks.

Problem 129. A garnet crystal nucleates and grows in a rock undergoing regional metamorphism. With respect to the deforming material in which it is embedded, the garnet behaves approximately as a rigid sphere that is carried along passively. If it nucleates in a surface of distinct composition or change of composition, called the *fiducial surface*, that surface may be engulfed by the growing garnet and appear within it as a surface of inclusions. (a) Sketch the fiducial surface both inside and outside a garnet that grew steadily in a rock undergoing uniform simple shear across planes parallel to the undeformed fiducial surface far from the garnet. (b) Show that the apparent angle of rotation Φ of the garnet is equal to $(1/2)\tan\Theta$, where Θ is the *angle of shear* (that is, the angle through which planes of material points originally perpendicular to the shear plane rotated during the deformation). Approach the problem by formulating a strain rate in a coordinate system with its origin far from the garnet and with the x_1 axis parallel to the rotation axis of the garnet, the x_2 axis coinciding with the fiducial plane, and the x_3 axis normal to this plane.

Problem 130. (a) Approach the problem of garnet growth and rotation by formulating Green's strain tensor $\mathfrak{E}_{ij}$ for the far field in original and final coordinate systems with the same orientation as in the previous problem; (b) find the principal directions of the tensor and the elongation for the original direction parallel to a_3; (c) sketch the traces on the a_2–a_3 (and x_2–x_3) plane of those original and final material planes intersecting along a line parallel to a_1 and x_1, which were mutually orthogonal before the strain and are so again after the strain, and identify in this sketch the angle θ from the x_3 coordinate to the long principal strain axis and the angle φ needed to restore the material a_1–a_2 plane to the x_1–x_2 orientation required by a simple-shear history; (d) compare your results with those of Problem 129; (e) sketch the traces, on the x_2–x_3 plane, of an original cube, its strained counterpart, and its strained and rotated counterpart, which accumulate to represent a simple-shear strain with the angle of shear $\Theta = 60°$ and find the numerical principal values, elongations, and directions of Green's tensor for that case.

Problem 131. Some rotated garnets show apparent rotations of 360° and more (the trace of the fiducial plane inside the garnet crystal appears to be wound more than once around the rotation axis). Assuming, as before, simple shear across planes parallel to the fiducial surfaces, what would be the orientation and axial ratios of initially spherical pebbles in the same rocks? Assume that the mechanical properties of the pebbles are the same as those of the matrix. Solve analytically and numerically for an apparent rotation of 360°, using the methods of analytical geometry, in an Eulerian reference frame.

Problem 132. Solve Problem 131 by tensor methods.

As we have seen, petrographic evidence in a rock may constitute a record of the history of relative rotations of constituent grains. Growing megacrysts that progressively incorporate a marker plane while they themselves rotate relatively to their matrix record that rotation. Another kind of petrographic evidence may record the cumulative strain of a rock. In the course of the strain (without overall rotation), individual members of a population of either rod- or plate-shaped marker grains (say hornblende needles or mica flakes) do rotate, and these rotations can serve as a record of the strain if the original distribution of rod or flake orientations is known (Owens, 1973).

The case is simplest when the original distribution of orientations of either of these marker grains was random and thus approximately uniform for all orientations in a large enough sample. March (1932) formulated this for a homogeneously strained material. An originally uniform angular distribution of rod-axis orientations becomes nonuniform, and the new distribution is characterized by the following equation in principal strain coordinates:

$$\varepsilon_i = {}^A\rho_i^{1/3} - 1 \ , \tag{7.6}$$

where the ε_i are the principal, finite, deviatoric strains, but stated conventionally (i.e., as the change in length divided by the original length, $\Delta l/{}^o l$) and where the ${}^A\rho_i$ (prescript A for axes) are the principal angular densities of long marker axes, normalized to make ${}^A\rho_1 {}^A\rho_2 {}^A\rho_3 = 1$. Under similar conditions, an originally uniform angular distribution of the orientations of the poles on plate-shaped, or tabular, markers (prescript P for poles) becomes:

$$\varepsilon_i = {}^P\rho_i^{-1/3} - 1 \ , \tag{7.7}$$

where the pole densities are also normalized to make ${}^P\rho_1 {}^P\rho_2 {}^P\rho_3 = 1$.

Problem 133. Let (S_{ij}) be a symmetric transformation matrix representing a finite homogeneous strain without dilatation, rotation, or translation. (a) Find Green's strain tensor $\mathfrak{E}_{ij}$ and the elongation in the direction l_i in terms of S_{ij}. (b) Apply the strain so defined to a material solid angle (cone), including a material line r_i, from the center of a unit sphere to a surface element on that sphere with area db. This surface element need not have a definite shape but should be compact and small. Express that area db as a fraction of the area of the whole unit sphere, and also the volume of the solid angle cone dv as a fraction of the volume of the whole unit sphere. (c) The transformation of the solid angle changes the length of the material line from $r = 1$ to R. How large is the surface element dB, subtended by the transformed solid angle on a sphere with the radius R? (d) How large is another surface element db', subtended by the transformed solid angle on the original unit sphere, expressed in terms of db and R, and expressed also in terms of db, r_i, and S_{ij}?

The results of this problem are the foundations of the March theory for rod-shaped markers. The following problem provides those for tabular markers.

Problem 134.   Define the orientation of a prestrain plane by the equation $r_i l_i = 0$, in which all vectors r_i are parallel to any plane normal to l_i. Apply to the space containing this plane the same homogeneous strain as in Problem 133 by means of the transformation matrix (S_{ij}). (a) Find the transformed vectors R_i, and let them define the orientation of the deformed planes, in terms of r_i and S_{ij}. (b) Find the normal L_i of the transformed plane and show that it has components proportional to those of a material line l_i, as in Problem 133, but subject to a strain that is the inverse of that effected by S_{ij}.

So far, we have not yet obtained the theories in the form of March's formulas [eqs. (7.1) and (7.2)].

Problem 135.   Assume that a population of marker bodies is embedded in the matrix of a material, with the mechanical properties of markers and matrix indistinguishable from each other; the markers are oriented at random, so that their angular distribution is virtually uniform.   Let this material be homogeneously strained and coordinate axes be the principal strain axes. (a) Assume that the markers are tabular, and show that March's formula (eq. 7.7) applies to the resulting nonuniform angular distribution of the poles on the plates. *Hints*: State the formula for principal elongations instead of principal strains, normalize the original pole density, and use equation (A.133.16). (b) Assume that the markers are rod-shaped, and show that March's formula [eq. (7.6)] describes the resulting nonuniform angular distribution of the rod axes.

The March method of measuring strain in a rock is not sensitive enough to measure a strain so small that it can be treated mathematically as an infinitesimally small strain, and neither is any other method that can be applied to a specimen or at an outcrop. Furthermore, almost all indicators of strain in a rock are cumulative, potentially resulting from a complex strain history. To make geological use of the strain or rotation evidence in a rock, it is therefore necessary to be able to factor it into two or more hypothetical strain increments, say, the strain due to the compaction of a sediment under an accumulating overburden and a later tectonically caused strain. Because both strains are large, they cannot be thought of as being simple summands (finite strains are not superposable); rather, the effect of successive large strain increments must be calculated by an appropriate multiplication of the transformation matrices representing them individually. That is the subject of the last chapter.

ॐ

8

STRAIN HISTORY AND POLAR

DECOMPOSITION

The effect of two consecutive strains (only two states enter into the calculation of a strain, the states before and after, independently of the actual strain path) can be calculated by premultiplying the transformation matrix of the first strain (its stretch tensor) with that of the second. Unless the two strains are *coaxial* (their principal directions coincide), however, the resulting cumulative transformation matrix represents not only a strain but also a rigid-body rotation; in that case the matrix is asymmetric. The method of *polar decomposition* allows one to interpret the combined transformation as if it had come about either by a strain followed by a rotation (*right polar decomposition*) or by a rotation followed by a strain (*left polar decomposition*).

Let $\mathbb{A}$ and $\mathbb{B}$ be two stretch tensors, or transformation matrices, representing each a strain without rotation; and let the strain $\mathbb{B}$ follow the strain $\mathbb{A}$. Then the combined transformation matrix $\mathbb{F}$ is:

$$\mathbb{B}\mathbb{A} = \mathbb{F} = \mathbb{R}\mathbb{U} = \mathbb{V}\mathbb{R} , \qquad (8.1)$$

where $\mathbb{F}$ results from premultiplication of the earlier stretch $\mathbb{A}$ with the later $\mathbb{B}$, where $\mathbb{R}\mathbb{U}$ is the "right" and $\mathbb{V}\mathbb{R}$ the "left" decomposition of $\mathbb{F}$, where $\mathbb{U}$ and $\mathbb{V}$ are two distinct stretch tensors, and where $\mathbb{R}$ is the transformation matrix for a rotation (elements of rotation matrices are indicated by the symbol a_{ij} elsewhere in this book). $\mathbb{F}$ is asymmetric and $\mathbb{R}$ differs from the identity matrix (δ_{ij}) except when $\mathbb{A}$ and $\mathbb{B}$ are coaxial. $\mathbb{U}$ and $\mathbb{V}$ have the same principal stretches and differ by orientation only (Fig. 8.1). In Problems 120 to 122, false approaches in the search for an appropriate decomposition of an asymmetric transformation were recognized by yielding impossible values for a rotation. Application of eq. (8.1) makes such a trial-and-error approach unnecessary.

Given a transformation $\mathbb{F}$, whether $\mathbb{A}$ and $\mathbb{B}$ are known or not, one can solve for the right polar stretch tensor $\mathbb{U}$:

$$\mathbb{U} = \sqrt{\mathbb{F}^{\mathsf{T}}\mathbb{F}} , \qquad (8.2)$$

where the square root of a matrix is understood to be the matrix that, when multiplied with itself, yields the radicand matrix. (For symmetric matrices like

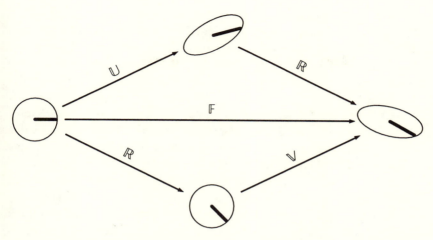

Figure 8.1. Polar decomposition. The transformation $\mathbb{F}$ (arrow from "before" to "after") can be interpreted as either the result of a hypothetical stretch transformation $\mathbb{U}$ followed by the rotation $\mathbb{R}$ (right polar) or that of the rotation $\mathbb{R}$ followed by the stretch $\mathbb{V}$ (left polar). One particular material line in the deformed body is marked by a heavy line.

$\mathbb{F}^T\mathbb{F}$ or $\mathbb{F}\mathbb{F}^T$, this root is most easily extracted by rotating the radicand to principal coordinates, extracting the positive square roots of its principal values, and then restoring the resulting matrices to the reference coordinates.) The solution for the left polar stretch tensor $\mathbb{V}$ is:

$$\mathbb{V} = \sqrt{\mathbb{F}\mathbb{F}^T} \ , \tag{8.3}$$

and the rotation matrix $\mathbb{R}$ is:

$$\mathbb{R} = \mathbb{F}\mathbb{U}^{-1} = \mathbb{V}^{-1}\mathbb{F} \ . \tag{8.4}$$

Because it is a rotation matrix, $\mathbb{R}$ describes a change of reference frame, not a rotation with respect to a fixed frame. As a corollary of eq. (8.1), the right and left compound stretches in terms of the incremental stretch tensors $\mathbb{A}$ and $\mathbb{B}$ are:

$$\mathbb{U} = \sqrt{\mathbb{A}^T\mathbb{B}^T\mathbb{B}\mathbb{A}} \ , \tag{8.5}$$

and:

$$\mathbb{V} = \sqrt{\mathbb{B}\mathbb{A}\,\mathbb{A}^T\mathbb{B}^T} \ . \tag{8.6}$$

Problem 136. Assume the following: An originally horizontal bed of sediment, after having been compacted vertically to 40% of its thickness, was then rigidly tilted so that its dip became 30° to the southwest and its strike 135°. In a certain domain, a tectonically caused strain with N, E, and down principal directions shortened the material by 50% north-south, leaving east-west lengths unchanged but lengthening up-down material lines so as to preserve postcompaction volume. As reference coordinates, use the geographic directions N, E, and down. Let $\mathbb{C}$ be the transformation matrix (and stretch tensor) for the compaction strain, as seen after tilting but before the tectonic event, with respect to geographic coordinates, and let $\mathbb{T}$ be the transformation matrix (and stretch tensor) for the tectonic strain with reference to the same coordinates. (a) Find the transformation matrix $\mathbb{F} = \mathbb{T}\mathbb{C}$ for the combined effect of compaction and tectonic strain. (b) Is $\mathbb{F}$ symmetric or not? (c) Perform the right and left polar decompositions of $\mathbb{F}$ into $\mathbb{U}$, $\mathbb{V}$, and $\mathbb{R}$. (d) Find the final bedding attitude and the angle between pre- and posttectonic bedding. *Note*: Intermediate numerical results must be carried to much higher precision than that expected in the final results. As a rule of thumb, three extra decimal places must be used and rounded only at the end.

The numerical solution of the preceding problem, and of several earlier ones, is laborious; wherever more than just a few such solutions are needed, it is necessary to use a computer. Algorithms for the numerical solution of equations in subscript or matrix notation are relatively easy to program, even for a computer of modest size. Before writing such a program, however, it is useful to have performed the calculation by hand, at least once.

Because this handbook does not aim for completeness, but merely attempts to equip the student for an independent mathematical formulation of problems, and because most real problems encountered by geologists tend to be numerically onerous, the set of problems may as well end here. By now, the readers and problem-solvers have probably become sufficiently acquainted with the power of tensor methods to have started thinking how they could apply them to their own research.

$\wp$

ANSWERS

Answer 1. Three vectors can form a triangle if their sum is the null vector:

$$A + B + C = 0 \ . \tag{A.1.1}$$

Answer 2. (a) Figure 1 shows a vector sum and a vector difference.

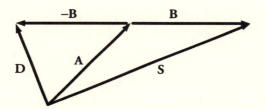

Figure A.2.1. S – sum $A + B$, D – difference
$A - B$.

(b) Define S as the vector sum of two vectors:

$$S \equiv A + B \ , \tag{A.2.1}$$

and D as the vector difference:

$$D \equiv A - B = A + (-B) \ . \tag{A.2.2}$$

Add eqs. (A.2.1) and (A.2.2):

$$2A = S + D \ , \tag{A.2.3}$$

and subtract eq. (A.2.2) from eq. (A.2.1):

$$2B = S - D \ . \tag{A.2.4}$$

Hence the result is:

$$A = (S + D)/2 \ ,$$
$$B = (S - D)/2 \ . \tag{A.2.5}$$

Answer 3. (a) If A is the magnitude of **A** then the unit vector **a** is parallel to **A**, given that:

$$\mathbf{a} = \mathbf{A}/A . \qquad\qquad (A.3.1)$$

Note: The unit vector is also called the *direction vector* considering that it characterizes the common direction of all parallel vectors.

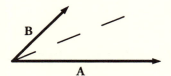

Figure A.3.1.   The dashed line bisects the angle formed by the vectors **A** and **B**.

(b) Let two different vectors **A** and **B** issue from the same point (Fig. A.3.1), then the vector sum **C** of the unit vectors **a** and **b**, parallel to **A** and **B**, is parallel to the bisector (Fig. A.3.2):

$$\mathbf{C} = \mathbf{A}/A + \mathbf{B}/B = \mathbf{a} + \mathbf{b} . \qquad\qquad (A.3.2)$$

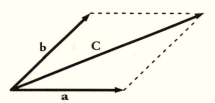

Figure A.3.2.   The vector sum **C** bisects the angle formed by the equal-length unit vectors **a** and **b**.

Answer 4. Prove that $\mathbf{X} = (\mathbf{A} + \mathbf{B})/3$.

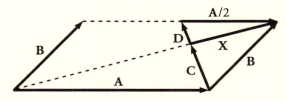

Figure A.4.1.   Trisection of the diagonal in a parallelogram.

By inspection of Fig. A.4.1:

$$C + D = B - A/2 , \qquad (A.4.1)$$

and:

$$C + X = B . \qquad (A.4.2)$$

Now postulate:

$$X = k(A + B) , \qquad (A.4.3)$$

and

$$D = nC , \qquad (A.4.4)$$

where k and n are scale factors. From eq. (A.4.4) follows that:

$$D + C = (n + 1)C . \qquad (A.4.5)$$

Substitute eq. (A.4.5) into eq. (A.4.1):

$$(n + 1)C = B - A/2 , \qquad (A.4.6)$$

and solve for C:

$$C = \frac{B - A/2}{n + 1} , \qquad (A.4.7)$$

eliminate X from eqs. (A.4.2) and (A.4.3):

$$k(A + B) = B - C , \qquad (A.4.8)$$

and substitute eq. (A.4.6) into (A.4.8):

$$k(A + B) = B \frac{n + 1}{n + 1} - B \frac{1}{n + 1} + A \frac{1}{2(n + 1)} . \qquad (A.4.9)$$

Simplify:

$$k(A + B) = B \frac{n}{n + 1} + A \frac{1}{2(n + 1)} . \qquad (A.4.10)$$

Because eq. (A.4.10) must hold for *any* A and B, the following two scalar equations must be true:

$$k = \frac{n}{n + 1} ; \quad k = \frac{1}{2(n + 1)} . \qquad (A.4.11)$$

Hence, by elimination of k from eqs. (A.4.11):

$$\frac{n}{n+1} = \frac{1}{2(n+1)} \ .$$

(A.4.12)

and multiplication of eq. (A.4.12) by $(n+1)$, we obtain n:

$$n = 1/2 \ .$$

(A.4.13)

Substitute eq. (A.4.12) into eq. (A.4.11) for the proof:

$$k = 1/3 \ \therefore$$

(A.4.14)

Alternative Answer 4. A "backhanded" proof consists in postulating that the factor k in eq. (A.4.3) be $1/3$ and proving the truth of equation (A.4.1), *using only this postulate.* Neither may one use the vector triangle formed by the vectors $\mathbf{A}/2$, $\mathbf{B}$, and $(\mathbf{C} + \mathbf{D})$ of the right-hand side of Fig. A.4.1. Restate eq. (A.4.3) as the postulate:

$$\mathbf{X} = (\mathbf{A} + \mathbf{B})/3 \ .$$

(A.4.15)

To obtain $\mathbf{C}$ in terms of $\mathbf{A}$ and $\mathbf{B}$, find by inspection of Fig. A.4.1:

$$\mathbf{C} = \mathbf{B} - \mathbf{X} \ ,$$

(A.4.16)

substitute eq. (A.4.15) into eq. (A.4.16):

$$\mathbf{C} = 3\mathbf{B}/3 - \mathbf{A}/3 - \mathbf{B}/3 \ ,$$

(A.4.17)

and simplify:

$$\mathbf{C} = 2\mathbf{B}/3 - \mathbf{A}/3 \ .$$

(A.4.18)

Calculate $\mathbf{D}$ in an analogous way; by inspection:

$$\mathbf{D} = \mathbf{X} - \mathbf{A}/2 \ ,$$

(A.4.19)

substitute eq. (A.4.15) into eq. (A.4.19):

$$\mathbf{D} = 2\mathbf{A}/6 + 2\mathbf{B}/6 - 3\mathbf{A}/6 \ ,$$

(A.4.20)

and simplify:

$$\mathbf{D} = 2\mathbf{B}/6 - \mathbf{A}/6 \ .$$

(A.4.21)

Add eqs. (A.4.18) and (A.4.21):

$$\mathbf{C} + \mathbf{D} = 4\mathbf{B}/6 + 2\mathbf{B}/6 - 2\mathbf{A}/6 - \mathbf{A}/6 \ ,$$

(A.4.22)

and simplify for comparison with eq. (A.4.1):

$$\mathbf{C} + \mathbf{D} = \mathbf{B} - \mathbf{A}/2 \ \therefore \tag{A.4.23}$$

Answer 5. (a) Three unequal and nonparallel vectors, **A**, **B**, and **C**, are coplanar if and only if:

$$\mathbf{A} \bullet (\mathbf{B} \times \mathbf{C}) = 0 \ . \tag{A.5.1}$$

$\mathbf{B} \times \mathbf{C}$ is parallel to the pole of the plane defined by the two vectors, and the equation ensures that **A** is perpendicular to that pole. (b) Because they fulfill the condition of eq. (A.5.1), the vectors **A**, **B**, and $s(\mathbf{A} + \mathbf{B})$ are coplanar if **A** and **B** are unequal and nonparallel ($\mathbf{A} \neq k\mathbf{B}$) and s and k are arbitrary scale factors.

Answer 6. Given that $\mathbf{A} \times \mathbf{B} = \mathbf{0}$ and $\mathbf{A} \bullet \mathbf{B} = 0$, suppose that neither $A = 0$ nor $B = 0$. Because their cross product is a null vector, **A** is either parallel or antiparallel to **B**. Hence:

$$\mathbf{A} = k\mathbf{B} \ , \tag{A.6.1}$$

and thus:

$$\cos\theta = \pm 1 \ , \tag{A.6.2}$$

where θ is the angle formed by the two vectors. It follows that:

$$\mathbf{A} \bullet \mathbf{B} = kA^2 = 0 \ ; \tag{A.6.3}$$

hence, either $k = 0$ or $A = 0$. Substitute either into eq. (A.6.1):

$$\mathbf{B} = \mathbf{0} \ , \tag{A.6.4}$$

hence:

$$B = 0 \ . \tag{A.6.5}$$

This is contradictory to the assumption, and the hypothesis is *false*. Thus the answer is: *No*.

Answer 7. (a) If three unequal and neither parallel nor antiparallel vectors are coplanar, then the cross product of any two of them must be perpendicular to the third. (b) An example would be the three vectors **A**, **B**, and $(\mathbf{A} + \mathbf{B})$, where $\mathbf{A} \neq k\mathbf{B}$ for an arbitrary k.

Answer 8. We label the three distinct position vectors by left superscripts (and adapt the convention that left superscripts have no mathematical function). Then:

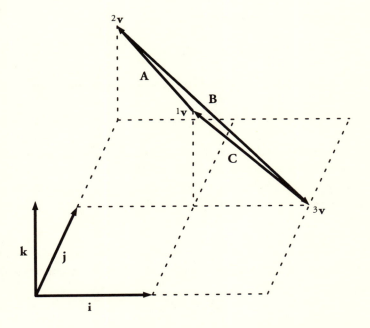

Figure A.8.1.   Vector triangle with vertices defined by position vectors $^{i}\mathbf{v}$.

$$\begin{aligned}
^{1}\mathbf{v} &= \mathbf{i} + \mathbf{j} + \mathbf{k} , \\
^{2}\mathbf{v} &= \quad\;\; 2\mathbf{j} + \mathbf{k} , \\
^{3}\mathbf{v} &= 2\mathbf{i} + \mathbf{j} ,
\end{aligned} \tag{A.8.1}$$

where $\mathbf{i}$, $\mathbf{j}$, and $\mathbf{k}$ are unit vectors along the coordinate axes x_1, x_2, and x_3. By inspection of Fig. A.8.1:

$$\mathbf{A} = {}^{2}\mathbf{v} - {}^{1}\mathbf{v} ; \quad \mathbf{B} = {}^{3}\mathbf{v} - {}^{2}\mathbf{v} ; \quad \mathbf{C} = {}^{1}\mathbf{v} - {}^{3}\mathbf{v} . \tag{A.8.2}$$

It follows that:

$$\begin{aligned}
\mathbf{A} &= -1\mathbf{i} + 1\mathbf{j} + 0\mathbf{k} , \\
\mathbf{B} &= \quad 2\mathbf{i} - 1\mathbf{j} - 1\mathbf{k} , \\
\mathbf{C} &= -1\mathbf{i} + 0\mathbf{j} + 1\mathbf{k} .
\end{aligned} \tag{A.8.3}$$

The sum of eqs. (A.8.3) thus is:

$$\Sigma = 0\mathbf{i} + 0\mathbf{j} + 0\mathbf{k} = \mathbf{0} \quad \therefore \tag{A.8.4}$$

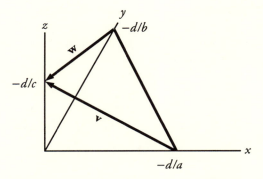

Figure A.9.1. Vectors **v** and **w** defined by coordinate intercepts.

Answer 9. The plane defined by the equation

$$ax + by + cz + d = 0 \tag{A.9.1}$$

has three intercepts (Fig. A.9.1) with the coordinate axes (at an intercept two of the coordinates are zero). They are:

$$ax + d = 0 ,$$
$$by + d = 0 ,$$
$$cz + d = 0 . \tag{A.9.2}$$

Hence at the intercepts:

$$x = -d/a , \quad y = -d/b , \quad z = -d/c . \tag{A.9.3}$$

Thus the vectors in the two vertical coordinate planes **v** and **w** are:

$$\mathbf{v} = \mathbf{i}\,d/a - \mathbf{k}\,d/c ,$$
$$\mathbf{w} = \mathbf{j}\,d/b - \mathbf{k}\,d/c . \tag{A.9.4}$$

Divide both vectors by the scalar d and form the cross product **c** of the scaled vectors:

$$\mathbf{c} = \mathbf{v} \times \mathbf{w} = \begin{vmatrix} \mathbf{i} & \mathbf{j} & \mathbf{k} \\ 1/a & 0 & -1/c \\ 0 & 1/b & -1/c \end{vmatrix} = \mathbf{i}\frac{1}{cb} + \mathbf{j}\frac{1}{ac} + \mathbf{k}\frac{1}{ab} . \tag{A.9.5}$$

Scale the cross product **c** to **C** by multiplying it with abc:

$$\mathbf{C} = \mathbf{i}\,a + \mathbf{j}\,b + \mathbf{k}\,c . \tag{A.9.6}$$

Answer 10. Choose mirrors oriented parallel to the coordinate planes; then reflected beams are oriented like the direct beams, except for a change of sign of the component perpendicular to the mirror plane. Let the beam enter the system of mirrors in the direction **V**, where:

$$\mathbf{V} = \mathbf{i}V_1 + \mathbf{j}V_2 + \mathbf{k}V_3 \ . \tag{A.10.1}$$

On reflection from the mirror at right angles to **i** the new beam is oriented in the **V'** direction:

$$\mathbf{V}' = \mathbf{i}(-V_1) + \mathbf{j}V_2 + \mathbf{k}V_3 \ , \tag{A.10.2}$$

and after reflection from all three mirrors, the reflected beam **V'''** is, for any order of reflections:

$$\mathbf{V}''' = -\mathbf{i}V_1 - \mathbf{j}V_2 - \mathbf{k}V_3 = -\mathbf{V} \ \therefore \tag{A.10.3}$$

Answer 11. Given:

$$\begin{aligned} \mathbf{A} &= \mathbf{i} + 2\mathbf{j} + 3\mathbf{k} \ , \\ \mathbf{B} &= 5\mathbf{i} + 7\mathbf{j} + 11\mathbf{k} \ . \end{aligned} \tag{A.11.1}$$

(a) The dot product of the two vectors is:

$$\mathbf{A} \bullet \mathbf{B} = 5 + 14 + 33 = 52 \ . \tag{A.11.2}$$

(b) Calculate the cross product:

$$\mathbf{A} \times \mathbf{B} = \begin{vmatrix} \mathbf{i} & \mathbf{j} & \mathbf{k} \\ 1 & 2 & 3 \\ 5 & 7 & 11 \end{vmatrix}$$

$$= \mathbf{i}(22 - 21) + \mathbf{j}(15 - 11) + \mathbf{k}(7 - 10) \ , \tag{A.11.3}$$

with the result:

$$\mathbf{A} \times \mathbf{B} = \mathbf{i} + 4\mathbf{j} - 3\mathbf{k} \ . \tag{A.11.4}$$

(c) In the inverse order, the cross product is a different vector:

$$\mathbf{B} \times \mathbf{A} = -\mathbf{A} \times \mathbf{B} = -\mathbf{i} - 4\mathbf{j} + 3\mathbf{k} \ . \tag{A.11.5}$$

Thus the vector **B** × **A** is antiparallel to **A** × **B**, but it has the same magnitude.

Answer 12. The three given vectors are:

$$
\begin{aligned}
{}^1\mathbf{v} &= \mathbf{i} + \mathbf{j} + \mathbf{k} \ , \\
{}^2\mathbf{v} &= \mathbf{i} \qquad\quad - \mathbf{k} \ , \\
{}^3\mathbf{v} &= \mathbf{i} - 2\mathbf{j} + \mathbf{k} \ .
\end{aligned}
\tag{A.12.1}
$$

(a) The cross product of the first two vectors is:

$$
{}^1\mathbf{v} \times {}^2\mathbf{v} =
\begin{vmatrix}
\mathbf{i} & \mathbf{j} & \mathbf{k} \\
1 & 1 & 1 \\
1 & 0 & -1
\end{vmatrix}
= -\mathbf{i} + 2\mathbf{j} - \mathbf{k} = -(\mathbf{i} - 2\mathbf{j} + \mathbf{k}) \ .
\tag{A.12.2}
$$

This cross product is antiparallel with ${}^3\mathbf{v}$. The cross product of the last two vectors is:

$$
{}^2\mathbf{v} \times {}^3\mathbf{v} =
\begin{vmatrix}
\mathbf{i} & \mathbf{j} & \mathbf{k} \\
1 & 0 & -1 \\
1 & -2 & -1
\end{vmatrix}
= -2\mathbf{i} - 2\mathbf{j} - 2\mathbf{k} = -2(\mathbf{i} + \mathbf{j} + \mathbf{k}) \ .
\tag{A.12.3}
$$

This cross product is antiparallel with ${}^1\mathbf{v}$. The cross product of the last with the first vector is:

$$
{}^3\mathbf{v} \times {}^1\mathbf{v} =
\begin{vmatrix}
\mathbf{i} & \mathbf{j} & \mathbf{k} \\
1 & -2 & 1 \\
1 & 1 & 1
\end{vmatrix}
= -3\mathbf{i} + 3\mathbf{k} = -3(\mathbf{i} - \mathbf{k}) \ .
\tag{A.12.4}
$$

This cross product is antiparallel with ${}^2\mathbf{v}$. (b) Hence the three vectors $\mathbf{v}$ are mutually perpendicular.

Answer 13. (a) In subscript notation, the three given vectors are:

$$
\begin{aligned}
[{}^1v_i] &= [\ 1 \quad 1 \quad 1\ ] \ , \\
[{}^2v_i] &= [\ 1 \quad 0 \quad -1\ ] \ , \\
[{}^3v_i] &= [\ 1 \quad -2 \quad 1\ ] \ .
\end{aligned}
\tag{A.13.1}
$$

To determine whether the vectors are orthogonal, use their dot products, D:

$$
D = A_i B_i \equiv A_1 B_1 + A_2 B_2 + A_3 B_3 \ ,
\tag{A.13.2}
$$

or explicitly:

$$^1D = \begin{bmatrix} ^2v_i \end{bmatrix} \bullet \begin{bmatrix} ^3v_i \end{bmatrix} = 1 + 0 - 1 = 0 ,$$

$$^2D = \begin{bmatrix} ^3v_i \end{bmatrix} \bullet \begin{bmatrix} ^1v_i \end{bmatrix} = 1 - 2 + 1 = 0 ,$$

$$^3D = \begin{bmatrix} ^1v_i \end{bmatrix} \bullet \begin{bmatrix} ^2v_i \end{bmatrix} = 1 + 0 - 1 = 0 . \qquad (A.13.3)$$

Note the convention that as left superscript for the resulting dot product one chooses the numeral *not* occurring among the left superscripts of the participating vectors. (b) Given the vectors $^1\mathbf{w} = \mathbf{i} + \mathbf{j} + 2\mathbf{k}$, $^2\mathbf{w} = -\mathbf{i} + x\mathbf{k}$, and $^3\mathbf{w} = 2\mathbf{i} + y\mathbf{j} + z\mathbf{k}$, ensure the orthogonality of $^1\mathbf{w}$ and $^2\mathbf{w}$ by dot-multiplying them and setting the product to zero:

$$\begin{bmatrix} ^1w_i \end{bmatrix} \bullet \begin{bmatrix} ^2w_i \end{bmatrix} = \begin{bmatrix} 1 & 1 & 2 \end{bmatrix} \bullet \begin{bmatrix} -1 & 0 & x \end{bmatrix}$$
$$= -1 + 2x = 0 , \qquad (A.13.4)$$

hence $x = 1/2$. Set the dot product of $^2\mathbf{w}$ with $^3\mathbf{w}$ to zero, substituting $1/2$ for x:

$$\begin{bmatrix} ^2w_i \end{bmatrix} \bullet \begin{bmatrix} ^3w_i \end{bmatrix} = \begin{bmatrix} -1 & 0 & 1/2 \end{bmatrix} \bullet \begin{bmatrix} 2 & y & z \end{bmatrix}$$
$$= -2 + z/2 = 0 , \qquad (A.13.5)$$

hence $z = 4$. Finally, set the dot product of $^3\mathbf{w}$ with $^1\mathbf{w}$ to zero, substituting 4 for z:

$$\begin{bmatrix} ^3w_i \end{bmatrix} \bullet \begin{bmatrix} ^1w_i \end{bmatrix} = \begin{bmatrix} 2 & y & 4 \end{bmatrix} \bullet \begin{bmatrix} 1 & 1 & 2 \end{bmatrix}$$
$$= 2 + y + 8 = 0 , \qquad (A.13.6)$$

hence $x = 1/2$, $y = -10$, $z = 4$.

Answer 14. (a) Let the equation $\mathbf{A} \bullet \mathbf{B} = \mathbf{A} \bullet \mathbf{C}$ be true for a given $\mathbf{A}$. Then:

$$B \cos {}^{AB}\theta = C \cos {}^{AC}\theta . \qquad (A.14.1)$$

Figure A.14.1 provides an example of different vectors $\mathbf{B}$ and $\mathbf{C}$ forming the same dot product with $\mathbf{A}$. Even in this simple case in which all three vectors lie in the same plane, it is demonstrably possible that $\mathbf{C} \neq \mathbf{B}$. In general, in order to form identical dot products with the vector $\mathbf{A}$, vectors $\mathbf{B}$, $\mathbf{C}$, etc., merely must have the same projection onto $\mathbf{A}$; that is, their heads must lie in the same plane orthogonal to $\mathbf{A}$. Hence, the answer is *no*. (b) A statement true for every $\mathbf{A}$ must be true for three different, nonparallel, and noncoplanar examples of $\mathbf{A}$; in that case, the three distinct planes orthogonal to the

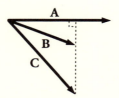

Figure A.14.1. Dot
products of **A** with
coplanar **B** and **C**.

different **A**s intersect in a single point, and the heads of **B** and **C** must both
necessarily lie at that point. Hence **B** = **C**, and the answer is *yes*. (c) In
subscript notation, the equation $\mathbf{A} \cdot \mathbf{B} = \mathbf{A} \cdot \mathbf{C}$ becomes:

$$a_i b_i = a_i c_i . \tag{A.14.2}$$

Subtract the right-hand from the left-hand side:

$$a_i (b_i - c_i) = 0 , \tag{A.14.3}$$

hence:

$$b_i - c_i = 0 . \tag{A.14.4}$$

Thus $b_i = c_i$ and **B** = **C**.

Answer 15. (a) To find a vector **N** of arbitrary magnitude, to be normal to the
plane defined by two vectors **V** and **W**, form their cross product in terms of
their components v_i and w_i:

$$\mathbf{N} = \begin{vmatrix} \mathbf{i} & \mathbf{j} & \mathbf{k} \\ v_1 & v_2 & v_3 \\ w_1 & w_2 & w_3 \end{vmatrix} . \tag{A.15.1}$$

Find the unit vector **n** parallel to **N**:

$$\mathbf{n} \equiv \mathbf{N}/N . \tag{A.15.2}$$

This vector has the components:

$$n_1 = (v_2 w_3 - v_3 w_2)/N ,$$
$$n_2 = (v_3 w_1 - v_1 w_3)/N ,$$
$$n_3 = (v_1 w_2 - v_2 w_1)/N . \tag{A.15.3}$$

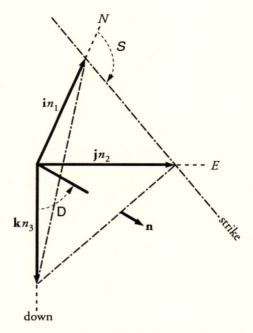

Figure A.15.1. A plane with strike S, dip D,
and pole **n**.

The same task is more succinctly stated in subscript notation with the
summation convention implied:

$$N_i = \epsilon_{ijk} v_j w_k \ ,$$ (A.15.4)

where ϵ_{ijk} is the alternating matrix . The unit normal has components:

$$n_i = N_i / \left(N_j \ N_j \right)^{1/2}.$$ (A.15.5)

(b) Use coordinates $(x_i) = ($ north east down $)$, a set of right-handed coordinates
commonly used in geology and geophysics. By inspection of the horizontal
triangle in Fig. A.15.1, the strike S is:

$$S = \pi - \tan^{-1} \left(n_2 / n_1 \right) = \pi - \tan^{-1} \left(N_2 / N_1 \right) \ .$$ (A.15.6)

The dip D is:

$$D = \cos^{-1} n_3 \ .$$ (A.15.7)

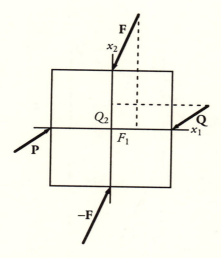

Figure A.16.1. Balance of forces.
Forces **F**, −**F**, **P**, and **Q** act on a rigid
cube (view from the top).

Answer 16. The diagram (Fig. A.16.1) shows the forces acting on the cube. To maintain the balance of forces (no acceleration, either linear or angular) requires the following:

$$\mathbf{P} = -\mathbf{Q} \ , \tag{A.16.1}$$

to prevent linear acceleration (if acting on a single point, these two forces cancel). To avoid angular acceleration by application of a torque, the following additional conditions on vector components must be met:

$$F_1 = Q_2 \ ,$$
$$F_3 = P_3 = Q_3 = 0 \ . \tag{A.16.2}$$

The edge length is irrelevant for the problem. (Inclusion of extraneous "information" in the statement of the problem is intentional and will be found in several more problems; in the progress of research, one of the necessary steps usually is the separation of relevant data from the remainder.)

Answer 17. (a) The magnitude of **u** (Fig. A.17.1) is:

$$u = \left(u_1^2 + u_2^2 + u_3^2\right)^{1/2} = \left(9 + 16 + 144\right)^{1/2} = 13 \ . \tag{A.17.1}$$

Note: This is the three-dimensional version of Pythagoras' theorem. The magnitude h of **h**, the horizontal projection of **u** is, by the conventional

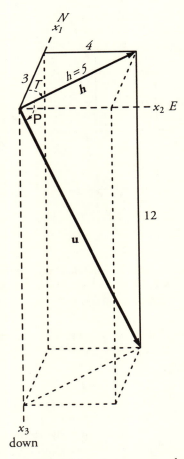

Figure A.17.1. A vector **u**, the vertical plane containing it, and its projection onto the horizontal plane.

form of Pythagoras:

$$h = \left(h_1^2 + h_2^2\right)^{1/2} = \left(9 + 16\right)^{1/2} = 5 \ . \tag{A.17.2}$$

Hence, by trigonometry, the trend T and plunge P of **u** are:

$$\mathsf{T} = \sin^{-1}\!\left\langle u_2 / \left(u_1^2 + u_2^2\right)^{1/2}\right\rangle = \sin^{-1}\left(4/5\right) \approx 53° \ , \tag{A.17.3}$$

and:

$$\mathsf{P} = \tan^{-1}\!\left\langle u_3 / \left(u_1^2 + u_2^2\right)^{1/2}\right\rangle = \tan^{-1}\left(12/5\right) \approx 67° \ . \tag{A.17.4}$$

(b) To find the angle θ formed by **u** with **v**, we first find the magnitude $v = 85$ by the method of eq. (A.17.1). The angle is most conveniently determined from the dot product of the two vectors:

$$\theta = \cos^{-1}\left(\frac{u_i v_i}{uv}\right) = \cos^{-1}\left(\frac{15 + 48 + 1008}{13 \cdot 85}\right)$$

$$= \cos^{-1}\left(\frac{1071}{1105}\right) \approx 14° \ . \tag{A.17.5}$$

Another, possibly more convenient, method of finding trend and plunge of a line is based on the standard conversion from cartesian to spherical coordinates (trend and the complement of plunge may be considered spherical coordinates on the unit sphere) and can be found in the Summary of Formulæ. (c) The direction cosines of **u** and **v** are the components of their unit, or direction, vectors:

$$\left[u_i/u\right] = \left[\ 3/13 \ \ 4/13 \ \ 12/13 \ \right] , \tag{A.17.6}$$

and:

$$\left[v_i/v\right] = \left[\ 5/85 \ \ 12/85 \ \ 84/85 \ \right] . \tag{A.17.7}$$

(d) The angle φ between x_1 (north, horizontal) and **u** is the arccosine of the dot product between unit vectors in these directions, or the arccosine of $[u_1/u]$ in eq. (A.17.6). Hence:

$$\varphi = \cos^{-1}\left(u_1/u\right) \approx 77° \ . \tag{A.17.8}$$

Answer 18. (a) Given is the position vector as a function of time $\mathbf{R}(t)$ defined by:

$$\mathbf{R} = \mathbf{i}\,a\cos\frac{2\pi t}{T} + \mathbf{j}\,a\sin\frac{2\pi t}{T} \ . \tag{A.18.1}$$

where a and T are constants. The magnitude R of **R** is found by inspection of Fig. A.18.1:

$$R = a\left(\cos^2\frac{2\pi t}{T} + \sin^2\frac{2\pi t}{T}\right)^{1/2}. \tag{A.18.2}$$

Hence (because for any angle α, $\cos^2\alpha + \sin^2\alpha = 1$):

$$R = a \ . \tag{A.18.3}$$

Also by inspection, the orientation $^R\theta$ of **R** is:

$$^R\theta = \tan^{-1}\left(R_2/R_1\right) \ . \tag{A.18.4}$$

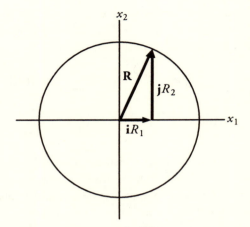

Figure A.18.1. The position vector **R** and
its component vectors.

Substitute for the components of **R** from eq. (A.18.1) to obtain:

$$^R\theta = \tan^{-1}\left(\sin\frac{2\pi t}{T}\bigg/\cos\frac{2\pi t}{T}\right) = \tan^{-1}\left(\tan\frac{2\pi t}{T}\right) = \frac{2\pi t}{T} \quad . \quad (A.18.5)$$

Thus a particle marked by the position vector **R** moves in a counterclockwise circular motion along a circle of radius a, T being the time needed for one revolution. Described is a positive rigid-body rotation about the x_3 axis at a constant rate. (b) The velocity **V** is the time derivative of **R**:

$$\mathbf{V} = \dot{\mathbf{R}} = \mathbf{i}\frac{dR_1}{dt} + \mathbf{j}\frac{dR_2}{dt}$$

$$= -\mathbf{i}\frac{2\pi t}{T}\sin\frac{2\pi t}{T} + \mathbf{j}\frac{2\pi t}{T}\cos\frac{2\pi t}{T} \quad . \quad (A.18.6)$$

By the same method as that used to find **R**, the magnitude V of **V** is:

$$V = 2\pi a/T \quad , \quad (A.18.7)$$

and the orientation of **V** is:

$$^V\theta = \tan^{-1}\left(-\cot\frac{2\pi t}{T}\right) = \frac{\pi}{2} + \frac{2\pi t}{T} \quad . \quad (A.18.8)$$

The velocity has a phase shift of $+\pi/2$ against the position. (c) The acceleration **A** is the time derivative of **V**:

$$\mathbf{A} = \dot{\mathbf{V}} = \ddot{\mathbf{R}} = \mathbf{i}\frac{dV_1}{dt} + \mathbf{j}\frac{dV_2}{dt}$$

$$= -\mathbf{i}\frac{4\pi^2 a}{T^2}\cos\frac{2\pi t}{T} - \mathbf{j}\frac{4\pi^2 a}{T^2}\sin\frac{2\pi t}{T} = -\frac{4\pi^2 a}{T^2}\mathbf{R} \ . \qquad \text{(A.18.9)}$$

The acceleration is thus antiparallel to the position. Its magnitude is:

$$A = 4\pi^2 a/T^2 = V^2/a \ , \qquad \text{(A.18.10)}$$

its orientation:

$$^A\theta = \tan^{-1}\left(\tan\frac{2\pi t}{T}\right) = \pi + \frac{2\pi t}{T} \ . \qquad \text{(A.18.11)}$$

The acceleration has $a + \pi/2$ phase shift against the velocity and hence a 1π phase shift against the position. It is directed toward the origin (caused by a centripetal force).

Answer 19. Given:

$$\mathbf{R} = \mathbf{A} + f(t)\mathbf{B} \ , \qquad \text{(A.19.1)}$$

since $\mathbf{A}$ is constant and thus:

$$\dot{\mathbf{A}} = \ddot{\mathbf{A}} = \mathbf{0} \ , \qquad \text{(A.19.2)}$$

it follows that:

$$\dot{\mathbf{R}} \times \ddot{\mathbf{R}} = f'f''(\mathbf{B} \times \mathbf{B}) = \mathbf{0} \ . \qquad \text{(A.19.3)}$$

$\dot{\mathbf{R}}$ and $\ddot{\mathbf{R}}$ are parallel or antiparallel.

Answer 20. Newton's law relates force $\mathbf{F}$, mass m, and velocity $\mathbf{V}$ as follows:

$$\mathbf{F} = \frac{d}{dt}(m\mathbf{V}) \ . \qquad \text{(A.20.1)}$$

With the mass constant for the present problem, the relation becomes:

$$\mathbf{F} = m\frac{d\mathbf{V}}{dt} \ . \qquad \text{(A.20.2)}$$

Suppose that the equation given for the rate of increase is correct:

$$\frac{d}{dt}(\tfrac{1}{2}m\mathbf{V}\cdot\mathbf{V}) = \mathbf{F}\cdot\mathbf{V} \ . \qquad \text{(A.20.3)}$$

Carry out the differentiation:

$$\frac{d}{dt}\left(\tfrac{1}{2}m\mathbf{V}\bullet\mathbf{V}\right)=\tfrac{1}{2}m\left(\mathbf{V}\bullet\frac{d\mathbf{V}}{dt}+\frac{d\mathbf{V}}{dt}\bullet\mathbf{V}\right). \qquad (A.20.4)$$

The two dot products on the right-hand side are identical (dot products are commutative). Hence:

$$\frac{d}{dt}\left(\tfrac{1}{2}m\mathbf{V}\bullet\mathbf{V}\right)=m\frac{d\mathbf{V}}{dt}\bullet\mathbf{V}. \qquad (A.20.5)$$

Substitution of eq. (A.20.1) into (A.20.5) yields eq. (A.20.3) and thus confirms the supposition.

Answer 21. By definition:

$$\nabla r^n=\mathbf{i}\frac{\partial r^n}{\partial x_1}+\mathbf{j}\frac{\partial r^n}{\partial x_2}+\mathbf{k}\frac{\partial r^n}{\partial x_3}. \qquad (A.21.1)$$

Given the position vector:

$$\mathbf{r}=\mathbf{i}x_1+\mathbf{j}x_2+\mathbf{k}x_3, \qquad (A.21.2)$$

the magnitude r of $\mathbf{r}$ is:

$$r=\left(x_1^2+x_2^2+x_3^2\right)^{1/2}. \qquad (A.21.3)$$

Differentiation is similar in all three directions; so differentiation with respect to x_1 is typical for all of them:

$$\frac{dr^n}{dx_1}=\frac{d\left(x_1^2+x_2^2+x_3^2\right)^{n/2}}{dx_1}$$

$$=\frac{2x_1 n}{2}\left(x_1^2+x_2^2+x_3^2\right)^{(n-2)/2}. \qquad (A.21.4)$$

Cancel factors of 2 and substitute eq. (A.21.3) into eq. (A.21.4) to obtain:

$$\frac{dr^n}{dx_1}=x_1 n r^{(n-2)}. \qquad (A.21.5)$$

After analogous differentiations with respect to x_2 and x_3, the result is:

$$\nabla r^n=\mathbf{i}x_1 n r^{(n-2)}+\mathbf{j}x_2 n r^{(n-2)}+\mathbf{k}x_3 n r^{(n-2)}. \qquad (A.21.6)$$

Substituting eq. (A.21.2) into eq. (A.21.6) yields the desired result:

$$\nabla r^n = n\, r^{(n-2)} \mathbf{r} \quad \therefore \qquad (A.21.7)$$

Answer 22. Let the vector $\mathbf{a}$ be constant and $\mathbf{r}$ be a position vector of the form:

$$\mathbf{r} = \mathbf{i}x_1 + \mathbf{j}x_2 + \mathbf{k}x_3 \; . \qquad (A.22.1)$$

(a) Then by definition:

$$
\begin{aligned}
\nabla(\mathbf{a} \bullet \mathbf{r}) &\equiv \operatorname{grad}(\mathbf{a} \bullet \mathbf{r}) \\
&\equiv \mathbf{i}\frac{\partial(a_1\,x_1 + a_2\,x_2 + a_3\,x_3)}{\partial x_1} + \mathbf{j} \; \cdots
\end{aligned}
\qquad (A.22.2)
$$

However, only variable terms differentiate to nonzero, as, for example:

$$\frac{d(a_1\,x_1 + a_2 x_2 + a_3\,x_3)}{d x_1} = a_1 \, , \; \text{etc.} \qquad (A.22.3)$$

Thus:

$$\nabla(\mathbf{a} \bullet \mathbf{r}) = \mathbf{i}a_1 + \mathbf{j}a_2 + \mathbf{k}a_3 = \mathbf{a} \; . \qquad (A.22.4)$$

General conclusion: The gradient of the dot product of a constant vector with a variable position vector is that same vector.

(b) Again by definition:

$$\mathbf{a} \times \mathbf{r} = \mathbf{i}(a_2\,x_3 - a_3\,x_2) + \mathbf{j}(a_3\,x_1 - a_1\,x_3) + \mathbf{k}(a_1\,x_2 - a_2\,x_1) \; . \qquad (A.22.5)$$

Then:

$$\nabla \times (\mathbf{a} \times \mathbf{r}) \equiv \operatorname{curl}(\mathbf{a} \times \mathbf{r})$$

$$
\equiv
\begin{vmatrix}
\mathbf{i} & \mathbf{j} & \mathbf{k} \\[6pt]
\dfrac{\partial}{\partial x_1} & \dfrac{\partial}{\partial x_2} & \dfrac{\partial}{\partial x_3} \\[10pt]
(a_2\,x_3 - a_3\,x_2) & (a_3\,x_1 - a_1\,x_3) & (a_1\,x_2 - a_2\,x_1)
\end{vmatrix}
$$

$$
= \mathbf{i}\left\langle \frac{\partial(a_1\,x_2 - a_2 x_1)}{\partial x_2} - \frac{\partial(a_3\,x_1 - a_1\,x_3)}{\partial x_3} \right\rangle - \mathbf{j} \; \cdots \; , \qquad (A.22.6)
$$

where, in each partial of the last expression, only the one term is nonzero that contains the variable with respect to which the term is differentiated (x_2 in the first, x_3 in the second partial). Thus:

$$\nabla \times (\mathbf{a} \times \mathbf{r}) = \mathbf{i}(a_1 + a_1) + \mathbf{j}(a_2 + a_2) + \mathbf{k}(a_3 + a_3)$$

$$= 2\mathbf{a} \ . \tag{A.22.7}$$

General conclusion: The curl of the cross product of a constant vector with a variable position vector is twice that same vector.

(c) Starting again from the definition, we find:

$$\nabla \bullet (\mathbf{a} \times \mathbf{r}) \equiv \frac{\partial(a_2 x_3 - a_3 x_2)}{\partial x_1} + \frac{\partial(a_3 x_1 - a_1 x_3)}{\partial x_2} + \frac{\partial(a_1 x_2 - a_2 x_1)}{\partial x_3}$$

$$= 0 + 0 + 0 \ . \tag{A.22.8}$$

General conclusion: The divergence of the cross product of a constant vector with a variable position vector is zero.

Answer 23. In the equation for the velocity at a point in the rotating body:

$$\mathbf{v} = {}^{\circ}\mathbf{v} + \mathbf{\Omega} \times \mathbf{r} \ , \tag{A.23.1}$$

where $\mathbf{\Omega}$, the angular velocity, is a true constant, and ${}^{\circ}\mathbf{v}$, the velocity of the origin is a pseudoconstant, in the sense that is not a function of either x, y, or z.

(a) By definition:

$$\text{curl } \mathbf{v} = \nabla \times \mathbf{v} = \nabla \times {}^{\circ}\mathbf{v} + \nabla \times (\mathbf{\Omega} \times \mathbf{r}) \ . \tag{A.23.2}$$

According to the results of Problem 22 (a) and (b), the result is:

$$\text{curl } \mathbf{v} = 2\mathbf{\Omega} \ . \tag{A.23.3}$$

(b) By definition:

$$\text{div } \mathbf{v} \equiv \nabla \bullet \mathbf{v} = \nabla \bullet {}^{\circ}\mathbf{v} + \nabla \bullet (\mathbf{\Omega} \times \mathbf{r}) \ . \tag{A.23.4}$$

According to the results of Problem 22 (c), the result is:

$$\text{div } \mathbf{v} = 0 \ . \tag{A.23.5}$$

(c) The velocity field $\mathbf{v}$ is solenoidal.

Answer 24. Hubble's law for the velocity $\mathbf{v}$ of distant galaxies is:

$$\mathbf{v} = H\mathbf{r} \ , \tag{A.24.1}$$

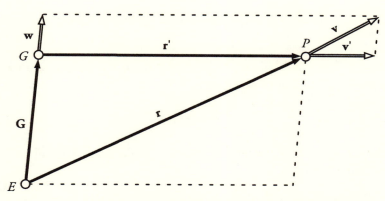

Figure A.24.1. Hubble's law from Earth and from a distant galaxy. E – Earth. G – a distant galaxy at the end of vector **G**. **r** – vector from Earth to another distant point P, **r'** – vector to the same point from the galaxy. **v**, **v'** – velocities at the distant point P, relative to Earth and to the galaxy. **w** – velocity of the galaxy relative to earth.

where H is the Hubble constant and **r** the position vector of a galaxy. (a) The divergence of the velocity field is:

$$\text{div } \mathbf{v} \equiv \nabla \cdot \mathbf{v} = H \nabla \cdot \mathbf{r} = 3H \ . \tag{A.24.2}$$

Note: One H for each dimension. (b) The independence of Hubble's law of the position of the observer is illustrated by Fig. A.24.1. By inspection of the triangle formed by **r** and the dashed lines parallel to **G** and **r'**:

$$\mathbf{r'} = \mathbf{r} - \mathbf{G} \ . \tag{A.24.3}$$

By Hubble's law, eq. (A.24.1), the velocity of the galaxy is:

$$\mathbf{w} = H\mathbf{G} \ . \tag{A.24.4}$$

By inspection of the small triangle near P, the velocity of P relative to G is:

$$\mathbf{v'} = \mathbf{v} - H\mathbf{G} \ . \tag{A.24.5}$$

Substitute eq. (A.24.1) into eq. (A.24.5):

$$\mathbf{v'} = H\mathbf{r} - H\mathbf{G} = H(\mathbf{r} - \mathbf{G}) \ , \tag{A.24.6}$$

and eq. (A.24.3) into eq. (A.24.6):

$$\mathbf{v'} = H\mathbf{r'} \ \therefore \tag{A.24.7}$$

Observations from the distant galaxy produce the same general law as those from Earth.

Answer 25. (a) A vector field is irrotational if either $\nabla \times \mathbf{v} = \mathbf{0}$ or $\mathbf{v} = \nabla\varphi$, with φ any scalar function. Given is:

$$\mathbf{V} = r^n \mathbf{r} \ , \tag{A.25.1}$$

where $\mathbf{r}$ is a position vector. If, according to the result of Problem 21, the following holds:

$$m \, r^{(m-2)} \mathbf{r} = \nabla r^m, \tag{A.25.2}$$

then, setting $m = n + 2$:

$$r^n \mathbf{r} = \frac{1}{n+2} \nabla r^{(n+2)}. \tag{A.25.3}$$

Hence (constants *may* be written inside the differential):

$$\mathbf{V} = \nabla \left\langle \frac{1}{n+2} \, r^{(n+2)} \right\rangle \ , \tag{A.25.4}$$

Thus $\mathbf{V}$ is the gradient of a scalar and therefore irrotational. (b) To find whether $\mathbf{V}$ is solenoidal, form its dot product with the del:

$$\nabla \cdot \mathbf{V} = \frac{\partial}{\partial x_1}(r^n x_1) + \frac{\partial}{\partial x_2}(r^n x_2) + \frac{\partial}{\partial x_3}(r^n x_3) \ , \tag{A.25.5}$$

or, explicitly:

$$\nabla \cdot \mathbf{V} = n \, r^{(n-1)}\frac{x_1^2}{r} + r^n + n \, r^{(n-1)}\frac{x_2^2}{r} + r^n + n \, r^{(n-1)}\frac{x_3^2}{r} + r^n . \tag{A.25.6}$$

Collect terms:

$$\nabla \cdot \mathbf{V} = n \, r^{(n-1)}\left(x_1^2 + x_2^2 + x_3^2\right)/r + 3 \, r^n . \tag{A.25.7}$$

Considering that the quantity in parentheses is r^2, this simplifies to:

$$\nabla \cdot \mathbf{V} = n \, r^n + 3 \, r^n = (n+3) r^n . \tag{A.25.8}$$

Thus $\nabla \cdot \mathbf{V} = 0$ holds and $\mathbf{V}$ is solenoidal if and only if $n = -3$.

Answer 26. (a) The body characterized by the given conductivity tensor K_{ij} is thermally anisotropic and is most conductive in the E–W horizontal direction. (b) The body conducts approximately 30% less well in the N–S horizontal, and a further 40% less well in the vertical direction. (c) The given thermal gradient indicates that temperature decreases most rapidly in the upward vertical direction. (d) According to the given equation, the heat flow is:

$$[q_i] = -\begin{bmatrix} 50 & 0 & 0 \\ 0 & 70 & 0 \\ 0 & 0 & 30 \end{bmatrix}\begin{bmatrix} 0 \\ 0 \\ 1.6 \end{bmatrix}$$

$$= -\begin{bmatrix} 0 & 0 & 30\times1.6 \end{bmatrix}$$

$$= \begin{bmatrix} 0 & 0 & -48 \end{bmatrix} \text{ J m}^{-2}\text{ s}^{-1}, \qquad (A.26.1)$$

where the first brackets in the first line enclose the nine components of the thermal conductivity tensor, in units of $\text{J m}^{-1}\text{ s}^{-1}\text{ K}^{-1}$, and the second brackets the temperature gradient vector in units of K m^{-1}. The heat flow is vertically upward (note the sign). (e) In the second case, in which the temperature gradient is:

$$[T_{,i}] = \begin{bmatrix} 1.0 & 1.0 & 1.0 \end{bmatrix} \text{ K m}^{-1}, \qquad (A.26.2)$$

According to Fig. A.26.1a, the trend and plunge of the temperature gradient grad T are:

$$\mathsf{T} = \sin^{-1}\left(1/\sqrt{2}\right) = 45°,$$
$$\mathsf{P} = \tan^{-1}\left(1/\sqrt{2}\right) \approx 35°, \qquad (A.26.3)$$

and of the maximum *decrease*, grad T (upward-pointing vectors are given negative "plunges"):

$$\mathsf{T}' = \mathsf{T}+180° = 225°,$$
$$\mathsf{P}' = -\mathsf{P} \approx -35°. \qquad (A.26.4)$$

(f) Substituting the new numerical values into eq. (A.26.1), we obtain:

$$[q_i] = -\begin{bmatrix} 50 & 70 & 30 \end{bmatrix} \text{ J m}^{-2}\text{ s}^{-1}. \qquad (A.26.5)$$

The direction cosines of the heat flow vector are:

$$[q_i/q] = -\begin{bmatrix} 5/\sqrt{83} & 7/\sqrt{83} & 3/\sqrt{83} \end{bmatrix}, \qquad (A.26.6)$$

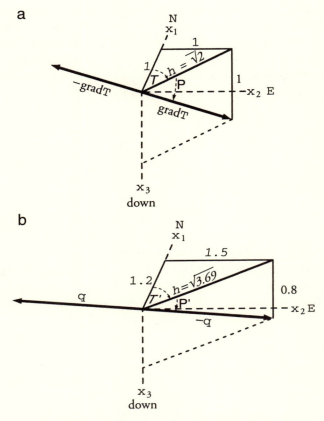

Figure A.26.1. Orientations, in the second case, of (a) the
temperature gradient and (b) the resulting heat flow.

and differ not only by the sign from those of the temperature gradient, which
are $\left[\, 1/\sqrt{3} \;\; 1/\sqrt{3} \;\; 1/\sqrt{3}\,\right]$. According to Fig. A.26.1b, the approximate trend and
plunge of the inverse vector of the heat flow are:

$$\mathsf{T'} = \sin^{-1}\!\left(70/\sqrt{70^2 + 50^2}\,\right) \approx 54°,$$

$$\mathsf{P'} = \tan^{-1}\!\left(70/\sqrt{70^2 + 50^2}\,\right) \approx 19°, \qquad (\text{A.26.7})$$

those of the heat flow itself:

$$\mathsf{T} = \quad \mathsf{T'} + 180° \approx 234°,$$

$$\mathsf{P} = -\mathsf{P'} \approx -19°. \qquad (\text{A.26.8})$$

Thus in an anisotropic material $\mathbf{q}$ and grad T are not necessarily antiparallel.
(g) Suppose that the actual thermal conductivity tensor is that of eq. (A.26.1)
and the actual temperature gradient that of eq. (A.26.2), yet suppose that only the
vertical component of that gradient had been measured and had erroneously
been combined with the assumption of both horizontal components being
zero; the erroneous gradient would have appeared to be:

$$\left[{}^*T_{,i} \right] = \left[\begin{array}{ccc} 0 & 0 & 1.0 \end{array} \right] \mathrm{K\ m^{-1}} . \tag{A.26.9}$$

Suppose further that this was combined with the incorrect assumption of an
isotropic thermal conductivity of the average value:

$$\left[{}^*K_{ij} \right] = \left[\begin{array}{ccc} 50 & 0 & 0 \\ 0 & 50 & 0 \\ 0 & 0 & 50 \end{array} \right] \mathrm{J\ m^{-1}\ s^{-1}\ K^{-1}} . \tag{A.26.10}$$

The calculated vertical component of the heat flow (the only component of
geophysical significance, sideways flow making no contributions to the
Earth's heat budget) would have been:

$$ {}^*q_3 = -50 \ \mathrm{J\ m^{-2}\ s^{-1}}, \tag{A.26.11}$$

instead of the actual:

$$ q_3 = -30 \ \mathrm{J\ m^{-2}\ s^{-1}}, \tag{A.26.12}$$

an overestimate of 67%, similar to many instances of actual errors caused by
oversimplified assumptions.

Answer 27. (a) By inspection of Fig. A.27.1, the given vector:

$$\left[q_i \right] = \left[\begin{array}{ccc} 1 & -\sqrt{3} & 2\sqrt{3} \end{array} \right] , \tag{A.27.1}$$

has the following magnitude, trend, and plunge:

$$ q = (1 + 3 + 12)^{1/2} = \sqrt{16} = 4 , $$

$$ T = \sin^{-1}(-\sqrt{3}/2) = -60° = 300° , $$

$$ P = \tan^{-1}(2\sqrt{3}/2) = \tan^{-1}\sqrt{3} = 60° . \tag{A.27.2}$$

(b) Given furthermore the tensor:

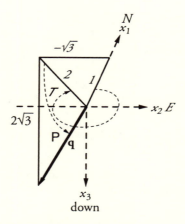

Figure A.27.1. Magnitude and
components of the vector **q**.

$$[S_{ij}] = \begin{bmatrix} 21 & 5\sqrt{3} & -2\sqrt{3} \\ 5\sqrt{3} & 31 & -6 \\ -2\sqrt{3} & -6 & 28 \end{bmatrix} , \tag{A.27.3}$$

and the relationship:

$$p_i = S_{ij} q_j , \tag{A.27.4}$$

calculate the components of p_i one by one:

$$\begin{aligned}
p_1 &= S_{11} q_1 + S_{12} q_2 + S_{13} q_3 \\
&= \quad 21 \quad\quad -15 \quad\quad -12 \quad = -6 , \\
p_2 &= S_{21} q_1 + S_{22} q_2 + S_{23} q_3 \\
&= \quad 5\sqrt{3} \quad -31\sqrt{3} \quad -12\sqrt{3} = -38\sqrt{3} , \\
p_3 &= S_{31} q_1 + S_{32} q_2 + S_{33} q_3 \\
&= -2\sqrt{3} \quad +6\sqrt{3} \quad +56\sqrt{3} = 60\sqrt{3} ,
\end{aligned} \tag{A.27.5}$$

to obtain:

$$\begin{aligned}
[p_i] &= \begin{bmatrix} -6 & -38\sqrt{3} & 60\sqrt{3} \end{bmatrix} \\
&= 2\sqrt{3} \begin{bmatrix} -\sqrt{3} & -19 & 30 \end{bmatrix} .
\end{aligned} \tag{A.27.6}$$

(c) The square of the magnitude is:

$$p^2 = 36 + 4332 + 10800 = 15168 , \tag{A.27.7}$$

and the magnitude:

$$p = 123 \ . \tag{A.27.8}$$

(d) By using the dot product $p_i \, q_i$:

$$^{pq}\theta \approx 18° \ . \tag{A.27.9}$$

(e) Given the different vector (simplified):

$$\left[q'_i \right] = 3 \left[\ 1 \quad \sqrt{3} \quad 2\sqrt{3} \ \right] , \tag{A.27.10}$$

the magnitude and orientation are, by the same methods as before:

$$q' = 12 \ ,$$
$$\mathbf{T'} = \sin^{-1}\left(\sqrt{3}/2 \right) = 60° \ ,$$
$$\mathbf{P'} = \tan^{-1}\sqrt{3} = 60° \ . \tag{A.27.11}$$

(f) In order to find the magnitude of $\mathbf{p'}$, one needs to calculate the vector itself first; it is:

$$\left[p'_i \right] = \left[\ 72 \quad 72\sqrt{3} \quad 144\sqrt{3} \ \right]$$
$$= 72 \left[\ 1 \quad \sqrt{3} \quad 2\sqrt{3} \ \right] = 24 \left[q'_i \right] , \tag{A.27.12}$$

from which the magnitude is:

$$p' = 288 \ . \tag{A.27.13}$$

(g) The vectors p_i and q_i differ only by a positive scalar factor [eq. (A.27.12)] and are thus parallel. They therefore lie in one of the principal directions of S_{ij}.

Answer 28. Given a tensor equation, such as:

$$p_i = T_{ij} \, q_j \ , \tag{A.28.1}$$

with a set of numerical values, say:

$$\left[p_i \right] = \left[\ 16 \quad 20 \quad 24 \ \right] , \quad \left[T_{ij} \right] = \begin{bmatrix} 6 & 1 & 8 \\ 7 & 5 & 3 \\ 2 & 9 & 4 \end{bmatrix} , \tag{A.28.2}$$

one can solve for q_i (a) by means of the system of linear equations:

$$16 = 6q_1 + q_2 + 8q_3 \ , \tag{A.28.3}$$

$$20 = 7q_1 + 5q_2 + 3q_3 \ , \tag{A.28.4}$$

$$24 = 2q_1 + 9q_2 + 4q_3 \ . \tag{A.28.5}$$

Like any other system, this can be solved directly, by addition or subtraction of equations, and substitution. Subtract eq. (A.28.3) from eq. (A.28.4) and eq. (A.28.4) from eq. (A.28.5):

$$4 = q_1 + 4q_2 - 5q_3 \ , \tag{A.28.6}$$

and:

$$4 = -5q_1 + 4q_2 + q_3 \ , \tag{A.28.7}$$

subtract eq. (A.28.7) from eq. (A.28.6), and solve for q_3:

$$q_3 = q_1 \ . \tag{A.28.8}$$

Substitute eq. (A.28.8) into eq. (A.28.6), and solve for q_2:

$$q_2 = q_1 + 1 \ . \tag{A.28.9}$$

Substitute eqs. (A.28.8) and (A.28.9) into eq. (A.28.3), and solve for q_1, substitute the result into eq. (A.28.8) for q_3 and into eq. (A.28.9) for q_2 to obtain:

$$q_1 = 1 \ , \ q_2 = 2 \ , \ q_3 = 1 \ , \tag{A.28.10}$$

which implies:

$$[q_i] = \begin{bmatrix} 1 & 2 & 1 \end{bmatrix} \ . \tag{A.28.11}$$

(b) For matrix inversion by Cramer's rule, the determinant of the tensor (here treated simply as a matrix) is needed:

$$\left| T_{ij} \right| = 360 \ . \tag{A.28.12}$$

The inverted tensor is:

$$\left[T_{ij}^{-1} \right] = \frac{1}{360} \begin{bmatrix} -7 & 68 & -37 \\ -22 & 8 & 38 \\ 53 & -52 & 23 \end{bmatrix} \ , \tag{A.28.13}$$

where the elements in the right-hand brackets are the transposed *cofactors* of $[T_{ij}]$. The inverted tensor can then be used in the equation:

$$q_i = T_{ij}^{-1} p_j , \qquad (A.28.14)$$

to solve in the usual way for q_i with the results already stated as eqs. (A.28.10) and (A.28.11).

Answer 29. Elements a_{ij} in a transformation matrix for the rotation of coordinate axes cannot be greater than $+1$ or smaller than -1 because, being projections of unit vectors onto axes, they are the cosines of angles.

Answer 30. (a) Using the trend and plunge angles of the new axes with reference to the old axes for the first row and third column, and the method of Problem 17 for the remainder, the set of direction angles (α_{ij}) from *old* to *new* axes is:

$$(\alpha_{ij}) = \begin{pmatrix} \alpha_{11} & \alpha_{12} & \alpha_{13} \\ \alpha_{21} & \alpha_{22} & \alpha_{23} \\ \alpha_{31} & \alpha_{32} & \alpha_{33} \end{pmatrix}$$

$$= \begin{pmatrix} 120° & 30° & 90° \\ 64.3° & 75.5° & 150° \\ 138.6° & 115.7° & 120° \end{pmatrix} . \qquad (A.30.1)$$

The angles formed by each of the new axes (rows, first subscripts) with each of the old axes (columns, second subscripts) form a three-by-three array. The set of cosines (a_{ij}) of these angles constitutes a *coordinate transformation matrix* or *rotation matrix* and may be looked at as an array of the direction cosines of the new axes in the old coordinate system (the three rows) or, alternatively, as an array of direction cosines of the old axes in the new coordinate system (the three columns). Figure A.30.1 shows the new axes x_i' in an equal-area projection oriented according to the old coordinate system x_i. When plotting, only one hemisphere is available, conventionally the lower in structural geology; for upward-pointing vectors, it is therefore necessary to plot the antiparallel axes, marking them as such. This makes it possible to inspect the handedness of the axes on the plot. In the present case the set of direction cosines, or transformation matrix, is:

$$(a_{ij}) = \begin{pmatrix} -1/2 & \sqrt{3}/2 & 0 \\ \sqrt{3}/4 & 1/4 & -\sqrt{3}/2 \\ -3/4 & -\sqrt{3}/4 & -1/2 \end{pmatrix} . \qquad (A.30.2)$$

Because each row of this matrix represents the direction vector of one of the new coordinate axes in the old coordinate system and each column the

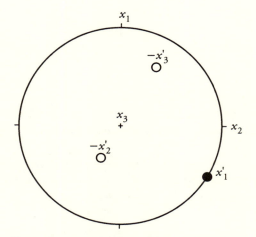

Figure A.30.1. New coordinate axes x_i' in an
equal-area projection oriented according to
the old coordinates x_i. Lower hemisphere.
Filled, positive branch of a new axis.

direction vector of one of the old coordinate axes in the new system, the vector
magnitude of every row or column must be unity, which is easily verified in
the present example. (b) The determinant of the transformation matrix of eq.
(A.30.2) is:

$$\left| a_{ij} \right| = +1 \; , \tag{A.30.3}$$

and (c) the new axes x_i' are right-handed like the old axes x_i. (d) The
transformation matrix from *new* to *newer* axes is:

$$\left(a'_{ij} \right) = \begin{pmatrix} -1 & 0 & 0 \\ 0 & -1 & 0 \\ 0 & 0 & -1 \end{pmatrix} . \tag{A.30.4}$$

This is an *inversion of axes*, (e) the determinant is:

$$\left| a'_{ij} \right| = -1 \; , \tag{A.30.5}$$

and (f) the *newer* axes x''_i are left-handed. (g) The transformation matrix from
old to newer axes $x_i \rightarrow x''_i$ is defined by:

$$a''_{ij} = -a_{ij} \; , \tag{A.30.6}$$

(h) where:

$$\left| a''_{ij} \right| = -1 \; . \tag{A.30.7}$$

(i) The transformation matrix from *new* to *old* coordinates is:

$$\left(\overset{*}{a}_{ij} \right) = \left(a_{j\,i} \right) = \left(a_{ij}^{-1} \right) , \tag{A.30.8}$$

thus the inversion of a transformation matrix rotating an orthogonal coordinate system is a simple transposition. (j) The determinant of ($\overset{*}{a}_{ij}$) is:

$$\left| \overset{*}{a}_{ij} \right| \equiv \left| a_{j\,i} \right| = \left| a_{ij} \right| = +1 , \tag{A.30.9}$$

transposition of a matrix does not affect its determinant. (k) The required transformation is effected by the matrix:

$$\left(\hat{a}_{ij} \right) = \begin{pmatrix} 1 & 0 & 0 \\ 0 & -1 & 0 \\ 0 & 0 & 1 \end{pmatrix} . \tag{A.30.10}$$

(l) The mirror plane for this reflection is the x_1–x_3 plane normal to x_2, passing through the coordinate origin. (m) The determinant of the reflection matrix is:

$$\left| \hat{a}_{ij} \right| = -1 , \tag{A.30.11}$$

which (n) shows that the coordinates $\hat{x}_i$ are left-handed.

Answer 31. No matter what is the azimuth of the horizontal projection of the *new* axis, the angle from the *old* downward axis to a *new* 30° obliquely upward one is 120°. The one desired direction-cosine element of the transformation matrix, a_{23}, is therefore:

$$a_{23} = \cos 120° = -1/2 . \tag{A.31.1}$$

All other information provided in the problem is therefore unnecessary for the solution.

Answer 32. (a) Either by intersection of the planes normal to the direction angles, the inverse of the solution of Problem 9, or by the method used for the solution of Problem 17, the new axes x_i' have the following trends and plunges:

$$^{x'_1}T = 144°, \quad ^{x'_1}P = -27°,$$
$$^{x'_2}T = 289°, \quad ^{x'_2}P = -58°,$$
$$^{x'_3}T = 226°, \quad ^{x'_3}P = 16°. \tag{A.32.1}$$

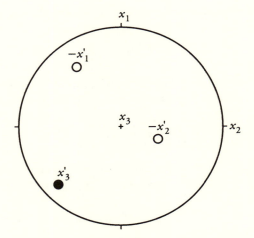

Figure A.32.1. Equal-area projection of
rotated axes. Lower hemisphere. Filled,
positive branch of an axis.

The corresponding set of direction angles from *old* axes i to *new* axes j (α_{ij})
is:

$$\left(\alpha_{ij}\right) \approx \begin{pmatrix} 136° & 58° & 117° \\ 80° & 120° & 148° \\ 132° & 134° & 74° \end{pmatrix}. \tag{A.32.2}$$

(b) Either by plotting on a stereographic or an equal-area projection
(Fig. A.32.1), or by calculating the *new* axes in terms of the *old* with the help of
the given transformation matrix and then forming the mutual cross products
of unit vectors along the *new* axes, it can be verified that these *new* axes are
orthogonal within the accuracy of the respective method.

Answer 33. In the following set of direction cosines consider the framed
elements as known:

$$\left(a_{ij}\right) = \begin{pmatrix} \boxed{a_{11}} & \boxed{a_{12}} & \boxed{a_{13}} \\ \boxed{a_{21}} & \boxed{a_{22}} & \boxed{a_{23}} \\ a_{31} & a_{32} & a_{33} \end{pmatrix}. \tag{A.33.1}$$

If handedness is to be preserved by the transformation, the missing row vector
should be the cross product of the first with the second, or:

$$a_{3i} = \epsilon_{ijk} a_{1j} a_{2k}. \tag{A.33.2}$$

Note that eq. (A.33.2) has both numerical and letter subscripts. The sequence of number subscripts (counting up from 3, modulo 3) is critical. Equation (A.33.2) represents the following explicit equations:

$$a_{31} = a_{12}\,a_{23} - a_{13}\,a_{22} \; ,$$
$$a_{32} = a_{13}\,a_{21} - a_{11}\,a_{23} \; ,$$
$$a_{33} = a_{11}\,a_{22} - a_{12}\,a_{21} \; . \tag{A.33.3}$$

What appears to be an alternative method, to solve for the missing element in each column by means of $a_{i1}\,a_{i1} = 1$, etc., is ambiguous insofar as it leaves the sign of the a_{3i} undetermined.

Answer 34. Given the tensor:

$$\left[S_{ij}\right] = \begin{bmatrix} 21 & 5\sqrt{3} & -2\sqrt{3} \\ 5\sqrt{3} & 31 & -6 \\ -2\sqrt{3} & -6 & 28 \end{bmatrix} , \tag{A.34.1}$$

and the rotation matrix:

$$\left(a_{ij}\right) = \begin{pmatrix} \sqrt{3}/2 & -1/2 & 0 \\ 1/4 & \sqrt{3}/4 & \sqrt{3}/2 \\ -\sqrt{3}/4 & -3/4 & 1/2 \end{pmatrix} , \tag{A.34.2}$$

rotation of the tensor according to the equation:

$$S'_{ij} = a_{ik}\,a_{jl}\,S_{kl} \; , \tag{A.34.3}$$

produces the result:

$$\left[S'_{ij}\right] = \begin{bmatrix} 16 & 0 & 0 \\ 0 & 24 & 0 \\ 0 & 0 & 40 \end{bmatrix} . \tag{A.34.4}$$

Each of the three *principal values*, forming the *diagonal* components of a tensor with all *nondiagonal* components zero, satisfies the cubic equation in determinant form:

$$\begin{vmatrix} (21 - S'_{ii}) & 5\sqrt{3} & -2\sqrt{3} \\ 5\sqrt{3} & (31 - S'_{ii}) & -6 \\ -2\sqrt{3} & -6 & (28 - S'_{ii}) \end{vmatrix} = 0 \;\; (\text{no sum on } i) \; . \tag{A.34.5}$$

This is the so-called *secular equation* for the tensor in eq. (A.34.1), except that the unknown in this cubic equation is customarily designated by the letter λ instead of S'_{ii}.

Answer 35. Rewrite the given equation:

$$\left[B_{ij}\right] = \begin{bmatrix} B & 0 & 0 \\ 0 & B & 0 \\ 0 & 0 & B \end{bmatrix} = B \begin{bmatrix} 1 & 0 & 0 \\ 0 & 1 & 0 \\ 0 & 0 & 1 \end{bmatrix} . \qquad \text{(A.35.1)}$$

The same statement can be made in subscript notation:

$$B_{ij} = B\delta_{ij} , \qquad \text{(A.35.2)}$$

where δ_{ij} is the Kronecker delta. Rotating B_{ij} by means of an arbitrary a_{ij} results in:

$$B'_{ij} = a_{ik}\, a_{jl}\, B_{kl} . \qquad \text{(A.35.3)}$$

By inspection of eq. (A.35.1) or by the definition of the Kronecker delta used in eq. (A.35.2), in all nonzero components of B_{ik}, $k = l$, hence, subscripts in eq. (A.35.3) may be redefined:

$$k \equiv m , \quad l \equiv m . \qquad \text{(A.35.4)}$$

Then eq. (A.35.3) becomes:

$$B'_{ij} = a_{im}\, a_{jm}\, B . \qquad \text{(A.35.5)}$$

However, the expression $a_{im}\, a_{jm}$ represents the dot product of two unit vectors, row vectors of (a_{jk}) representing new axes x_i and x_j. This dot product being zero unless $i = j$ and unity if $i = j$, it has the exact properties of the Kronecker delta δ_{ij}:

$$a_{ik}\, a_{jk} = \delta_{ij} . \qquad \text{(A.35.6)}$$

This is called the *orthogonality relation* of (a_{ij}). Substitute eq. (A.35.6) into eq. (A.35.5) and the result into eq. (A.35.1) to obtain:

$$B'_{ij} = B\delta_{ij} = B_{ij} \; \therefore \qquad \text{(A.35.7)}$$

The significance of this identity is that the B_{ij}, given in eq. (A.35.1), does not change with an arbitrary rotation, that it is *invariant to rotation*.

Answer 36. The sum T_{ii} of diagonal components of the tensor T_{ij}, also called its *trace*, can be shown to be invariant by subjecting it to an arbitrary rotation:

$$T'_{ii} = a_{ij} a_{ik} T_{jk} .$$ (A.36.1)

Use the orthogonality relation of the columns of (a_{ij}), $a_{ij} a_{ik} = \delta_{jk}$, to restate eq. (A.36.1) as:

$$T'_{ii} = \delta_{jk} T_{jk} ,$$ (A.36.2)

and the substitution property of the Kronecker delta to obtain:

$$T'_{ii} = T_{jj} \; \therefore$$ (A.36.3)

Answer 37. The solution is obvious, the dot product being a scalar and hence independent of orientation. Thus the following serves merely to show a general method of proving invariance. Given that, after an arbitrary rotation by the matrix (a_{ij}), the vectors p_i and q_i become:

$$p'_i = a_{ij} p_j ,$$
$$q'_i = a_{ij} q_i ,$$ (A.37.1)

the dot product is:

$$p'_i q'_i = a_{ij} a_{ik} p_j q_k .$$ (A.37.2)

Use first the orthogonality relation $a_{ij} a_{ik} = \delta_{jk}$, then the substitution property $\delta_{jk} q_k = q_j$, and obtain:

$$p'_i q'_i = p_j q_j \; \therefore$$ (A.37.3)

Note that we could equally well have substituted the subscript j in the first vector of the right-hand side by k, rather than substituting the subscript k of the second vector by j; as a matter of fact, we could quite arbitrarily have used any other repeated letter as the dummy subscript. By convention, one avoids alphabetic gaps between subscript letters if possible.

Answer 38. We form the dot product of p_i with itself to find the square of the magnitude:

$$p^2 = p_i p_i .$$ (A.38.1)

Using the result of the previous problem, we obtain:

$$p^2 = p_j p_j = p'_j p'_j \; \therefore$$ (A.38.2)

Answer 39. The array:

$$\langle \chi_i \rangle = \langle p_1/q_1 \quad p_2/q_2 \quad p_3/q_3 \rangle \,, \tag{A.39.1}$$

is not a vector for the following reason. Suppose that the magnitudes p and q are both finite and that originally all six components of p_i and q_i are nonzero. Now assume that, with a particular rotation, one of the components of q_i' becomes zero while all of the components of p_i' are finite. That makes one of the three ratios χ_i' and thus also the magnitude χ infinite. Hence, rotation could change the "magnitude" of the array from finite to infinite, a property of the array that is incompatible with the invariance of vector magnitudes.

Answer 40. Given that:

$$T_{ij}' = a_{im} a_{jk} T_{mk} = \pm a_{im} a_{jk} T_{km} \,, \tag{A.40.1}$$

and considering that the factors of a multiplication are commutative, eq. (A.40.1) may be rewritten as:

$$T_{ij}' = a_{jk} a_{im} T_{km} = \pm T_{ji}' \,. \tag{A.40.2}$$

General conclusion: After an arbitrary rotation, a symmetric tensor of the second rank remains symmetric, and an antisymmetric tensor remains antisymmetric.

Answer 41. Given:

$$p_i = S_{ij} q_j \,, \tag{A.41.1}$$

where:

$$[S_{ij}] = \begin{bmatrix} S_1 & 0 & 0 \\ 0 & S_2 & 0 \\ 0 & 0 & S_3 \end{bmatrix} \,, \tag{A.41.2}$$

and:

$$S_1 > S_2 > S_3 \,. \tag{A.41.3}$$

(a) Yes, with all nondiagonal components zero, the tensor is referred to its principal axes. (b) To find the maximum magnitude p, we first take its square:

$$p^2 \equiv p_i p_i \equiv p_1^2 + p_2^2 + p_3^2 = S_1^2 q_1^2 + S_2^2 q_2^2 + S_3^2 q_3^2 \,, \tag{A.41.4}$$

where:

$$q^2 \equiv q_1^2 + q_2^2 + q_3^2 = \text{const.} \,, \tag{A.41.5}$$

and:

$$u = f(q_i) = S_1^2 q_1^2 + S_2^2 q_2^2 + S_3^2 q_3^2, \tag{A.41.6}$$

subject to the constraint:

$$0 = \varphi(q_i) = q_1^2 + q_2^2 + q_3^2 - q^2. \tag{A.41.7}$$

The first derivatives of the function f and the constraint function φ are:

$$0 = 2S_1^2 q_1 + 2S_2^2 q_2 + 2S_3^2 q_3, \tag{A.41.8}$$

$$0 = 2q_1 + 2q_2 + 2q_3. \tag{A.41.9}$$

Multiply eq. (A.41.9), the derivative of the constraint equation, by the Lagrange multiplier λ, add it to the derivative of function f, eq. (A.41.8), and obtain the following equations:

$$(S_1^2 + \lambda)q_1 = 0,$$
$$(S_2^2 + \lambda)q_2 = 0,$$
$$(S_3^2 + \lambda)q_3 = 0. \tag{A.41.10}$$

Equations (A.41.10) simultaneously with eq. (A.41.5) yield the following solutions:

i) $q_1 = q$, $q_2 = 0$, $q_3 = 0$; $\lambda = -S_1^2$,

ii) $q_1 = 0$, $q_2 = q$, $q_3 = 0$; $\lambda = -S_2^2$,

iii) $q_1 = 0$, $q_2 = 0$, $q_3 = q$; $\lambda = -S_3^2$. $\qquad$ (A.41.11)

Thus the maximum p^2 is the one with the greatest S_i^2, hence the one along x_1 because of the inequality (A.41.3). (c) In this case:

$$[p_i] = [\ S_1 q \quad 0 \quad 0 \ ], \tag{A.41.12}$$

hence p is parallel to q, and both are parallel to x_1; the coordinates are the principal coordinates of the tensor. (d) In principal coordinates the representation quadric is:

$$S_1 x_1^2 + S_2 x_2^2 + S_3 x_3^2 = 1, \tag{A.41.13}$$

or, in subscript notation:

$$S_i x_i^2 = 1, \tag{A.41.14}$$

Note that the assemblage of principal values of the second-rank tensor $[S_{ij}]$ can, with caution, be treated as the pseudo-vector $[S_i]$. (e) Apply the method of normal lines and tangent planes to find the radius vector x_i parallel to the desired unit vector u_i so that:

$$x_i = x u_i \, , \tag{A.41.15}$$

where x is the length of the radius vector $\mathbf{x}$, or in principal coordinates the length $S_i^{-1/2}$ of the appropriate half-axis of the quadric:

$$x_i = S_i^{-1/2} u_i \quad (\text{no sum on } i) \, . \tag{A.41.16}$$

Note: Subscript equations involving pseudovectors like $[S_i]$ commonly require the no-sum specification. Without it, eq. (A.41.16) would be an improper equation, having a free subscript on the left-hand-side term but what looks like dummy subscripts on the right-hand side. (f) The desired outward normal unit vector n_i where the radius vector x_i touches the quadric surface is:

$$n_i = N_i / N \, , \tag{A.41.17}$$

where:

$$N_i = u_{,i}\big|_{°x_i} \, , \tag{A.41.18}$$

and where in turn the function $u(x_i)$ is characterized by eq. (A.41.16) as:

$$u(x_i) = S_i x_i^2 \quad (\text{no sum on } i) \, . \tag{A.41.19}$$

In principal coordinates, $°x_i$ can stand for x_1, x_2, or x_3. Differentiate eq. (A.41.19) and obtain:

$$u_{,i} = 2 S_i x_i \quad (\text{no sum on } i) \, . \tag{A.41.20}$$

Then normalize as in eq. (A.41.17), factors of 2 cancelling, to:

$$n_i = S_i x_i / \left(S_j^2 x_j^2 \right)^{1/2} \quad (\text{no sum on } i) \, . \tag{A.41.21}$$

Note that the sum over j in this equation is not revoked. (g) To find p_i, substitute u_i for q_i in eq. (A.41.1) and obtain for principal coordinates:

$$p_i = S_i u_i \quad (\text{no sum on } i) \, . \tag{A.41.22}$$

(h) To find the angle between p_i and n_j, solve eq. (A.41.15) for u_i and substitute the result into eq. (A.41.22):

$$p_i = S_i\, x_i/x \quad (\text{no sum on } i)\,.\tag{A.41.23}$$

The vector p_i thus differs from n_i [eq. (A.41.21)] only by a scalar factor of proportionality, and they are parallel. Note: This also constitutes a more thoroughly reasoned answer to question (a).

Answer 42. Postulate that the following be true:

$$T_{ij} = S_{ij} + A_{ij}\,,\tag{A.42.1}$$

where:

$$S_{ij} = S_{ji}\,,\tag{A.42.2}$$

and:

$$A_{ij} = -A_{ji}\,,\tag{A.42.3}$$

and thus:

$$T_{ji} = S_{ji} + A_{ji} = S_{ij} - A_{ij}\,.\tag{A.42.4}$$

Add eq. (A.42.1) to (A.42.4) in its second form, and solve for S_{ij}:

$$S_{ij} = \tfrac{1}{2}\big(T_{ij} + T_{ji}\big)\,.\tag{A.42.5}$$

Subtract the same equations from each other and solve for A_{ij}:

$$A_{ij} = \tfrac{1}{2}\big(T_{ij} - T_{ji}\big)\,.\tag{A.42.6}$$

Hence the premise is confirmed and:

$$\mathbb{S} + \mathbb{A} = \mathbb{T}\,.\tag{A.42.7}$$

Note: Open font letters represent tensors or other matrices in matrix notation, that is, without explicit subscripts.

Answer 43. Given the *asymmetric* tensor:

$$\big[T_{ij}\big] = \begin{bmatrix} 1 & 2 & 3 \\ 4 & 5 & 6 \\ 7 & 8 & 9 \end{bmatrix},\tag{A.43.1}$$

the *symmetric* tensor is:

$$\big[S_{ij}\big] = \begin{bmatrix} 1 & 3 & 5 \\ 3 & 5 & 7 \\ 5 & 7 & 9 \end{bmatrix},\tag{A.43.2}$$

and the *antisymmetric* tensor:

$$[A_{ij}] = \begin{bmatrix} 0 & -1 & -2 \\ 1 & 0 & -1 \\ 2 & 1 & 0 \end{bmatrix}. \tag{A.43.3}$$

Answer 44. (a) The transformation matrix is:

$${}^{2}_{\theta}(a_{ij}) = \begin{pmatrix} \cos\theta & 0 & -\sin\theta \\ 0 & 1 & 0 \\ \sin\theta & 0 & \cos\theta \end{pmatrix}. \tag{A.44.1}$$

This is a positive rotation of the coordinate system about its x_2 axis through the angle θ. The same matrix thus may also indicate a *physical rotation* through $-\theta$ with respect to stationary coordinates. (b) The other two positive coordinate rotations about coordinate axes are:

$${}^{1}_{\theta}(a_{ij}) = \begin{pmatrix} 1 & 0 & 0 \\ 0 & \cos\theta & \sin\theta \\ 0 & -\sin\theta & \cos\theta \end{pmatrix}, \tag{A.44.2}$$

and:

$${}^{3}_{\theta}(a_{ij}) = \begin{pmatrix} \cos\theta & \sin\theta & 0 \\ -\sin\theta & \cos\theta & 0 \\ 0 & 0 & 1 \end{pmatrix}. \tag{A.44.3}$$

(c) A rotation about x_2 through $180°$ is effected by:

$${}^{2}_{180°}(a_{ij}) = \begin{pmatrix} -1 & 0 & 0 \\ 0 & 1 & 0 \\ 0 & 0 & -1 \end{pmatrix}. \tag{A.44.4}$$

(d) Let the transformation matrix of eq. (A.44.4) act on the tensor ${}_{0°}[S_{ij}]$:

$${}_{180°}[S_{ij}] = \begin{bmatrix} S_{11} & -S_{12} & S_{13} \\ -S_{12} & S_{22} & -S_{23} \\ S_{13} & -S_{23} & S_{33} \end{bmatrix}. \tag{A.44.5}$$

(e) This could be taken as a symmetry operation with x_2 as a dyad axis. In a monoclinic crystal with x_2 parallel to the dyad axis (see Fig. A.47.1), tensor properties like that of eq. (A.44.5) must be identical before and after such a symmetry operation. (f) To uphold Neumann's principle, the crystal must have tensor properties such that:

$$_{180^\circ}[S_{ij}] = {_{0^\circ}}[S_{ij}] = \begin{bmatrix} S_{11} & 0 & S_{13} \\ 0 & S_{22} & 0 \\ S_{13} & 0 & S_{33} \end{bmatrix}. \tag{A.44.6}$$

Tensor properties with symmetry *higher* than that of the tensor in eq. (A.44.6) are equally compatible with Neumann's principle because the symmetry operation of eq. (A.44.4) still leads to tensor identity. Tensor properties with *lower* symmetry are incompatible with the principle.

Answer 45. Tensor properties in trigonal crystals oriented with their threefold axis parallel to x_3 must remain unchanged by a rotation through ±120°. Positive rotation about the x_3 axis through an angle θ is effected by the rotation matrix:

$$_{\theta}(a_{ij}) = \begin{pmatrix} \cos\theta & \sin\theta & 0 \\ -\sin\theta & \cos\theta & 0 \\ 0 & 0 & 1 \end{pmatrix}. \tag{A.45.1}$$

Let θ be +120°:

$$_{+120^\circ}(a_{ij}) = \begin{pmatrix} -1/2 & \sqrt{3}/2 & 0 \\ -\sqrt{3}/2 & -1/2 & 0 \\ 0 & 0 & 1 \end{pmatrix}. \tag{A.45.2}$$

or let it be −120°:

$$_{-120^\circ}(a_{ij}) = \begin{pmatrix} -1/2 & -\sqrt{3}/2 & 0 \\ \sqrt{3}/2 & -1/2 & 0 \\ 0 & 0 & 1 \end{pmatrix}. \tag{A.45.3}$$

Perform the positive rotation according to:

$$S'_{ij} = a_{ik}\, a_{jl}\, S_{kl}\,, \tag{A.45.4}$$

then the rotated tensor is, component by component:

$$S_{11} = S'_{11} = {}^1\!/_4\, S_{11}\boxed{-\sqrt{3}/_4\, S_{12} - \sqrt{3}/_4\, S_{21}} + {}^3\!/_4\, S_{22}\,. \tag{A.45.5}$$

To produce identity of S_{11} and S'_{11}, the framed terms must either cancel or must both be zero, and S_{11} must equal S_{22}:

$$S_{12} = S'_{12} = \sqrt{3}/_4\, S_{11}\boxed{+\, {}^1\!/_4\, S_{12} - 3/4\, S_{21}} - \sqrt{3}/_4\, S_{22}\,. \tag{A.45.6}$$

Considering that $S_{11} = S_{22}$ and that therefore the unframed terms of the right-hand side of eq. (A.45.6) necessarily cancel, considering further that the framed terms cannot cancel if those in eq. (A.45.5) do, to produce identity of S_{12} and S'_{12}, both the framed terms and hence S_{12} and S_{21} must be zero:

$$S_{13} = S'_{13} = {}^{-1}/2\, S_{13} + \sqrt{3}/2\, S_{23} . \qquad (A.45.7)$$

Unless a particular proportionality between S_{13} and S_{23} exists, both must be zero:

$$S_{21} = S'_{21} = \sqrt{3}/4\, S_{11} \boxed{-3/4 S_{12} + {}^1/4 S_{21}} - \sqrt{3}/4\, S_{22} . \qquad (A.45.8)$$

From eq. (A.45.6), the framed terms are individually zero, and from eq. (A.45.5) the unframed right-hand side terms cancel. Hence we prove a second time that S_{21} must be zero:

$$S_{22} = S'_{22} = {}^1/4\, S_{11} + \sqrt{3}/4\, S_{12} + \sqrt{3}/4\, S_{21} + {}^3/4\, S_{22} . \qquad (A.45.9)$$

By the same arguments as for eq. (A.45.5), we find redundantly that $S_{11} = S_{22}$:

$$S_{23} = S'_{23} = {}^{-\sqrt{3}}/2\, S_{13} - {}^1/2 S_{23} . \qquad (A.45.10)$$

Considering that two contradictory proportionalities between S_{13} and S_{23} would be required to validate both eqs. (A.45.7) and (A.45.10), both S_{13} and S_{23} must be zero:

$$S_{31} = S'_{31} = {}^{-1}/2 S_{31} + \sqrt{3}/2\, S_{32} , \qquad (A.45.11)$$

$$S_{32} = S'_{32} = {}^{-\sqrt{3}}/2 S_{31} - {}^1/2 S_{32} . \qquad (A.45.12)$$

By the arguments used for eqs. (A.45.7) and (A.45.10), S_{31} and S_{32} must also be zero. Finally:

$$S'_{33} = S_{33} . \qquad (A.45.13)$$

Equations (A.45.8), (A.45.11), and (A.45.12) are unnecessary if the tensor is specified to be symmetric. All arguments used for the positive rotation apply equally for the negative rotation with the matrix of eq. (A.45.3). No property of a crystal is known at present that would require an asymmetric tensor for its description. To allow the symmetry operations:

$$_{\pm 120°}\left[S'_{ij}\right] = \left[S_{ij}\right] , \qquad (A.45.14)$$

it is necessary and sufficient that the tensor be:

$$
[S_{ij}] = \begin{bmatrix} S_{11}=S_{22} & S_{12}=0 & S_{13}=0 \\ S_{21}=0 & S_{22}=S_{11} & S_{23}=0 \\ S_{31}=0 & S_{32}=0 & S_{33} \end{bmatrix}. \tag{A.45.15}
$$

The tensor of eq. (A.45.15) has axial symmetry with x_3 for the unique axis, which implies that it is invariant to rotation about the x_3 axis. The conventional form of writing such a tensor is:

$$
[S_{ij}] = \begin{bmatrix} S_1 & 0 & 0 \\ 0 & S_1 & 0 \\ 0 & 0 & S_3 \end{bmatrix}, \tag{A.45.16}
$$

as, for example, in Nye (1964, p. 43).

Answer 46. Considering that in the cubic system the crystallographic axes a, b, and c are orthogonal and conventionally oriented in the x_1, x_2, and x_3 directions of cartesian coordinates, and that their lengths a, b, and c are equal, the unit vector in the [UVW] direction is:

$$
[u_i] = [U\,V\,W]/(U^2 + V^2 + W^2)^{1/2}. \tag{A.46.1}
$$

Thus for [UVW] = [1 1 1] it is:

$$
[u_i] = \begin{bmatrix} 1/\sqrt{3} & 1/\sqrt{3} & 1/\sqrt{3} \end{bmatrix}. \tag{A.46.2}
$$

Answer 47. In the monoclinic system **a** is conventionally taken to be parallel to x_1, **b** parallel to x_2, and **c** to lie in the x_1–x_3 plane, the mirror plane, where **a**, **b**, and **c** are vectors in the directions of the crystallographic axes with generally different magnitudes (relative unit-cell dimensions) a, b, and c. Inspection of Fig. A.47.1 shows that in cartesian coordinates:

$$
[c_i] = \begin{bmatrix} c\cos\beta & 0 & c\sin\beta \end{bmatrix}. \tag{A.47.1}
$$

This allows calculation of the unit vector in the direction of [UVW]:

$$
[u_i] = \frac{[Ua + Wc\cos\beta \qquad Vb \qquad Wc\sin\beta]}{\langle V^2 b_i b_i + (Ua_i + Wc_i)(Ua_i + Wc_i)\rangle^{1/2}}, \tag{A.47.2}
$$

or:

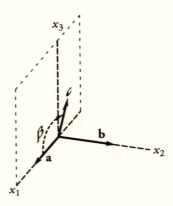

Figure A.47.1. Orientations
of crystallographic axes in the
monoclinic system.

$$[u_i] = \frac{[Ua + Wc\cos\beta \qquad Vb \qquad Wc\sin\beta]}{\langle U^2 a_i a_i + V^2 b_i b_i + Wc_i(Ua_i + Wc_i)\rangle^{1/2}} , \qquad (A.47.3)$$

or:

$$[u_i] = \frac{[Ua + Wc\cos\beta \qquad Vb \qquad Wc\sin\beta]}{(U^2 a^2 + V^2 b^2 + W^2 c^2 + 2UaWc\cos\beta)^{1/2}} , \qquad (A.47.4)$$

where Wc contributes components to $\mathbf{u}$ both in the x_1 and the x_3 directions. In the x_1 direction this component is added to Ua; in the x_3 direction it is the only contributor. The normalizing constant consists of the square root of the sum of the squares of the orthogonal components of $\mathbf{u}$.

Answer 48. The given matter tensor K_{ij} is invariant to all transformations by rotation (see Problem 35) and therefore possesses all symmetry elements a crystal can possibly have, and it thus contains no information about the crystal; the property would be valid for any crystal.

Answer 49. (a) Of the body diagonals of a conventionally oriented cubic crystal choose the one that traverses the {001} cube from the all-negative to the all-positive corner. The unit vector pointing in the all-positive direction of this body diagonal is:

$$[l_i] = \begin{bmatrix} 1/\sqrt{3} & 1/\sqrt{3} & 1/\sqrt{3} \end{bmatrix} . \qquad (A.49.1)$$

The components of this unit vector are direction cosines. Unit vectors along the other body diagonals (and senses) have components of the same magnitude

with all possible combinations of signs. (b) Positive rotation through 120°
about $[l_i]$ carries axes so that:

$$x_1' = x_2, \quad x_2' = x_3, \quad x_3' = x_1, \qquad \text{(A.49.2)}$$

and therefore the transformation matrix effecting this rotation (see Problem
30) is:

$$\left(a_{ij} \right) = \begin{pmatrix} 0 & 1 & 0 \\ 0 & 0 & 1 \\ 1 & 0 & 0 \end{pmatrix}. \qquad \text{(A.49.3)}$$

(c) Under this and similar transformations, in which each row and each
column of the transformation matrix contains only a single nondiagonal and
nonzero element of the value +1 or −1, second-rank tensors are invariant only
if they are *isotropic*, that is, invariant under any rotation. This is the
requirement imposed on second-rank tensor properties of cubic crystals by
Neumann's Principle.

Answer 50. Given the temperature gradient:

$$\left[T_{,i} \right] = \begin{bmatrix} 1.0 & 1.0 & 1.0 \end{bmatrix} \text{ K m}^{-1}, \qquad \text{(A.50.1)}$$

the heat flow:

$$\left[q_i \right] = -\begin{bmatrix} 50 & 70 & 30 \end{bmatrix} \text{ J m}^{-2} \text{ s}^{-1}, \qquad \text{(A.50.2)}$$

and the relationship:

$$q_i = -K_{ij} \, T_{,j}, \qquad \text{(A.50.3)}$$

then K, the magnitude of K_{ij}, is the component of flow in the direction of the
gradient, divided by the magnitude of the gradient. Thus:

$$K = \frac{-q_i \, T_{,i}}{\left(T_{,j} \, T_{,j} \right)^{1/2} \left(T_{,k} \, T_{,k} \right)^{1/2}}, \qquad \text{(A.50.4)}$$

where the second factor of the numerator divided by the appropriate
normalizing constant, the first square root of a dot product in the
denominator, is the unit vector in the direction of the gradient; the second,
similar, square root in the denominator is the (scalar) magnitude of that
gradient. The complete numerator, being a dot product, makes the whole
fraction a scalar. Note that, to avoid confusion, different dummy subscripts
were chosen to indicate each of the three distinct dot products in this equation.
Equation (A.50.4) simplifies to:

$$K = -q_i \, T_{,i} / \left(T_{,j} \, T_{,j} \right) . \tag{A.50.5}$$

For the given numerical values and direction, this becomes:

$$K = 150/3 = 50 \, \text{J m}^{-1} \, \text{s}^{-1} \, \text{K}^{-1} . \tag{A.50.6}$$

One of the infinite number of tensors to have that magnitude in this direction happens to be:

$$\left[K_{ij} \right] = \begin{bmatrix} 50 & 0 & 0 \\ 0 & 70 & 0 \\ 0 & 0 & 30 \end{bmatrix} \text{J m}^{-1} \, \text{s}^{-1} \, \text{K}^{-1} . \tag{A.50.7}$$

(See Problem 26.) Since the tensor of eq. (A.50.7) is in its principal coordinates, one can solve for K by means of:

$$K = l_i^2 \, K_i , \tag{A.50.8}$$

where l_i is the unit vector (the set of direction cosines) for the direction in which K is sought. For the direction of the temperature gradient given in eq. (A.50.1):

$$\left[l_i \right] = \left[\ 1/\sqrt{3} \ \ 1/\sqrt{3} \ \ 1/\sqrt{3} \ \right] . \tag{A.50.9}$$

and hence, as in eq. (A.50.6):

$$K = 50 \, \text{J m}^{-1} \, \text{s}^{-1} \, \text{K}^{-1} . \tag{A.50.10}$$

Answer 51. Inspection of the list of measured properties:

Temperature gradient $(10^4 \, \text{K m}^{-1})$	Magnitude of conductivity in the direction of the gradient $(\text{J m}^{-1} \, \text{s}^{-1} \, \text{K}^{-1})$
[1.1 0.0 0.0]	8
[0.0 1.7 0.0]	16
[0.0 0.0 1.0]	8
[0.8 0.8 0.0]	24
[0.0 0.7 0.7]	20
[0.7 0.0 0.7]	12

immediately yields:

$$K_{11} = 8 \, \text{J m}^{-1} \, \text{s}^{-1} \, \text{K}^{-1} ,$$
$$K_{22} = 16 \, \text{J m}^{-1} \, \text{s}^{-1} \, \text{K}^{-1} ,$$
$$K_{33} = 8 \, \text{J m}^{-1} \, \text{s}^{-1} \, \text{K}^{-1} . \tag{A.51.1}$$

These values are independent of the temperature gradient values stated in the first column of the table. We are interested only in the directions in which the temperature gradient (and presumably heat flow) were measured to calculate the conductivity magnitudes stated in the second column of the table. (The magnitude of a second-rank tensor, relating a dependent to an independent vector, is the component of the dependent vector in the direction of the independent vector, divided by the magnitude of the independent vector.) For the fourth line of the table of data, the direction vector of the applied temperature gradient is:

$$[l_i] = \left[\begin{array}{ccc} \sqrt{2}/2 & \sqrt{2}/2 & 0 \end{array} \right] . \tag{A.51.2}$$

Then by the definition of magnitude, $K = K_{ij} l_i l_j$, the measurement implies:

$$24 \text{ J m}^{-1} \text{ s}^{-1} \text{ K}^{-1} = {}^1\!/_2 (K_{11} + K_{12} + K_{21} + K_{22}) , \tag{A.51.3}$$

which, by substitution of known numerical values, together with the knowledge that the tensor K_{ij} is symmetric, hence $K_{21} = K_{12}$, allows one to solve for K_{12}:

$$\begin{aligned} K_{12} = K_{21} &= {}^1\!/_2 (-8 + 48 - 16) \text{ J m}^{-1} \text{ s}^{-1} \text{ K}^{-1} \\ &= 12 \text{ J m}^{-1} \text{ s}^{-1} \text{ K}^{-1} . \end{aligned} \tag{A.51.4}$$

Similarly, from the last line of the data:

$$\begin{aligned} K_{13} = K_{31} &= {}^1\!/_2 (-8 + 24 - 8) \text{ J m}^{-1} \text{ s}^{-1} \text{ K}^{-1} \\ &= 4 \text{ J m}^{-1} \text{ s}^{-1} \text{ K}^{-1} , \end{aligned} \tag{A.51.5}$$

and from line 5:

$$\begin{aligned} K_{23} = K_{32} &= {}^1\!/_2 (-16 + 40 - 8) \text{ J m}^{-1} \text{ s}^{-1} \text{ K}^{-1} \\ &= 8 \text{ J m}^{-1} \text{ s}^{-1} \text{ K}^{-1} . \end{aligned} \tag{A.51.6}$$

Thus the thermal conductivity tensor for the specimen is:

$$[K_{ij}] = \begin{bmatrix} 8 & 12 & 4 \\ 12 & 16 & 8 \\ 4 & 8 & 8 \end{bmatrix} \text{ J m}^{-1} \text{ s}^{-1} \text{ K}^{-1} . \tag{A.51.7}$$

Answer 52. For greater ease of answering both questions, use the knowledge from Problem 34 that the rotation matrix:

$$\left(a_{ij}\right) = \begin{pmatrix} \sqrt{3}/2 & -1/2 & 0 \\ 1/4 & \sqrt{3}/4 & \sqrt{3}/2 \\ -\sqrt{3}/4 & -3/4 & 1/2 \end{pmatrix} , \tag{A.52.1}$$

rotates the given tensor to its principal coordinates, so as to make it:

$$\left[{}^{P}S_{ij}\right] = \begin{bmatrix} 16 & 0 & 0 \\ 0 & 24 & 0 \\ 0 & 0 & 40 \end{bmatrix} . \tag{A.52.2}$$

(a) To find S in the direction of the first vector, A_i:

$$\left[A_i\right] = \begin{bmatrix} 1 & -\sqrt{3} & 2\sqrt{3} \end{bmatrix} , \tag{A.52.3}$$

with the direction $[l_i] = [A_i/A]$:

$$\left[l_i\right] = \begin{bmatrix} 1/4 & -\sqrt{3}/4 & \sqrt{3}/2 \end{bmatrix} , \tag{A.52.4}$$

the direction vector must be transformed to the principal coordinates of S_{ij} by means of a_{ij}:

$$\left[{}^{P}l_i\right] = \left[a_{ij}\, l_j\right] = \begin{bmatrix} \sqrt{3}/4 & 5/8 & 3\sqrt{3}/8 \end{bmatrix} . \tag{A.52.5}$$

Then the magnitude in this direction is:

$$S = {}^{P}l_i^2\, S_i = 3 + \frac{75 + 135}{8} = \frac{117}{4} = 29.25 . \tag{A.52.6}$$

(b) By similar methods the vector B_i:

$$\left[B_i\right] = \begin{bmatrix} 1 & \sqrt{3} & 2\sqrt{3} \end{bmatrix} , \tag{A.52.7}$$

has the direction:

$$\left[l_i'\right] = \begin{bmatrix} 1/4 & \sqrt{3}/4 & \sqrt{3}/2 \end{bmatrix} , \tag{A.52.8}$$

or in the principal coordinates of $[S_{ij}]$:

$$\left[{}^{P}l_i'\right] = \begin{bmatrix} 0 & 1 & 0 \end{bmatrix} . \tag{A.52.9}$$

With the direction vector in the $^Px_2$ direction, the magnitude thus is a principal value:

$$S' = 24 .$$ (A.52.10)

Answer 53. For the given tensor:

$$[T_{ij}] = \begin{bmatrix} 6 & 1 & 8 \\ 7 & 5 & 3 \\ 2 & 9 & 4 \end{bmatrix} ,$$ (A.53.1)

T in the direction of x_2 is:

$$T = T_{22} = 5 .$$ (A.53.2)

In the direction of $[v_i] = [1\ 1\ 1]$ the direction vector is:

$$[l_i] = \begin{bmatrix} 1/\sqrt{3} & 1/\sqrt{3} & 1/\sqrt{3} \end{bmatrix} .$$ (A.53.3)

In this direction the magnitude T' is:

$$T' = (6 + 1 + 8 + 7 + 5 + 3 + 2 + 9 + 4)/3 = 15 .$$ (A.53.4)

Answer 54. For a test of invariance to rotation about x_3 rotate the given tensor:

$$[S_{ij}] = \begin{bmatrix} S_1 & 0 & 0 \\ 0 & S_1 & 0 \\ 0 & 0 & S_3 \end{bmatrix} ,$$ (A.54.1)

by the general rotation matrix for a positive rotation through the angle θ of the coordinates about that axis:

$$(a_{ij}) = \begin{pmatrix} \cos\theta & \sin\theta & 0 \\ -\sin\theta & \cos\theta & 0 \\ 0 & 0 & 1 \end{pmatrix} .$$ (A.54.2)

This can be done either immediately or, more conveniently, by decomposing S_{ij} as follows:

$$[S_{ij}] = \begin{bmatrix} S_1 & 0 & 0 \\ 0 & S_1 & 0 \\ 0 & 0 & S_1 \end{bmatrix} + \begin{bmatrix} 0 & 0 & 0 \\ 0 & 0 & 0 \\ 0 & 0 & S_3 - S_1 \end{bmatrix}$$

$$= \begin{bmatrix} {}^iS_{ij} \end{bmatrix} + \begin{bmatrix} \overset{*}{S}_{ij} \end{bmatrix} .$$ (A.54.3)

The first tensor of the decomposition, $^iS_{ij}$, is invariant to any rotation (see Problem 35). If we omit products with zero components of the tensor $\overset{*}{S}_{ij}$, its rotation about x_3 simplifies to:

$$\overset{*}{S}'_{ij} = a_{i3}\,a_{j3}\,\overset{*}{S}_{ij} = a_{i3}\,a_{j3}(S_3 - S_1) \ . \tag{A.54.4}$$

Because for the particular rotation matrix of eq. (A.54.2):

$$a_{i3} = \begin{cases} 0 \ , \ \text{if} \ i = 1, 2 \\ 1 \ , \ \text{if} \ i = 3 \end{cases} , \tag{A.54.5}$$

eq. (A.54.4) can be expanded to:

$$\overset{*}{S}'_{ij} = a_{ik}\,a_{jl}\,\overset{*}{S}_{kl} = \overset{*}{S}_{ij} \ . \tag{A.54.6}$$

In view of the invariance of $^iS'_{ij} = {}^iS_{ij}$, recombination of the decomposed tensors after the rotation yields:

$$S'_{ij} = {}^iS'_{ij} + \overset{*}{S}'_{ij} = {}^iS_{ij} + \overset{*}{S}_{ij} = S_{ij} \ \therefore \tag{A.54.7}$$

Answer 55. (a) Show that:

$$|A_{pq}| = \epsilon_{ijk}\,A_{i1}\,A_{j2}\,A_{k3} = \epsilon_{ijk}\,A_{1i}\,A_{2j}\,A_{3k} \ . \tag{A.55.1}$$

Note that the middle and the right-hand expressions in eq. (A.55.1) each contains one set of three-letter dummy subscripts and one of numerical subscripts. In the first part of eq. (A.55.1) the columns of A_{ij} have the numerical subscripts, in the second part the rows. A square matrix that has as columns the rows of another, in the same order, is called the transpose of that other matrix. Considering that the alternating matrix is defined as:

$$\epsilon_{ijk} \equiv \begin{cases} 1 \ , \ \text{if} \ i, j, k = 1, 2, 3 \ \text{ or } \ 2, 3, 1 \ \text{ or } \ 3, 1, 2 \\ -1 \ , \ \text{if} \ i, j, k = 3, 2, 1 \ \text{ or } \ 2, 1, 3 \ \text{ or } \ 1, 3, 2 \ , \\ 0 \ , \ \text{if} \ i = j \ \text{ or } \ i = k \ \text{ or } \ j = k \end{cases} \tag{A.55.2}$$

writing out each nonzero term in the two expressions of eq. (A.55.1) produces the exact result of the ordinary technique of calculating the determinant, except for the order in which the terms appear and the order of multiplication; by the conventional technique and according to the first and second right-hand expressions, the determinant is:

$$|A_{pq}| = \quad A_{11} A_{22} A_{33} - \langle A_{11} A_{23} A_{32} \rangle$$
$$- \underline{A_{12} A_{21} A_{33}} + [A_{12} A_{23} A_{31}]$$
$$+ (A_{13} A_{21} A_{32}) - \{A_{13} A_{22} A_{31}\}$$

$$= \quad A_{11} A_{22} A_{33} + [A_{12} A_{23} A_{31}]$$
$$+ (A_{13} A_{21} A_{32}) - \{A_{13} A_{22} A_{31}\}$$
$$- \underline{A_{12} A_{21} A_{33}} - \langle A_{11} A_{23} A_{32} \rangle$$

$$= \quad A_{11} A_{22} A_{33} + (A_{21} A_{32} A_{13})$$
$$+ [A_{31} A_{12} A_{23}] - \{A_{31} A_{22} A_{13}\}$$
$$- \underline{A_{21} A_{12} A_{33}} - \langle A_{11} A_{32} A_{23} \rangle . \qquad \text{(A.55.3)}$$

Corresponding terms in the three versions of eq. (A.55.3) are marked by similar bracketing.

General rule: The determinant of a 3 × 3 (or any other square) matrix equals that of its transpose. (b) Note that in eq. (A.55.2) which defines the alternating matrix elements with numerical subscripts in the order 1, 2, 3 and the cyclic permutations of that order are positive, elements with subscripts in the order 3, 2, 1 and the cyclic permutations of that inverse order are negative. Note also that exchange of any two of the numerical subscripts changes a sequence from the increasing (1, 2, 3) to the decreasing order (3, 2, 1). The determinant $|D_{pq}|$ formed by interchanging columns 1 and 2 of a matrix (A_{rs}) is, therefore:

$$|D_{pq}| = \epsilon_{ijk} A_{i2} A_{j1} A_{k3} = -\epsilon_{ijk} A_{j1} A_{i2} A_{k3} = -|A_{rs}|, \quad \text{(A.55.4)}$$

or similarly for other interchanges of rows or columns.

General rule: If any two columns or rows of a 3 × 3 matrix are exchanged, the determinants of the resulting and the original matrix have opposite signs. As a consequence, if two columns or rows of a matrix are identical then:

$$|A_{rs}| = -|A_{rs}|, \qquad \text{(A.55.5)}$$

and thus:

$$|A_{rs}| = 0 \quad \therefore \qquad \text{(A.55.6)}$$

(c) Prove:

$$|A_{ij}| \epsilon_{pqr} = \epsilon_{ijk} A_{pi} A_{qj} A_{rk} . \qquad \text{(A.55.7)}$$

On the right-hand side of this equation, we have replaced the numerical subscripts of the second right-hand expression of eq. (A.55.1) by letter subscripts p, q, and r, hence by the defining eq. (A.55.2):

$$\epsilon_{ijk} A_{pi} A_{qj} A_{rk} = \begin{cases} |A_{ij}|, & \text{if } p, q, r = 1, 2, 3 \\ & \text{or } 2, 3, 1 \text{ or } 3, 1, 2 \\ -|A_{ij}|, & \text{if } p, q, r = 3, 2, 1 \\ & \text{or } 2, 1, 3 \text{ or } 1, 3, 2 \\ 0, & \text{if } p = q \\ & \text{or } p = r \text{ or } q = r \end{cases} \quad . \quad \text{(A.55.8)}$$

This in turn corresponds exactly to the definition of $|A_{ij}| \epsilon_{pqr}$, and eq. (A.55.7) is true. The statement:

$$|A_{ij}| \epsilon_{pqr} = \epsilon_{ijk} A_{ip} A_{jq} A_{kr} , \qquad \text{(A.55.9)}$$

can be proved analogously. (d) Given that $|A_{ij}|$ and $|B_{ij}|$ are determinants and that:

$$C_{ij} = A_{ik} B_{jk} , \qquad \text{(A.55.10)}$$

show that:

$$|C_{ij}| = |A_{pq}||B_{rs}| . \qquad \text{(A.55.11)}$$

By eq. (A.55.10), $|C_{ij}| = |A_{ik} B_{jk}|$ and by eq. (A.55.1):

$$|A_{ik} B_{jk}| = \epsilon_{ijk} A_{il} B_{1l} A_{jm} B_{2m} A_{kn} B_{3n} . \qquad \text{(A.55.12)}$$

Rearranging the order of multiplication, this is also:

$$|A_{ik} B_{jk}| = \epsilon_{ijk} A_{il} A_{jm} A_{kn} B_{1l} B_{2m} B_{3n} . \qquad \text{(A.55.13)}$$

By means of eq. (A.55.1) this in turn may be written as:

$$|A_{ik} B_{jk}| = |A_{pq}| \epsilon_{lmn} B_{1l} B_{2m} B_{3n} , \qquad \text{(A.55.14)}$$

and further as:

$$|C_{ij}| = |A_{ik} B_{jk}| = |A_{pq}||B_{rs}| \quad \therefore \qquad \text{(A.55.15)}$$

Answer 56. Given the identity of eq. (A.55.11) as long as eq. (A.55.10) holds, and given the orthogonality of the rotation matrix (a_{ij}):

$$a_{ij} a_{kj} = \delta_{ik} , \qquad \text{(A.56.1)}$$

it follows that:

$$|a_{ij} a_{kj}| = |a_{pq}||a_{rs}| = |\delta_{ik}| = 1 , \qquad \text{(A.56.2)}$$

and, since $|a_{pq}| = |a_{rs}|$, also that:

$$|a_{pq}|^2 = 1 \; . \tag{A.56.3}$$

Thus:

$$|a_{pq}| = \pm 1 \; \therefore \tag{A.56.4}$$

Answer 57. Postulate that in principal coordinates the magnitude surface of a symmetric tensor of the second rank S_{ij} is:

$$\left(S_1 x_1^2 + S_2 x_2^2 + S_3 x_3^2\right)^2 = \left(x_1^2 + x_2^2 + x_3^2\right)^3 . \tag{A.57.1}$$

The radius vector r to a point x_i is:

$$r = \left(x_1^2 + x_2^2 + x_3^2\right)^{1/2}. \tag{A.57.2}$$

In principal coordinates the magnitude $S = S_{ij} l_i l_j$ becomes:

$$S = S_i l_i^2 , \tag{A.57.3}$$

where the unit vector l_i in the direction from the origin to the position x_i is:

$$l_i = x_i / \left(x_1^2 + x_2^2 + x_3^2\right)^{1/2}. \tag{A.57.4}$$

Thus:

$$S = r = \left(x_1^2 + x_2^2 + x_3^2\right)^{1/2} = \frac{S_1 x_1^2 + S_2 x_2^2 + S_3 x_3^2}{x_1^2 + x_2^2 + x_3^2} . \tag{A.57.5}$$

Squaring both sides of the right-hand equation yields:

$$x_1^2 + x_2^2 + x_3^2 = \frac{\left(S_1 x_1^2 + S_2 x_2^2 + S_3 x_3^2\right)^2}{\left(x_1^2 + x_2^2 + x_3^2\right)^2} . \tag{A.57.6}$$

To verify the correctness of the postulate eq. (A.57.1), multiply eq. (A.57.6) with the denominator:

$$\left(S_1 x_1^2 + S_2 x_2^2 + S_3 x_3^2\right)^2 = \left(x_1^2 + x_2^2 + x_3^2\right)^3 \; \therefore \tag{A.57.7}$$

The magnitude ovaloid [strictly, eq. (A.57.6) also includes the origin] is a third order surface and inconvenient for calculations. The representation quadric $S_{ij} x_i x_j = 1$ is a more suitable surface; its radius vector $\mathbf{r}$ is related to the magnitude S by $S = 1/r^2$.

Answer 58. (a) The secular equation for the given tensor is:

$$\begin{vmatrix} (21-\lambda) & -3\sqrt{6} & -5 \\ -3\sqrt{6} & (14-\lambda) & 3\sqrt{6} \\ -5 & 3\sqrt{6} & (21-\lambda) \end{vmatrix} = 0 \ , \tag{A.58.1}$$

or:

$$\lambda^3 - 56\,\lambda^2 + 896\,\lambda - 4096 = 0 \ . \tag{A.58.2}$$

(b) The results can be guessed, considering that they must all be even integers. Begin by guessing a first solution, say $\lambda = 16$ (this being the cube root of the last coefficient), and then continuing to guess that the coefficients might be:

$$\lambda^3 - \left(\frac{7 \times 16}{2}\right)\lambda^2 + \left(\frac{7 \times 16^2}{2}\right)\lambda - 16^3 = 0 \ . \tag{A.58.3}$$

The roots then are:

$$^1\lambda = 8, \quad ^2\lambda = 16, \quad ^3\lambda = 32 \ . \tag{A.58.4}$$

(c) The shortest principal axis of the representation ellipsoid has the orientation $^{max}l_i$ of the largest principal value of the given tensor, $^{max}S_i = 32$; hence according to eq. (3.50):

$$^{max}\lambda \,{}^{max}l_i = S_{ij}\,{}^{max}l_j \ . \tag{A.58.5}$$

Rearrange this equation, dropping the identifying left superscripts:

$$S_{ij}\,l_j - \lambda\,l_i = 0 \ . \tag{A.58.6}$$

For $i = 1$, this yields:

$$(S_{11} - \lambda)\,l_1 + S_{12}\,l_2 + S_{13}\,l_3 = 0 \ , \tag{A.58.7}$$

or:

$$-11l_1 - 3\sqrt{6}\,l_2 - 5l_3 = 0 \ , \tag{A.58.8}$$

and similarly for $i = 3$:

$$S_{31}\,l_1 + S_{32}\,l_2 + (S_{33} - \lambda)\,l_3 = 0 \ , \tag{A.58.9}$$

or:

$$-5l_1 + 3\sqrt{6}\,l_2 - 11l_3 = 0 \ . \tag{A.58.10}$$

[The equation for $i = 2$ is not independent of eq. (A.58.7)]. Add eqs. (A.58.8) and (A.58.10) and solve for l_1:

$$l_1 = -l_3 \ . \tag{A.58.11}$$

Substitute eq. (A.58.11) into eq. (A.58.8) and also solve for l_1:

$$l_1 = -3\sqrt{6}\,l_2/6 = -\sqrt{6}\,l_2/2 \ . \tag{A.58.12}$$

Because $[\,l_i\,]$ is a unit vector and thus $l_1^2 + l_2^2 + l_3^2 = 1$:

$$6l_2^2/4 + l_2^2 + 6l_2^2/4 = 1 \ , \tag{A.58.13}$$

or:

$$l_2^2 = \tfrac{1}{4} \ . \tag{A.58.14}$$

Thus:

$$\left[\,^{\max}l_i\,\right] = \left[\,\mp\sqrt{6}/4 \quad \pm 1/2 \quad \pm\sqrt{6}/4\,\right] , \tag{A.58.15}$$

and (d) by similar methods:

$$\left[\,^{\text{int}}l_i\,\right] = \left[\,\pm\sqrt{2}/2 \quad 0 \quad \pm\sqrt{2}/2\,\right] , \tag{A.58.16}$$

and either by similar methods or as the cross product of the first two direction vectors:

$$\left[\,^{\min}l_i\,\right] = \left[\,\pm\sqrt{2}/4 \quad \pm\sqrt{3}/2 \quad \mp\sqrt{2}/4\,\right] . \tag{A.58.17}$$

(e) The secular equation for the tensor in principal coordinates is:

$$\begin{vmatrix} (32-\lambda) & 0 & 0 \\ 0 & (16-\lambda) & 0 \\ 0 & 0 & (8-\lambda) \end{vmatrix} = 0 \ , \tag{A.58.18}$$

or:

$$(32-\lambda)(16-\lambda)(8-\lambda) = 0 \ , \tag{A.58.19}$$

or:

$$\lambda^3 - 56\lambda^2 + 896\lambda - 4096 = 0 \ . \tag{A.58.20}$$

Equations (A.58.20) and (A.58.2) are identical; they are the secular equation of the same tensor referred to differently oriented coordinates. This shows that the coefficients of the secular equation are independent of coordinate orientation; they are *invariant* to rotation.

Answer 59. The results are (a):

$$\delta_{ij}\,T_{kj} = T_{ki} \ , \tag{A.59.1}$$

by subscript substitution, a second-rank tensor (or any other square matrix) with two free subscripts. (b):

$$\delta_{pq}\,\delta_{rq}\,S_{pr} = \delta_{rp}\,S_{pr} = S_{rr}\,, \tag{A.59.2}$$

by repeated substitution, a scalar with a pair of dummy subscripts. (c):

$$\delta_{ij}\,\delta_{mn}\,A_{jm}\,B_{np}\,C_{pq} = \delta_{mn}\,A_{im}\,B_{np}\,C_{pq} = A_{in}\,B_{np}\,C_{pq}\,, \tag{A.59.3}$$

by substituting twice, a second-rank tensor. (d):

$$\delta_{pq}\,\delta_{qr}\,\delta_{rs}\,T_{st} = \delta_{pr}\,\delta_{rs}\,T_{st} = \delta_{pr}\,T_{rt} = T_{pt}\,, \tag{A.59.4}$$

by substituting three times, a second-rank tensor. (e):

$$\delta_{ij}\,\delta_{ji} = \delta_{ii} = 1 + 1 + 1 = 3\,, \tag{A.59.5}$$

by substitution and numerical evaluation, assuming that $i, j = 1, 2, 3$, as is usual in the context of this book.

Answer 60. Rotation of the expressions by the rotation matrix (a_{ij}) yields (a):

$$a_{im}\,a_{jn}\,AB_{mn}\,, \tag{A.60.1}$$

a tensor of the second rank. (b):

$$a_{it}\,a_{jk}\,a_{kv}\,a_{lw}\,B_{tu}\,C_{vw}\,, \tag{A.60.2}$$

a tensor of the fourth rank. (c):

$$p_k\,q_k\,, \tag{A.60.3}$$

a scalar. (d):

$$a_{jk}\,p_m\,A_{mk}\,, \tag{A.60.4}$$

a vector. Note that only j is a free subscript. (e):

$$a_{im}\,a_{jn}\,A_{mp}\,B_{np}\,, \tag{A.60.5}$$

a second-rank tensor. Note that m, n, and p are dummy subscripts.

Answer 61. The results are (a):

$$\delta_{pq}\,\delta_{rs}\,T_{ps} = T_{qr}\,, \tag{A.61.1}$$

(b):

$$\delta_{23} T_{13} = 0 \ , \tag{A.61.2}$$

because $\delta_{23} = 0$. (c):

$$(\delta_{ii})^2 = 3^2 = 9 \ , \tag{A.61.3}$$

(d):

$$\delta_{ii}^2 \equiv 1^2 + 1^2 + 1^2 = 3 \ . \tag{A.61.4}$$

Answer 62. To prove the statement "If A_{ij} and B_{ij} are tensors, then $C_{jk} = A_{ij} B_{ik}$ is also a tensor," rotate C_{jk} and verify that the general rotation rule for second-rank tensors holds:

$$
\begin{aligned}
C'_{ik} &= A'_{ij} B'_{jk} \\
&= a_{ip} \boxed{a_{jq} a_{jr}} a_{ks} A_{pq} B_{rs} \\
&= a_{ip} a_{ks} \boxed{\delta_{qr}} A_{pq} B_{rs} \\
&= a_{ip} a_{ks} \boxed{A_{pq} B_{qs}} \\
&= a_{ip} a_{ks} C_{ps} \ \therefore
\end{aligned}
\tag{A.62.1}
$$

By virtue of the orthogonality relations of the rotation matrix, the framed matrices in the second line can be replaced by the Kronecker delta in the third, and by virtue of the substitution property of the Kronecker delta, the subscript r of B_{rs} in the third line can be replaced by q in the fourth. Because this q plays the role of a dummy subscript, it can be replaced by any letter subscript that does not occur in the same equation, say, j as in the initial statement; this allows substitution for the framed product in the fourth line, with insertion of the appropriate free subscripts p and s, by means of the definition of C_{jk}.

Answer 63. Prove that a matrix (D_{ij}) is a tensor, given:

$$D_{im} = A_{ij} B_{jk} C_{km} \ , \tag{A.63.1}$$

where $[A_{ij}], [B_{ij}],$ and $[C_{ij}],$ are tensors. Note that j and k are dummy subscripts. For the proof, rotate all three tensors of eq. (A.63.1):

$$
\begin{aligned}
D'_{ps} &= A'_{pq} B'_{qr} C'_{rs} \\
&= a_{pi} \boxed{a_{qj} a_{qk}} \boxed{a_{rl} a_{rm}} a_{sn} A_{ij} B_{kl} C_{mn} \\
&= a_{pi} a_{sn} \delta_{jk} \delta_{lm} A_{ij} B_{kl} C_{mn} \\
&= a_{pi} a_{sn} A_{ij} B_{jl} C_{ln} \ .
\end{aligned}
\tag{A.63.2}
$$

Note that the framed pairs of rotation matrices are replaced by Kronecker deltas in the next line and that those serve to substitute subscripts of the tensors in the line below. Because they have the same pairings of dummy subscript as in eq. (A.63.1), that equation can be substituted into the last line of eq. (A.63.2), thus showing that D_{ij} are indeed the components of a tensor of the second rank which follows the rotation rule:

$$D'_{ps} = a_{pi} a_{sn} D_{in} \quad \therefore \tag{A.63.3}$$

Answer 64. To prove the statement "If $p'_i = a_{ij} p_j$ then $p_j = a_{ij} p'_i$," first multiply both sides of the first equation of the statement by a_{ik}, then use the orthogonality relations (framed transformation matrices) to introduce the Kronecker delta, and then use its substitution property:

$$a_{ik} p'_i = \boxed{a_{ik} a_{ij}}\, p_j$$
$$= \delta_{kj} p_j$$
$$= p_k \quad \therefore \tag{A.64.1}$$

This result may appear surprising, but remember that $\cos(-\theta) = \cos\theta$.

Answer 65. Rotate the vector:

$$\left[p_i \right] = \left[\; \sqrt{3}\,(\sqrt{2} - 1) \quad \sqrt{2} - 1 \quad 1 \; \right], \tag{A.65.1}$$

by the operation $p'_i = a_{ij} p_j$ with the rotation matrix:

$$\left(a_{ij} \right) = \begin{pmatrix} 1/2 & \sqrt{3}/2 & 0 \\ -\sqrt{6}/4 & \sqrt{2}/4 & \sqrt{2}/2 \\ \sqrt{6}/4 & -\sqrt{2}/4 & \sqrt{2}/2 \end{pmatrix}, \tag{A.65.2}$$

and obtain:

$$\left[p'_i \right] = \left[\; \sqrt{3}\,(\sqrt{2} - 1) \quad \sqrt{2} - 1 \quad 1 \; \right] = \left[p_i \right]. \tag{A.65.3}$$

The rotation represented by the matrix of eq. (A.65.2) has an axis parallel to the vector of eq. (A.65.1).

Answer 66. (a) The given relationship between two vectors:

$$V_{ij} = -p_i q_j + p_j q_i, \tag{A.66.1}$$

may be written out fully as:

$$[V_{ij}] = \begin{bmatrix} 0 & p_2 q_1 - p_1 q_2 & p_3 q_1 - p_1 q_3 \\ p_1 q_2 - p_2 q_1 & 0 & p_3 q_2 - p_2 q_3 \\ p_1 q_3 - p_3 q_1 & p_2 q_3 - p_3 q_1 & 0 \end{bmatrix} , \quad \text{(A.66.2)}$$

which can (b) be restated as:

$$[V_{ij}] = \begin{bmatrix} 0 & -r_3 & r_2 \\ r_3 & 0 & -r_1 \\ -r_2 & r_1 & 0 \end{bmatrix} , \quad \text{(A.66.3)}$$

where:

$$[r_i] = \begin{bmatrix} p_2 q_3 - p_3 q_2 & p_3 q_1 - p_1 q_3 & p_1 q_2 - p_2 q_1 \end{bmatrix} . \quad \text{(A.66.4)}$$

Subscript notation simplifies eq. (A.66.4) to:

$$r_i = \epsilon_{ijk} p_j q_k . \quad \text{(A.66.5)}$$

This is the cross product of $\mathbf{p}$ and $\mathbf{q}$. It is an axial vector; as such, it does not transform as simply as a polar vector does but as:

$$r_i' = |a_{km}| a_{ij} r_j . \quad \text{(A.66.6)}$$

The determinant that enters as a factor into this transformation rule is unnecessary as long as handedness is preserved in the transformation. The relationship between r_i and V_{ij} shows that the vector $[r_i]$ is a degenerate second-rank tensor with only three independent components (so are all axial vectors). (c) The antisymmetric matrix of eq. (A.66.3) must be a second-rank tensor because the coefficients of its secular equation are invariant to coordinate rotation. The secular equation of $[V_{ij}]$ is:

$$\begin{vmatrix} -\lambda & -r_3 & r_2 \\ r_3 & -\lambda & -r_1 \\ -r_2 & r_1 & -\lambda \end{vmatrix} = 0 , \quad \text{(A.66.7)}$$

or:

$$\lambda^3 + \lambda \left(r_1^2 + r_2^2 + r_3^2 \right) = 0 , \quad \text{(A.66.8)}$$

where the quantity in parentheses is the square of the magnitude r and thus invariant to rotation. It follows that $[V_{ij}]$ is a tensor of the second rank. (d) Equation (A.66.8) can be divided by λ:

$$\lambda^2 = -\left(r_1^2 + r_2^2 + r_3^2 \right) . \quad \text{(A.66.9)}$$

Thus no real solution exists for λ, the square root of a negative constant.

Answer 67. For p_i and q_i to be parallel, real, nonzero, finite vectors, their components must be proportional:

$$p_i = \lambda q_i \ . \tag{A.67.1}$$

Expressing p_i in terms of A_{ij} and q_j, this becomes:

$$A_{ij} q_j = \lambda q_i \ . \tag{A.67.2}$$

This system of three equations may be solved for the three components of q_i provided a solution exists, that is, provided that:

$$\left| A_{ij} - \lambda \delta_{ij} \right| = 0 \ , \tag{A.67.3}$$

which is the secular equation for the tensor $[A_{ij}]$ and implies:

$$\lambda^3 - {}^1I \lambda^2 + {}^2I \lambda - {}^3I = 0 \ , \tag{A.67.4}$$

where the coefficients ${}^iI$, which must be independent of the coordinates and are usually designated as the first, second, and third invariants of the tensor $[A_{ij}]$, are defined as:

$$^1I \equiv A_{ii} \ ,$$

$$^2I \equiv \begin{vmatrix} A_{22} & A_{23} \\ A_{32} & A_{33} \end{vmatrix} + \begin{vmatrix} A_{11} & A_{13} \\ A_{31} & A_{33} \end{vmatrix} + \begin{vmatrix} A_{11} & A_{12} \\ A_{21} & A_{22} \end{vmatrix} \ ,$$

$$^3I \equiv \left| A_{ij} \right| \ . \tag{A.67.5}$$

In this particular case the invariants are:

$$^1I = 0 \ ,$$

$$^2I = A_{23}^2 + A_{13}^2 + A_{12}^2 \ ,$$

$$^3I = 0 \ . \tag{A.67.6}$$

Note that for ${}^2I$ each of the products within the parentheses is negative. Omitting the zero terms in the secular equation (A.67.4), we thus have:

$$\lambda^3 + \left(A_{23}^2 + A_{13}^2 + A_{12}^2 \right) \lambda = 0 \ . \tag{A.67.7}$$

One of the solutions is ${}^1\lambda = 0$; divide the eq. (A.67.7) by λ and obtain:

$$\lambda^2 = -\left(A_{23}^2 + A_{13}^2 + A_{12}^2\right) , \qquad (A.67.8)$$

a quadratic equation with the solutions:

$$^{2,3}\lambda = \pm\langle-\left(A_{23}^2 + A_{13}^2 + A_{12}^2\right)\rangle^{1/2} = 0 , \qquad (A.67.9)$$

where the second equality is derived by the following argument: The quantity in the parentheses can only be positive or zero, and thus the square root must be either imaginary or zero (and for the quantity in parentheses to be zero, A_{12}, A_{13}, and A_{23}, and hence each of the nine components of A_{ij}, must be zero). Thus, if eq. (A.67.1) holds, p_i can be parallel to q_i only for $\lambda = 0$. It follows that, for p_i and q_i to be parallel, either (for all three λ):

q_i is finite and $p_i = 0$, or

p_i is finite and q_i is infinite,

or (for $^2\lambda$ or $^3\lambda$):

q_i is real and p_i is imaginary, or

p_i is real and q_i is imaginary.

Hence, real, nonzero, finite p_i and q_i cannot be parallel if they are related to each other by A_{ij}.

General rule: An antisymmetric tensor of the second rank has no principal directions.

Alternative approach to Problem 67. The postulated relationship between the vectors $[p_i]$ and $[q_i]$ and the tensor $[A_{ij}]$ is:

$$p_i = A_{ij} q_j . \qquad (A.67.10)$$

It can be shown that in eq. (A.67.10), as long as $A_{ij} = -A_{ji}$, $[p_i]$ is at right angles to $[q_i]$ for all $[q_j]$. To prove this, multiply both sides of eq. (A.67.10) with q_i:

$$p_i q_i = A_{ij} q_j q_i . \qquad (A.67.11)$$

This is the dot product of $[p_i]$ with $[q_i]$. Since all terms on the right-hand side cancel, the dot product is always zero and the vectors are always mutually orthogonal and cannot satisfy the condition for a principal direction that they be parallel. Note that $[A_{ij}]$ can be written in terms of the axial vector $[r_i]$ (see Problem 66):

$$\left[A_{ij}\right] = \begin{bmatrix} 0 & -r_3 & r_2 \\ r_3 & 0 & -r_1 \\ -r_2 & r_1 & 0 \end{bmatrix} . \qquad \text{(i)}$$

By trying various **r** and **q** that have components on only one coordinate axis, it is possible to verify that the following must be true for the set of axis-parallel vectors and, because the finding must be independent of the choice of coordinate axes, for all vectors **r** and **q**:

$$p_i = \epsilon_{ijk} r_j q_k .$$ (ii)

Thus **p** is the cross product **r** × **q**, and therefore orthogonal not only to **q** but also to **r**. Comparing the roles of A_{ij} in eq. (A.67.10) and of $\epsilon_{ijk} r_j$ in eq. (ii), it is possible to state:

$$A_{ik} = \epsilon_{ijk} r_j .$$ (iii)

Answer 68. For the tensor:

$$\left[S_{ij} \right] = \begin{bmatrix} S_{11} & 0 & 0 \\ 0 & S_{22} & S_{23} \\ 0 & S_{23} & S_{33} \end{bmatrix} ,$$ (A.68.1)

the secular equation is:

$$(S_{11} - \lambda) \left\langle (S_{22} - \lambda)(S_{33} - \lambda) - S_{23}^2 \right\rangle = 0 .$$ (A.68.2)

If $\lambda = S_{11}$, then the leading term in parentheses becomes zero and the equation is true, no matter what the value of the brackets. Thus one of the roots of the equation, and hence a principal value of S_{ij}, is S_{11}. For a more formal proof, one may assume that the given tensor governs a relationship between vectors $p_i = S_{ij} q_j$ and investigate the full set of conditions for x_1 being a principal axis of a tensor relating p_i to q_i. They are:

$$p_1 = \lambda q_1 ,$$ (i)

$$p_1 = S_{1j} q_j ,$$ (ii)

$$q_1 = C x_1 .$$ (iii)

Conditions (ii) and (iii) are jointly satisfied by:

$$[p_i] = \begin{bmatrix} S_{11} q_1 & 0 & 0 \end{bmatrix} .$$ (A.68.3)

Hence p_i is proportional to q_i. The same statement in subscript notation is:

$$p_i = S_{11} q_i ,$$ (A.68.4)

and fulfills condition (i).

Answer 69. (a) The determinant to be investigated is explicitly:

$$\begin{vmatrix} S_{11} & S_{12} \\ S_{12} & S_{22} \end{vmatrix} = S_{11}S_{22} - S_{12}^2 \,. \tag{A.69.1}$$

The missing (or implied rotation) axis for the appropriate Mohr circle construction is x_3; hence $R = 3$, and therefore, modulo 3, $a = 1$ and $b = 2$. The vertical line from the abscissa to the point for $S_{ab} = S_{12}$ must therefore be erected (or dropped if negative) from the point for $S_{bb} = S_{22}$. Figure A.69.1 is sketched for positive S_{11}, S_{12}, and S_{22}. By application of the Pythagorean theorem to the right triangle in Fig. A.69.1, below the radius r:

$$\langle {}^1\!/_2(S_{22} - S_{11})\rangle^2 + S_{12}^2 = r^2 \,. \tag{A.69.2}$$

To solve for $S_{11}S_{22}$ in eq. (A.69.1), we use the binomial theorem:

$$(a \pm b)^2 = a^2 \pm 2ab + b^2 \,, \tag{i}$$

subtract the two versions of eq. (i) with different signs from each other:

$$(a + b)^2 - (a - b)^2 = 4ab \,, \tag{ii}$$

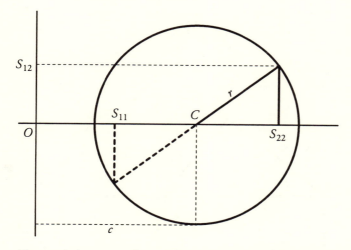

Figure A.69.1.   Mohr circle construction for a positive rotation of axes about x_3, $R = 3$. O – origin of Mohr space coordinates with the abscissa as axis of diagonal, ordinate of off-diagonal components. C – Mohr circle center. r – radius. c – distance OC (same conventions for all Mohr circle constructions).

and solve for ab:

$$ab = 1/4 (a + b)^2 - 1/4 (a - b)^2 . \tag{iii}$$

With the help of eq. (iii), we can write:

$$S_{11} S_{22} = \langle 1/2 (S_{11} + S_{22}) \rangle^2 - \langle 1/2 (S_{11} - S_{22}) \rangle^2 . \tag{A.69.3}$$

Substituting eq. (A.69.3) into eq. (A.69.1) yields:

$$\begin{vmatrix} S_{11} & S_{12} \\ S_{12} & S_{22} \end{vmatrix} = \langle 1/2 (S_{11} + S_{22}) \rangle^2 - \langle 1/2 (S_{11} - S_{22}) \rangle^2 - S_{12}^2 . \tag{A.69.4}$$

Substitute eq. (A.69.2) for the last two terms of eq. (A.69.4):

$$\begin{vmatrix} S_{11} & S_{12} \\ S_{12} & S_{22} \end{vmatrix} = \langle 1/2 (S_{11} + S_{22}) \rangle^2 - r^2 . \tag{A.69.5}$$

Inspection of Fig. A.69.1 shows that the line $OC \equiv c$ has the length:

$$c = 1/2 (S_{11} + S_{22}) , \tag{A.69.6}$$

and that neither c nor r changes with rotation about x_3. Substitution of eq. (A.69.6) into eq. (A.69.5) results in an equation in which the right-hand side, and hence the determinant, is invariant to such a rotation:

$$\begin{vmatrix} S_{11} & S_{12} \\ S_{12} & S_{22} \end{vmatrix} = c^2 - r^2 \therefore \tag{A.69.7}$$

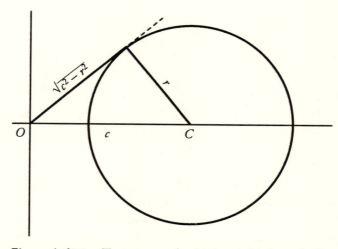

Figure A.69.2. Tangent on the Mohr circle through the origin.

(b) A geometric interpretation exists for the square root of the difference on the right-hand side of eq. (A.69.7) if the Mohr circle does not enclose the coordinate origin of Mohr space, or algebraically, if the difference is not negative. Inspection of Fig. A.69.2 shows that the distance between the origin and the tangent point, measured along the tangent on the Mohr circle through the origin, has the length $(c^2 - r^2)^{1/2}$. It is one of the orthogonal sides of the right triangle that has r as the other orthogonal side and c, the distance from O to C, as its hypotenuse.

Answer 70.

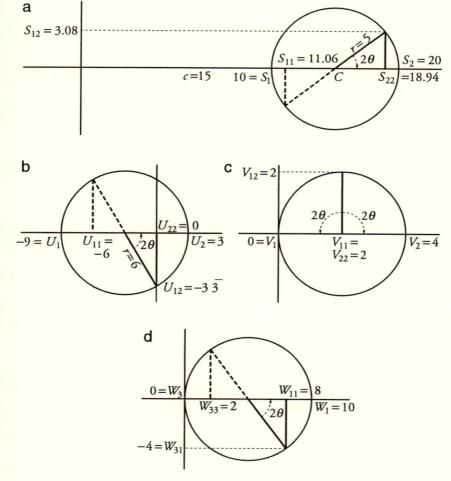

Figure A.70.1. Four Mohr circle constructions.

(a) Figure A.70.1a (for conventions, see Fig. A.69.1), for $R = 3$, $a = 1$, $b = 2$, shows:

$$c = \frac{1}{2}(S_{11} + S_{22}) = 15 , \tag{A.70.1}$$

$$r^2 = \frac{1}{4}(S_{22} - S_{11})^2 + S_{12}^2$$

$$= 7.88^2/4 + 3.08^2 \approx 25 , \tag{A.70.2}$$

$$r \approx 5 , \tag{A.70.3}$$

$$S_1 = c - r = 15 - 5 = 10 ,$$

$$S_2 = c + r = 15 + 5 = 20 ,$$

$$S_3 = 43 , \tag{A.70.4}$$

$$2\theta = -\cos^{-1}\left\langle (S_{22} - c)/r \right\rangle$$

$$= -\cos^{-1}(3.94/5) \approx -38° , \tag{A.70.5}$$

$$\theta \approx -19° . \tag{A.70.6}$$

Note that the sign of θ is for the rotation *from* given *to* principal coordinates.
(b) Figure A.70.1b , also for $R = 3$, $a = 1$, $b = 2$, shows:

$$c = -6/2 = -3 , \tag{A.70.7}$$

$$r^2 = 36/4 + 27 = 36 , \tag{A.70.8}$$

$$U_1 = -3 - 6 = -9 ,$$

$$U_2 = -3 + 6 = 3 ,$$

$$U_3 = 10 , \tag{A.70.9}$$

$$2\theta = \cos^{-1}(1/2) = 60° , \tag{A.70.10}$$

$$\theta = 30° . \tag{A.70.11}$$

(c) Similarly, still for $R = 3$, $a = 1$, $b = 2$, Fig. A.70.1c indicates:

$$V_1 = 0 , \quad V_2 = 4 , \quad V_3 = 9 , \quad \text{or}$$

$$V_1 = 4 , \quad V_2 = 0 , \quad V_3 = 9 , \tag{A.70.12}$$

$$2\theta = \pm 90° , \tag{A.70.13}$$

$$\theta = \pm 45° . \tag{A.70.14}$$

General rule: any two-dimensional symmetric tensor with equal diagonal components is brought to its principal orientation by a rotation of 45° in either direction; its principal values equal the value of the original diagonal components plus and minus that of the original off-diagonal components.

(d) Figure A.70.1d, for $R = 2$, $a = 3$, $b = 1$, indicates the following solutions:

$$W_1 = 10 \ , \ W_2 = 12 \ , \ W_3 = 0 \ , \tag{A.70.15}$$
$$2\theta = \cos^{-1}(3/5) \approx 53.1° \ , \tag{A.70.16}$$
$$\theta \approx 26.6° \ . \tag{A.70.17}$$

Answer 71. Rotate the given tensor:

$$[S_{ij}] = \begin{bmatrix} -1 & 3 & 8 \\ 3 & 10 & 6 \\ 8 & 6 & 2 \end{bmatrix} \ , \tag{A.71.1}$$

about the x_3 axis by:

$$S'_{ij} = a_{ik} \, a_{jl} \, S_{kl} \ , \tag{A.71.2}$$

where:

$$(a_{ij}) = \begin{pmatrix} \cos\theta & \sin\theta & 0 \\ -\sin\theta & \cos\theta & 0 \\ 0 & 0 & 1 \end{pmatrix} \ . \tag{A.71.3}$$

Hence:

$$
\begin{aligned}
S'_{22} &= a_{21} \, a_{21} \, S_{11} + a_{21} \, a_{22} \, S_{12} + a_{22} \, a_{21} \, S_{21} + a_{22} \, a_{22} \, S_{22} \\
&= S_{11} \sin^2\theta + 2 S_{12} \sin\theta \cos\theta + S_{22} \cos^2\theta \ .
\end{aligned} \tag{A.71.4}
$$

Substitute:

$$S'_{22} = 0 \ , \tag{A.71.5}$$

into eq. (A.71.4) and divide the result by $\cos^2\theta$:

$$S_{11} \tan^2\theta - 2 S_{12} \tan\theta + S_{22} = 0 \ . \tag{A.71.6}$$

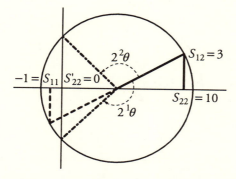

Figure A.71.1. Mohr construction to
obtain zero S_{22}.

This is a quadratic equation in $\tan\theta$. By standard methods:

$$\tan\theta = \frac{S_{12} \pm \left(S_{12}^2 - S_{11}S_{22}\right)^{1/2}}{S_{11}} = -3 \pm \sqrt{19} \approx \begin{cases} -7.36 \\ 1.36 \end{cases} \qquad \text{(A.71.7)}$$

Hence:

$$^1\theta \approx -82.26° \left(-82° \, 16'\right) ,$$
$$^2\theta \approx \quad 53.65° \left(\, 53° \, 38'\right) . \qquad \text{(A.71.8)}$$

The rotation about x_3 moves the positive x_1 toward the positive x_2. In Mohr space, Fig. A.71.1, $R = 3$, $a = 1$, $b = 2$, and the radius is:

$$r^2 = S_{12}^2 + \left\langle S_{22} - (S_{11} + S_{22})/2 \right\rangle^2$$
$$= 9 + \langle 10 - 4.5 \rangle^2 = 39.25 ,$$
$$r = 6.265 . \qquad \text{(A.71.9)}$$

Answer 72. (a) Given is a tensor:

$$\left[S_{ij} \right] = \begin{bmatrix} -1 & 0 & 0 \\ 0 & 7 & -3 \\ 0 & -3 & -1 \end{bmatrix} , \qquad \text{(A.72.1)}$$

and a rotation through $45°$ about the x_1 axis requires $R = 1$, $a = 2$, $b = 3$. Inspection of Fig. A.72.1 shows that right triangles above and below the abscissa (say, those to the right of the circle center) are congruent in Mohr space: Both have the radius r as hypotenuse, and one has S_{23}, the other S'_{23} as one of the orthogonal sides.

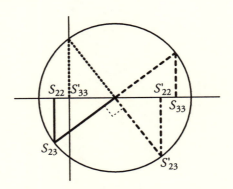

Figure A.72.1. Rotation through $45°$
by Mohr circle construction.

General rule: Because a 45° rotation about a coordinate axis is represented by a 90° rotation in Mohr space, this congruence holds for all 45° rotations.

Because of the known abscissa of the center of the Mohr circle and because of the congruence:

$$C = (S_{22} + S_{33})/2 = 3 ,$$
$$S'_{22} = C + S_{23} = 0 ,$$
$$S'_{33} = C - S_{23} = 6 ,$$
$$S'_{23} = S_{33} - C = -4 . \tag{A.72.2}$$

Hence:

$$[S'_{ij}] = \begin{bmatrix} -1 & 0 & 0 \\ 0 & 0 & -4 \\ 0 & -4 & 6 \end{bmatrix} . \tag{A.72.3}$$

(b) Because:

$$r = \sqrt{3^2 + 4^2} = 5 , \tag{A.72.4}$$

the same tensor in principal coordinates is:

$$[S'_{ij}] = \begin{bmatrix} -1 & 0 & 0 \\ 0 & 8 & 0 \\ 0 & 0 & -2 \end{bmatrix} , \tag{A.72.5}$$

and the angle θ of rotation from original to principal coordinates is:

$$-2\theta = \cos^{-1} \langle (C - S_{33})/r \rangle$$
$$= \cos^{-1} (4/5)$$
$$\approx 36.8° ,$$
$$\theta \approx -18.4° . \tag{A.72.6}$$

Answer 73. (a) Given the tensor:

$$[U_{ij}] = \begin{bmatrix} -8 & 0 & 5\sqrt{3} \\ 0 & 5\sqrt{3} & 0 \\ 5\sqrt{3} & 0 & 2 \end{bmatrix} , \tag{A.73.1}$$

a rotation of 30° about x_2 with $R = 2$, $a = 3$, $b = 1$, is represented in Mohr space by Fig. A.73.1. The rotation matrix is:

$$\left({}^{+30°}a_{ij} \right) = \begin{pmatrix} \sqrt{3}/2 & 0 & -1/2 \\ 0 & 1 & 0 \\ 1/2 & 0 & \sqrt{3}/2 \end{pmatrix} . \tag{A.73.2}$$

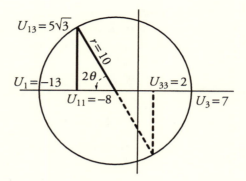

Figure A.73.1. Rotation through 30°, and
to principal coordinates, about the x_2 axis.

The rotated tensor happens to be in principal coordinates:

$$\left[^{+30°}U_{ij}\right] = \left[^{P}U_{ij}\right] = \begin{bmatrix} -13 & 0 & 0 \\ 0 & 5\sqrt{3} & 0 \\ 0 & 0 & 7 \end{bmatrix}. \tag{A.73.3}$$

Given:

$$\left[V_{ij}\right] = \begin{bmatrix} -8 & -2\sqrt{3} & 0 \\ -2\sqrt{3} & -4 & 0 \\ 0 & 0 & 2\sqrt{3} \end{bmatrix}, \tag{A.73.4}$$

and $R = 3$, $a = 1$, $b = 2$, Fig. A.73.2 shows in Mohr space the rotation through $\theta = 30°$ about x_3. The rotation matrix is:

$$\left(a_{ij}\right) = \begin{pmatrix} \sqrt{3}/2 & -1/2 & 0 \\ 1/2 & \sqrt{3}/2 & 0 \\ 0 & 0 & 1 \end{pmatrix}, \tag{A.73.5}$$

and the rotated tensor is:

$$\left[^{+30°}V_{ij}\right] = \begin{bmatrix} -4 & -2\sqrt{3} & 0 \\ -2\sqrt{3} & -8 & 0 \\ 0 & 0 & 2\sqrt{3} \end{bmatrix}. \tag{A.73.6}$$

(b) The principal values and orientation of $[U_{ij}]$ have already been found. Inspection of Fig. A.73.2 and the symmetry in Mohr space of the rotated and unrotated tensors of eqs. (A.73.1) and (A.73.6) show that the angle θ for

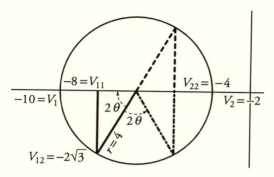

Figure A.73.2. Rotation through 30° about the
x_3 axis.

rotation of $[V_{ij}]$ to principal coordinates is $-30°$. This rotation makes x_2 move toward x_1, and in principal coordinates the tensor is:

$$[^P V_{ij}] = \begin{bmatrix} -10 & 0 & 0 \\ 0 & -2 & 0 \\ 0 & 0 & 2\sqrt{3} \end{bmatrix} . \tag{A.73.7}$$

Answer 74. Figure A.74.1a shows the rotation to the nearest principal coordinates of the given tensor:

$$[W_{ij}] = \begin{bmatrix} 12.52 & 0 & -7.12 \\ 0 & 7.00 & 0 \\ -7.12 & 0 & 1.48 \end{bmatrix} . \tag{A.74.1}$$

$R = 2$, $a = 3$, $b = 1$, and the radius r of the Mohr circle is:

$$r = \sqrt{5.52^2 + 7.12^2} \approx 9.01 . \tag{A.74.2}$$

The same tensor, in the principal coordinates nearest the coordinates in which it was given, is:

$$[^P W_{ij}] \approx \begin{bmatrix} 16.01 & 0 & 0 \\ 0 & 7.00 & 0 \\ 0 & 0 & -2.01 \end{bmatrix} . \tag{A.74.3}$$

The angle of rotation to principal coordinates is:

$$\theta = \tfrac{1}{2} \tan^{-1}\left(\frac{7.12}{5.52}\right) \approx +26.1° . \tag{A.74.4}$$

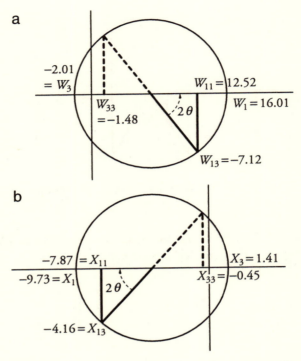

Figure A.74.1. Rotations to nearest principal coordinates.

The x_3 axis moves toward x_1. Further given (Fig. A.74.1b) is:

$$[X_{ij}] = \begin{bmatrix} -7.87 & 0 & -4.16 \\ 0 & -4.16 & 0 \\ -4.16 & 0 & -0.45 \end{bmatrix}. \qquad (A.74.5)$$

The axes are the same as for the previous tensor, and the Mohr circle radius is:

$$r = \sqrt{3.71^2 + 4.16^2} \approx 5.57 . \qquad (A.74.6)$$

In principal coordinates the tensor is:

$$[{}^{P}X_{ij}] \approx \begin{bmatrix} -9.73 & 0 & 0 \\ 0 & -4.16 & 0 \\ 0 & 0 & 1.41 \end{bmatrix}, \qquad (A.74.7)$$

and the angle of rotation to principal coordinates is:

$$\theta = {}^{-1}/_2 \tan^{-1}\left(\frac{4.16}{3.71}\right) \approx -24.1°. \tag{A.74.8}$$

The x_1 axis moves toward x_3.

Answer 75. In order to rotate the given tensor:

$$\left[S_{ij}\right] = \begin{bmatrix} -3 & 0 & 0 \\ 0 & 4 & -6 \\ 0 & -6 & -10 \end{bmatrix}, \tag{A.75.1}$$

about the x_1 axis ($R=1$, $a=1$, $b=2$) by the smallest angle to yield the component $S_{22}=6$, we calculate the radius r of the Mohr circle (see Fig. A.75.1) and the new triangle with r as hypotenuse and a side of length 9:

$$r = \sqrt{7^2+6^2} = \sqrt{49+36} = \sqrt{81+4} = \sqrt{9^2+2^2}, \tag{A.75.2}$$

The rotation θ is:

$$\theta = {}^{-1}/_2 \langle \tan^{-1}(6/7) - \tan^{-1}(2/9)\rangle \approx -14°, \tag{A.75.3}$$

a rotation that moves the x_3 axis moves toward x_2. The rotated tensor is:

$$\left[S'_{ij}\right] = \begin{bmatrix} -3 & 0 & 0 \\ 0 & 6 & -2 \\ 0 & -2 & -12 \end{bmatrix}. \tag{A.75.4}$$

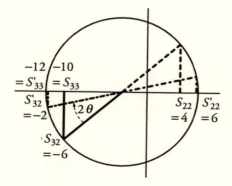

Figure A.75.1. Rotation to produce a specified tensor component.

Answer 76. Given the stress:

$$[\sigma_{ij}] = \begin{bmatrix} 1 & 3 & 2 \\ 3 & -9 & -6 \\ 2 & -6 & 16 \end{bmatrix} \text{ kPa ,} \qquad (A.76.1)$$

use Cauchy's formula:

$$p_i = \sigma_{ij} \, l_j \, , \qquad (A.76.2)$$

where l_j is the unit vector normal to the plane on which the traction p_i acts. To find the force F_i acting on an area A, use:

$$F_i = A \, p_i \, , \qquad (A.76.3)$$

The unit outward normal l_j is obtained by normalizing the given vector v_j:

$$[l_i] = \frac{1}{\sqrt{3}} [\, 1 \quad 1 \quad 1 \,] \, , \qquad (A.76.4)$$

and the force is:

$$[F_i] = [A \, \sigma_{ij} \, l_j] = \frac{1}{\sqrt{3}} [\, 6 \quad -12 \quad 12 \,] \text{ kN .} \qquad (A.76.5)$$

Note that $A = 1 \, \text{m}^2$. (a) To find the magnitude N of the normal component of the force F_i acting on A, we determine the length of the projection of the force vector onto the unit normal:

$$N = F_i \, l_i = 2 \text{kN} \, , \qquad (A.76.6)$$

(Note the different fonts used for the variable N and the unit N.) To find the force vector of the normal component, the magnitude must be multiplied with the outward normal unit vector l_i:

$$N_i = l_i \, F_j \, l_j \, . \qquad (A.76.7)$$

Thus numerically N_i is:

$$[N_i] = \frac{1}{\sqrt{3}} [\, 2 \quad 2 \quad 2 \,] \text{ kN .} \qquad (A.76.8)$$

(b) Because the total force F_i on the area A must be the sum of the normal and of the tangential forces, and of those forces only, the tangential or shear force T_i is:

$$T_i = F_i - N_i \, , \qquad (A.76.9)$$

or numerically:

$$[T_i] = \frac{1}{\sqrt{3}} [4 \quad -14 \quad 10] \text{ kN} .$$

(A.76.10)

Answer 77. (a) The secular equation for the given stress tensor is:

$$\begin{vmatrix} 4 - \lambda & -2\sqrt{3} & 6 \\ -2\sqrt{3} & 3 - \lambda & -3\sqrt{3} \\ 6 & -3\sqrt{3} & 9 - \lambda \end{vmatrix} = 0 .$$

(A.77.1)

Evaluating a determinant can be simplified by applying manipulation rules for determinants. The goal is to bring as many zeroes as possible into the top row because each zero in that row eliminates two terms from the evaluation. *A determinant can be multiplied by a given factor by multiplying one row or one column by that factor.* Multiply both sides of eq. (A.77.1) by a factor of 3 by applying this rule to row 1 of the determinant. *A determinant is unaltered if a multiple of one row (or column) is added to another row (or column).* Apply that rule by subtracting twice row 3 from row 1. The combined result of these two steps is:

$$\begin{vmatrix} -3\lambda & 0 & 2\lambda \\ -2\sqrt{3} & 3 - \lambda & -3\sqrt{3} \\ 6 & -3\sqrt{3} & 9 - \lambda \end{vmatrix} = 0 .$$

(A.77.2)

The λ in an off-diagonal element of this determinant is undesirable; therefore, multiply both sides of eq. (A.77.2) by a factor of 3 by applying the appropriate rule to column 3, then add twice column 1 to column 3 and obtain:

$$\begin{vmatrix} -3\lambda & 0 & 0 \\ -2\sqrt{3} & 3 - \lambda & -13\sqrt{3} \\ 6 & -3\sqrt{3} & 39 - 3\lambda \end{vmatrix} = 0 .$$

(A.77.3)

In this form the determinant is easy to evaluate as:

$$\lambda^2 (\lambda - 16) = 0 .$$

(A.77.4)

Divide eq. (A.77.4) by λ^2, with the implication that $^1\lambda = {}^2\lambda = 0$:

$$^3\lambda = 16 \text{MPa} .$$

(A.77.5)

It follows that the stress is uniaxial and has the magnitude of 16 megapascals. (b) Let X_i be the radius vector of the representation quadric for the stress, then for the direction of the unique principal stress:

$$\sigma_{ij} X_j = {}^3\lambda X_i .$$

(A.77.6)

Divide both sides of eq. (A.77.6) by the magnitude X of $[X_i]$, drop the identifying left superscript for λ, and rearrange:

$$\sigma_{ij} l_j - \lambda l_i = 0 .$$

(A.77.7)

With $[\sigma_{ij}]$ being:

$$[\sigma_{ij}] = \begin{bmatrix} 4 & -2\sqrt{3} & 6 \\ -2\sqrt{3} & 3 & -3\sqrt{3} \\ 6 & -3\sqrt{3} & 9 \end{bmatrix} \text{ MPa} ,$$

(A.77.8)

solve for the three components of l_i; from the first row of $[\sigma_{ij}]$ and eq. (A.77.7) obtain:

$$(4 - 16) l_1 - 2\sqrt{3} l_2 + 6 l_3 = 0 ,$$

(A.77.9)

and multiply by 3:

$$-36 l_1 - 6\sqrt{3} l_2 + 18 l_3 = 0 .$$

(A.77.10)

From the third row:

$$6 l_1 - 3\sqrt{3} l_2 + (9 - 16) l_3 = 0 ,$$

(A.77.11)

and multiply by (−2):

$$-12 l_1 + 6\sqrt{3} l_2 + 14 l_3 = 0 .$$

(A.77.12)

Add eqs. (A.77.10) and (A.77.12):

$$-48 l_1 + 32 l_3 = 0 ,$$

(A.77.13)

or:

$$l_1 = 2 l_3 / 3 .$$

(A.77.14)

Substitute eq. (A.77.14) into eq. (A.77.10):

$$-24 l_3 - 6\sqrt{3} l_2 + 18 l_3 = 0 ,$$

(A.77.15)

divide by 6:

$$(-4 + 3) l_3 - \sqrt{3} l_2 = 0 ,$$

(A.77.16)

or:

$$l_2 = -l_3 / \sqrt{3} .$$

(A.77.17)

Now express all three components l_i in terms of l_3 and set the sum of their squares equal to unity:

$$1 = (4/9 + 3/9 + 9/9)\, l_3^2 = 16/9\, l_3^2 \ . \tag{A.77.18}$$

Hence:

$$l_3 = \pm\sqrt{9/16} = \pm 3/4 \ . \tag{A.77.19}$$

Substitute eq. (A.77.19) into eqs. (A.77.14) and (A.77.17), and assemble the unit vector parallel to the unique tensile component of $[\,\sigma_{ij}\,]$:

$$\left[\, l_i \,\right] = \left[\ 1/2 \quad -\sqrt{3}/4 \quad 3/4\ \right] \ . \tag{A.77.20}$$

Answer 78. In principal coordinates, the deviator of a general stress is:

$$\left[\,{}^P\!\Delta_{ij}\,\right] = \begin{bmatrix} (\sigma_1 + p) & 0 & 0 \\ 0 & (\sigma_2 + p) & 0 \\ 0 & 0 & (\sigma_3 + p) \end{bmatrix}, \tag{A.78.1}$$

where p, the pressure, is the negative of the mean normal stress:

$$p = -(\sigma_1 + \sigma_2 + \sigma_3)/3 \ . \tag{A.78.2}$$

Note that the algebraic addition of p in eq. (A.78.1) represents a physical subtraction because of the opposite sign conventions for normal stress (tension positive) and hydrostatic pressure (compression positive). Thus, for principal coordinates, we can state the contents of eq. (A.78.1) in subscript notation:

$$^P\!\Delta_{ij} = {}^P\!\sigma_{ij} + p\,\delta_{ij} \ . \tag{A.78.3}$$

Considering that, like all tensors of the second rank with identical principal values, δ_{ij} is invariant to rotation, this may be generalized for arbitrary orientations:

$$\Delta_{ij} = \sigma_{ij} + p\,\delta_{ij} \ . \tag{A.78.4}$$

In subscript notation eq. (A.78.2) becomes:

$$p = -\sigma_{kk}/3 \ . \tag{A.78.5}$$

Substituting eq. (A.78.5) into eq. (A.78.4) yields the proof:

$$\Delta_{ij} = \sigma_{ij} - \delta_{ij}\,\sigma_{kk}/3 \quad \therefore \tag{A.78.6}$$

Answer 79. In terms of principal values λ, (a) a uniaxial stress $(\sigma_2 = \sigma_3 = 0)$ is:

$$\lambda^2 (\lambda - \sigma) = 0 , \tag{A.79.1}$$

or:

$$\lambda^3 - \lambda^2 \sigma = 0 . \tag{A.79.2}$$

(b) A biaxial stress $(\sigma_3 = 0)$ is:

$$\lambda (\lambda - \sigma_1) (\lambda - \sigma_2) = 0 , \tag{A.79.3}$$

or:

$$\lambda^3 - \lambda^2 (\sigma_1 + \sigma_2) + \lambda \sigma_1 \sigma_2 = 0 . \tag{A.79.4}$$

(c) A triaxial stress $(\sigma_1 \neq \sigma_2 \neq \sigma_3 \neq \sigma_1 \neq 0, \sigma_2 \neq 0, \sigma_3 \neq 0)$ is:

$$(\lambda - \sigma_1) (\lambda - \sigma_2) (\lambda - \sigma_3) = 0 , \tag{A.79.5}$$

or:

$$\lambda^3 - \lambda^2 (\sigma_1 + \sigma_2 + \sigma_3) + \lambda (\sigma_1 \sigma_2 + \sigma_2 \sigma_3 + \sigma_3 \sigma_1) + \sigma_1 \sigma_2 \sigma_3 = 0 . \tag{A.79.6}$$

(d) A pressure $(\sigma_1 = \sigma_2 = \sigma_3 = -p)$ is:

$$(\lambda + p) (\lambda + p) (\lambda + p) = 0 , \tag{A.79.7}$$

or:

$$(\lambda + p)^3 = 0 , \tag{A.79.8}$$

or:

$$\lambda^3 + 3p \lambda^2 + 3p^2 \lambda + p^3 = 0 . \tag{A.79.9}$$

(e) A pure shear stress $(\sigma_2 = -\sigma_1, \sigma_3 = 0)$ is:

$$\lambda (\lambda - \sigma) (\lambda + \sigma) = 0 , \tag{A.79.10}$$

or:

$$\lambda^3 - \lambda \sigma^2 = 0 . \tag{A.79.11}$$

Note the difference between eqs. (A.79.2) and (A.79.11).

Answer 80. (a) In terms of the nondiagonal components τ_{ij} of the stress tensor σ_{ij}, the body torque G_i is:

$$G_i = \epsilon_{ijk} \tau_{jk} . \tag{A.80.1}$$

or explicitly:

$$\tau_{32} - \tau_{23} + G_1 = 0 ,$$
$$\tau_{13} - \tau_{31} + G_2 = 0 ,$$
$$\tau_{21} - \tau_{12} + G_3 = 0 . \tag{A.80.2}$$

Since the stress is antisymmetric and $\tau_{ij} = -\tau_{ji}$:

$$\begin{bmatrix} G_i \end{bmatrix} = 2\begin{bmatrix} \tau_{23} & \tau_{31} & \tau_{12} \end{bmatrix} . \tag{A.80.3}$$

Note the factor of two which replaces the implied summation in eq. (A.80.1). (b) Axial vectors do not change sign upon inversion. To test G_i for axiality, invert it by applying to the components τ_{ij} in eq. (A.80.3) the transformation matrix for inversion, the negative Kronecker delta [see eq. (A.30.4)]:

$$\begin{bmatrix} G_i' \end{bmatrix} = 2\begin{bmatrix} \delta_{22}\,\delta_{33}\,\tau_{23} & \delta_{33}\,\delta_{11}\,\tau_{31} & \delta_{11}\,\delta_{22}\,\tau_{12} \end{bmatrix} . \tag{A.80.4}$$

The negative signs of the pairs of Kronecker deltas cancel, and therefore:

$$\begin{bmatrix} G_i' \end{bmatrix} = 2\begin{bmatrix} \tau_{23} & \tau_{31} & \tau_{12} \end{bmatrix} = \begin{bmatrix} G_i \end{bmatrix} . \tag{A.80.5}$$

This result also holds for a reflection, which is the cumulative effect of an inversion and a rotation. (c) Since in the antisymmetric stress tensor $\sigma_{11} = \sigma_{22} = \sigma_{33} = 0$:

$$p = -\sigma_{ii}/3 = 0 . \tag{A.80.6}$$

Answer 81. Given the homogeneous, asymmetric stress:

$$\begin{bmatrix} \sigma_{ij} \end{bmatrix} = \begin{bmatrix} 10 & 30 & 20 \\ 30 & -90 & -60 \\ 30 & 40 & 50 \end{bmatrix} \text{kPa} , \tag{A.81.1}$$

(a) the body torque per unit volume has the components:

$$G_1 = \sigma_{23} - \sigma_{32} = -60 - 40 = -100 \text{ kPa} ,$$
$$G_2 = \sigma_{31} - \sigma_{13} = 30 - 20 = 10 \text{ kPa} ,$$
$$G_3 = \sigma_{12} - \sigma_{21} = 30 - 30 = 0 \text{ kPa} . \tag{A.81.2}$$

Thus:

$$\begin{bmatrix} G_i \end{bmatrix} = \begin{bmatrix} -100 & 10 & 0 \end{bmatrix} \text{kPa} . \tag{A.81.3}$$

(b) The hydrostatic pressure is:

$$p = -\sigma_{ii}/3 = 10 \text{ kPa} . \tag{A.81.4}$$

This is a small pressure, approximately one-hundredth of the standard atmospheric pressure at sea level.

Answer 82. (a) In a medium that is not accelerating, the body force is:

$$\sigma_{ij,j} + F_i = 0 \ . \tag{A.82.1}$$

Given the stress field:

$$\left[\sigma_{ij}\right] = \begin{bmatrix} x_1+1 & x_1+x_2 & x_1+x_3 \\ x_2+x_1 & x_2+2 & x_2+x_3 \\ x_3+x_1 & x_3+x_2 & x_3+3 \end{bmatrix} \text{Pa} \ , \tag{A.82.2}$$

the components of the body force are:

$$-F_1 = \sigma_{1j,j} \equiv \frac{\partial \sigma_{11}}{\partial x_1} + \frac{\partial \sigma_{12}}{\partial x_2} + \frac{\partial \sigma_{13}}{\partial x_3} \ ,$$

$$-F_2 = \sigma_{2j,j} \equiv \frac{\partial \sigma_{21}}{\partial x_1} + \frac{\partial \sigma_{22}}{\partial x_2} + \frac{\partial \sigma_{23}}{\partial x_3} \ ,$$

$$-F_3 = \sigma_{3j,j} \equiv \frac{\partial \sigma_{31}}{\partial x_1} + \frac{\partial \sigma_{32}}{\partial x_2} + \frac{\partial \sigma_{33}}{\partial x_3} \ . \tag{A.82.3}$$

(b) At the point $[x_i] = [1 \ 2 \ 4]$m, and also at all other points in the medium of eq. (A.82.2) independently of the numerical values of x_i, the components of the body force per m^3 are:

$$F_1 = -(1+1+1) = -3 \text{ N} \ ,$$
$$F_2 = -(1+1+1) = -3 \text{ N} \ ,$$
$$F_3 = -(1+1+1) = -3 \text{ N} \ , \tag{A.82.4}$$

and the force vector is:

$$\left[F_i\right] = -3\begin{bmatrix} 1 & 1 & 1 \end{bmatrix} \text{N} \ . \tag{A.82.5}$$

Answer 83. The equation:

$$\sigma_{ij,j} = 0 \ , \tag{A.83.1}$$

implies that the total force on any small region of the medium subject to the stress σ_{ij} is zero, or, in other words, that all forces balance. This is trivially true for homogeneous stress, but also nontrivially for many analytically solved cases of inhomogeneous stress, such as those referring to ideally elastic bodies free of body forces, or neglecting body forces if they are insignificantly small. It is not true for "dynamic" stress solutions, such as the wave equations.

Answer 84. The secular equation can be expanded so that each of its elements is a difference:

$$\begin{vmatrix} \sigma_{11}-\lambda & \sigma_{12}-0 & \sigma_{13}-0 \\ \sigma_{12}-0 & \sigma_{22}-\lambda & \sigma_{23}-0 \\ \sigma_{13}-0 & \sigma_{23}-0 & \sigma_{33}-\lambda \end{vmatrix} = 0 \ . \qquad\qquad (A.84.1)$$

Since each element of the first row of this determinant is a sum of two terms, the total determinant equals the sum of the two determinants formed by the summands of that row or column together with the remainder of the determinant:

$$\begin{vmatrix} \sigma_{11}-\lambda & \sigma_{12}-0 & \sigma_{13}-0 \\ \sigma_{12}-0 & \sigma_{22}-\lambda & \sigma_{23}-0 \\ \sigma_{13}-0 & \sigma_{23}-0 & \sigma_{33}-\lambda \end{vmatrix} =$$

$$\begin{vmatrix} \sigma_{11} & \sigma_{12} & \sigma_{13} \\ \sigma_{12}-0 & \sigma_{22}-\lambda & \sigma_{23}-0 \\ \sigma_{13}-0 & \sigma_{23}-0 & \sigma_{33}-\lambda \end{vmatrix} + \begin{vmatrix} -\lambda & 0 & 0 \\ \sigma_{12}-0 & \sigma_{22}-\lambda & \sigma_{23}-0 \\ \sigma_{13}-0 & \sigma_{23}-0 & \sigma_{33}-\lambda \end{vmatrix} . \qquad (A.84.2)$$

Decompose the second row of each of the two separate determinants, and then the third row of each of the four resulting new determinants, according to the following scheme in which, for each decomposed row, $|\sigma|$ indicates a determinant with only components of σ_{ij} and $|\lambda|$ one with only λ:

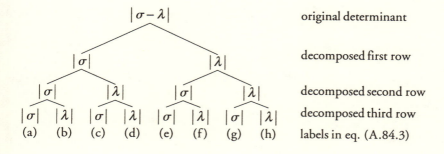

	original determinant
$|\sigma-\lambda|$	
	decomposed first row
$|\sigma|$ $|\lambda|$	
	decomposed second row
$|\sigma|$ $|\lambda|$ $|\sigma|$ $|\lambda|$	
	decomposed third row
$|\sigma|$ $|\lambda|$ $|\sigma|$ $|\lambda|$ $|\sigma|$ $|\lambda|$ $|\sigma|$ $|\lambda|$	labels in eq. (A.84.3)
(a) (b) (c) (d) (e) (f) (g) (h)	

Reordering the decomposed determinants (labels in parentheses) makes the secular equation:

$$
-\begin{vmatrix} \lambda & 0 & 0 \\ 0 & \lambda & 0 \\ 0 & 0 & \lambda \end{vmatrix}
$$
$$
\text{(h)}
$$

$$
+\begin{vmatrix} \sigma_{11} & \sigma_{12} & \sigma_{13} \\ 0 & \lambda & 0 \\ 0 & 0 & \lambda \end{vmatrix}
+\begin{vmatrix} \lambda & 0 & 0 \\ \sigma_{12} & \sigma_{22} & \sigma_{23} \\ 0 & 0 & \lambda \end{vmatrix}
+\begin{vmatrix} \lambda & 0 & 0 \\ 0 & \lambda & 0 \\ \sigma_{13} & \sigma_{23} & \sigma_{33} \end{vmatrix}
$$
$$
\text{(d)} \qquad\qquad \text{(f)} \qquad\qquad \text{(g)}
$$

$$
-\begin{vmatrix} \sigma_{11} & \sigma_{12} & \sigma_{13} \\ \sigma_{12} & \sigma_{22} & \sigma_{23} \\ 0 & 0 & \lambda \end{vmatrix}
-\begin{vmatrix} \lambda & 0 & 0 \\ \sigma_{12} & \sigma_{22} & \sigma_{23} \\ \sigma_{13} & \sigma_{23} & \sigma_{33} \end{vmatrix}
-\begin{vmatrix} \sigma_{11} & \sigma_{12} & \sigma_{13} \\ 0 & \lambda & 0 \\ \sigma_{13} & \sigma_{23} & \sigma_{33} \end{vmatrix}
$$
$$
\text{(b)} \qquad\qquad \text{(e)} \qquad\qquad \text{(c)}
$$

$$
+\begin{vmatrix} \sigma_{11} & \sigma_{12} & \sigma_{13} \\ \sigma_{12} & \sigma_{22} & \sigma_{23} \\ \sigma_{13} & \sigma_{23} & \sigma_{33} \end{vmatrix} = 0 \ . \tag{A.84.3}
$$
$$
\text{(a)}
$$

This simplifies to:

$$
\begin{array}{c} -\lambda^3 + \lambda^2\, \sigma_{ii} \\ \text{(h)} \quad \text{(d)(f)(g)} \end{array}
$$

$$
-\lambda \left(\begin{vmatrix} \sigma_{11} & \sigma_{12} \\ \sigma_{12} & \sigma_{22} \end{vmatrix} + \begin{vmatrix} \sigma_{22} & \sigma_{23} \\ \sigma_{23} & \sigma_{33} \end{vmatrix} + \begin{vmatrix} \sigma_{33} & \sigma_{13} \\ \sigma_{13} & \sigma_{11} \end{vmatrix} \right)
$$
$$
\text{(b)} \qquad\qquad \text{(e)} \qquad\qquad \text{(c)}
$$

$$
+\left| \sigma_{ij} \right| = 0 \ , \tag{A.84.4}
$$
$$
\text{(a)}
$$

and further to:

$$
-\lambda^3 + {}^1\!I\, \lambda^2 - {}^2\!I\, \lambda + {}^3\!I = 0 \ , \tag{A.84.5}
$$

or, in the standard form:

$$
\lambda^3 - {}^1\!I\, \lambda^2 + {}^2\!I\, \lambda - {}^3\!I = 0 \ \therefore \tag{A.84.6}
$$

Answer 85. The transformed tensor is:

$$
T'_{ij} = a_{ik}\, a_{jl}\, T_{kl} \ , \tag{A.85.1}
$$

and its determinant:

$$
\left| T'_{ij} \right| = \left| a_{ik}\, a_{jl}\, T_{kl} \right| \ . \tag{A.85.2}
$$

The rule that the product of determinants $C = A B$ is itself the determinant of the matrix product $\mathbb{C}$ formed from two matrices $\mathbb{A}$ and $\mathbb{B}$ by multiplying each element in the i th row of $\mathbb{A}$ with the corresponding element in the j th row of $\mathbb{B}$ and adding the products to obtain the j th column of $\mathbb{C}$ is applicable when, in subscript notation, the matrix product contains a dummy subscript. In eq. (A.85.2) one dummy subscript is l. This allows us to restate the equation as:

$$\left| T'_{ij} \right| = \left| a_{jl} \right| \left| a_{ik} T_{kl} \right| . \tag{A.85.3}$$

Considering that $\left| a_{ij} \right| = \pm 1$, this implies:

$$\left| T'_{ij} \right| = \pm \left| a_{ik} T_{kl} \right| . \tag{A.85.4}$$

The rule of determinants of matrix products can be applied once more because in eq. (A.85.4) k is a dummy subscript:

$$\left| T'_{ij} \right| = \pm \left| a_{ik} \right| \left| T_{kl} \right| . \tag{A.85.5}$$

Because the determinant of the same transformation matrix must be either $+1$ or -1, the ambiguity of sign vanishes, and we confirm the postulate:

$$\left| T'_{ij} \right| = \left| T_{kl} \right| \quad \therefore \tag{A.85.6}$$

Answer 86. Since the secular equation in terms of the customary (numbered) invariants is:

$$\lambda^3 - {}^1I\,\lambda^2 + {}^2I\,\lambda - {}^3I = 0 , \tag{A.86.1}$$

with the particular invariants having the values of $4, -11$, and -30, this yields:

$$\lambda^3 - 4\lambda^2 - 11\lambda + 30 = 0 . \tag{A.86.2}$$

This cubic equation in λ is easily solved by trial and error (all λ must be small, they are probably integer, at least one of them must be even, and at least one odd):

$$\lambda = 5, 2, -3 . \tag{A.86.3}$$

The check for such a set of trial solutions, but also for algebraic solutions to beware of mistakes, is to substitute each of the solutions into eq. (A.86.2):

$$\begin{aligned}
\lambda = 5 : & \quad 125 - 100 - 55 + 30 = 0 , \\
\lambda = 2 : & \quad 8 - 16 - 22 + 30 = 0 , \\
\lambda = -3 : & \quad -27 - 36 + 33 + 30 = 0 .
\end{aligned}$$

Thus, in the customary descending order of the principal values, the principal stresses are:

$$\sigma_1 = 5, \quad \sigma_2 = 2, \quad \sigma_3 = -3 \text{ units .} \tag{A.86.4}$$

Answer 87. Given the stress:

$$[\sigma_{ij}] = \begin{bmatrix} -5 & 2 & -3 \\ 2 & 2 & 1 \\ -3 & 1 & -1 \end{bmatrix} \text{ kPa ,} \tag{A.87.1}$$

and a direction defined by the vector:

$$[v_i] = [\,1 \quad 3 \quad 5\,] \text{ m ,} \tag{A.87.2}$$

for a unit direction vector (set of direction cosines), normalize the vector v_i:

$$[l_i] = {}^1\!/\!\sqrt{35}[v_i] = {}^1\!/\!\sqrt{35}[\,1 \quad 3 \quad 5\,] \text{ m .} \tag{A.87.3}$$

The traction (force per unit area) on the plane normal to v_i, $P_i = \sigma_{ij}\,l_j$ is:

$$[P_i] = {}^1\!/\!\sqrt{35}\,[(-5+6-15) \ (\,2+6+5) \ (-3+3-5)\,]$$
$$= {}^1\!/\!\sqrt{35}\,[-14 \quad 13 \quad -5\,] \text{ kPa .} \tag{A.87.4}$$

The magnitude N of the normal component of P_i is $N = P_i\,l_i$, or the scalar length of the projection of P_i onto l_i. To obtain the vectorial normal component of the traction, this magnitude is applied as a scalar factor to the vector l_i:

$$[N_i] = \left[\left[P_j\,l_j\right]l_i\right]$$
$$= \boxed{{}^1\!/\!\sqrt{35}\,(-14+39-25)}\ {}^1\!/\!\sqrt{35}[\,1 \quad 3 \quad 5\,]$$
$$= [\,0 \quad 0 \quad 0\,] \text{ kPa ,} \tag{A.87.5}$$

where the framed term represents the magnitude N of the normal force N_i, which happens to be zero. The tangential traction is $T_i = P_i - N_i$, the difference between the total and normal tractions. In the present case N_i is a null vector, and the tangential traction is $T_i = P_i$:

$$[T_i] = {}^1\!/\!\sqrt{35}\,[-14 \quad 13 \quad -5\,] \text{ kPa .} \tag{A.87.6}$$

Answer 88. (a) Given the stress in a cube, $-1 \le x_1 \le 1, -1 \le x_2 \le 1, -1 \le x_3 \le 1$:

$$[\sigma_{ij}] = \begin{bmatrix} -3x_1^2 x_2 & x_1^2 x_3 & 6x_1 x_2 x_3 \\ x_1^2 x_3 & 6x_1 x_2 x_3 & -4x_1 x_3^2 \\ 6x_1 x_2 x_3 & -4x_1 x_3^2 & -3x_2 x_3^2 \end{bmatrix}, \quad \text{(A.88.1)}$$

the body force can be calculated as:

$$F_i = -\sigma_{ij,j}, \quad \text{(A.88.2)}$$

or explicitly for the force component F_1:

$$F_1 = -\left\langle \frac{\partial(-3x_1^2 x_2)}{\partial x_1} + \frac{\partial(-3x_1^2 x_3)}{\partial x_2} + \frac{\partial(6x_1 x_2 x_3)}{\partial x_3} \right\rangle, \quad \text{(A.88.3)}$$

and similarly for F_2 and F_3. The differentiation yields:

$$F_1 = -(-6x_1 x_2 \ + \ 0 \ + \ 6x_1 x_2) = 0 \ ,$$
$$F_2 = -(\ 2x_1 x_3 + 6x_1 x_3 - 8x_1 x_3) = 0 \ ,$$
$$F_3 = -(\ 6x_2 x_3 \ - \ 0 \ - \ 6x_2 x_3) = 0 \ ; \quad \text{(A.88.4)}$$

thus no body force is found to act on the material in the cube. (b) At the surface where $x_3 = -1$, that is, at the surface normal to x_3, the normal traction has only one nonzero component:

$$N_3 = -3x_2 x_3^2 = -3x_2 = \sigma_{33} \ . \quad \text{(A.88.5)}$$

Where x_2 has a positive value, this normal traction (also commonly, but misleadingly, referred to as "*normal stress*") pulls toward positive x_3, that is, inward into the cube, and is thus compressive. Note that the even power of x_3 in eq. (A.88.5) leaves the sign of the traction unchanged despite the negative sign of the value for x_3. (c) The *shear* (or tangential) *traction* (commonly called "*shear stress*") on the same surface has only two nonzero components ($T_3 = 0$); the first is:

$$T_1 = 6x_1 x_2 x_3 = -6x_1 x_2 = -\sigma_{31} \ . \quad \text{(A.88.6)}$$

Note that x_3 occurs here with an odd power and that its sign, therefore, *does* modify the sign of the resulting expression. Where both x_1 and x_2 have positive values the component T_1 of the tangential traction pulls toward the negative direction of x_1. Note also that the change of sign of the traction with the sign change of x_3 does *not* affect the stress component σ_{31}, which refers equally to the tractions on both the positive and negative sides of the cube, a reason not to use the term "stress" for "traction." The second shear traction component is:

$$T_2 = -4x_1 x_3^2 = -4x_1 = \sigma_{32} , \tag{A.88.7}$$

a tangential traction pulling toward negative x_2 wherever x_1 has a positive value ($x_3^2 = +1$).

Answer 89. Given the stress:

$$\left[\sigma_{ij}\right] = \begin{bmatrix} -4 & -2\sqrt{3} & 6 \\ -2\sqrt{3} & 1 & -\sqrt{3} \\ 6 & -\sqrt{3} & 3 \end{bmatrix} \text{ kPa} , \tag{A.89.1}$$

the secular equation is:

$$\lambda^3 - 64\lambda = 0 , \tag{A.89.2}$$

with the solutions:

$$^1\lambda = 0 , \; \left(^{2,3}\lambda\right)^2 = 64$$

$$\lambda = 0, \; 8, -8 . \tag{A.89.3}$$

Thus eq. (A.89.1) represents a pure shear stress with $\sigma = 8\,\text{kPa}$.

Answer 90. (a) The previous answer can be generalized: If both the first and the third invariants of a symmetric tensor of the second rank are zero, the secular equation is:

$$\lambda\left(\lambda^2 + {}^2I\right) = 0 , \tag{A.90.1}$$

with the roots:

$$\lambda = 0 , \; +\sqrt{-{}^2I} , \; -\sqrt{-{}^2I} . \tag{A.90.2}$$

(b) Hence, a stress σ_{ij} with invariants ${}^1I = {}^3I = 0$ is a pure shear stress with the magnitude:

$$\sigma = \sqrt{-{}^2I} . \tag{A.90.3}$$

Answer 91. Given the stress:

$$\left[\sigma_{ij}\right] = \begin{bmatrix} (x_2 + 1) & x_3 & x_1 \\ x_3 & (x_1 + 1) & x_2 \\ x_1 & x_2 & (x_3 + 1) \end{bmatrix} \text{ Pa} , \tag{A.91.1}$$

(a) the body force $F_i = -\sigma_{ij,j}$ is:

$$F_1 = 0 ,$$
$$F_2 = 0 ,$$
$$F_3 = -(1 + 1 + 1) ,$$
$$[F_i] = [\ 0 \quad 0 \quad -3\ ]\ \text{N} , \qquad\qquad (A.91.2)$$

and (b), since σ_{ij} is symmetric, the body torque, $G_i = \epsilon_{ijk}\sigma_{jk}$, is necessarily:

$$[G_i] = [\ 0 \quad 0 \quad 0\ ]\ \text{N} . \qquad\qquad (A.91.3)$$

(c) For a plane with the outward normal:

$$[l_i] = {-1}/{\sqrt{3}}\, [\ 1 \quad 1 \quad 1\ ] , \qquad\qquad (A.91.4)$$

the total force $P_i = \sigma_{ij} l_j$ is:

$$[P_i] = {-1}/{\sqrt{3}}\, \left[(x_1 + x_2 + x_3 + 1)(x_1 + x_2 + x_3 + 1)(x_1 + x_2 + x_3 - 1) \right]\ \text{N} . \quad (A.91.5)$$

To find the traction, this force per unit area ($1\ \text{m}^3$), simply change the units in eq. (A.91.5) to Pa. On the face that has l_i (pointing obliquely backward and downward and left) as its outward normal (Fig. A.91.1), the variables x_i must satisfy the equation:

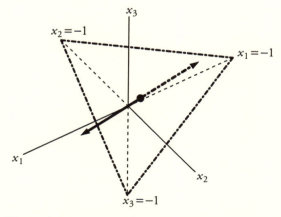

Figure A.91.1. Face of a regular octahedron with its outward normal unit vector (heavy dash-dot line) pointing away from, and traction vector (heavy line) pointing obliquely toward, the viewer.

$$x_1 + x_2 + x_3 = -1 \ . \tag{A.91.6}$$

For convenience, rearrange eq. (A.91.6) and multiply it by the factor $(-1/\sqrt{3}\ )$:

$$0 = -1/\sqrt{3}\ (x_1 + x_2 + x_3 + 1)\ , \tag{A.91.7}$$

then solve for each of the components of P_i in eq. (A.91.5) simultaneously with eq. (A.91.7). Subtracting eq. (A.91.7) from P_1 and P_2 yields:

$$P_1 = P_2 = 0\ ; \tag{A.91.8}$$

subtracting it from P_3, one obtains:

$$P_3 = 2/\sqrt{3}\ . \tag{A.91.9}$$

Hence, the traction acting uniformly over the face is:

$$[P_i] = [\ 0 \quad 0 \quad 2\ ]/\sqrt{3}\ \text{Pa}\ . \tag{A.91.10}$$

The normal component of this traction is:

$$[N_i] = l_j\, P_j\, [l_i] = 2[\ 1 \quad 1 \quad 1\ ]/3\sqrt{3}\ \text{Pa}\ , \tag{A.91.11}$$

with the magnitude:

$$N = \sqrt{\frac{3 \times 4}{3 \times 9}} = \sqrt{\frac{4}{9}} = 2/3\ \text{Pa}\ . \tag{A.91.12}$$

A positive force of $2/3\,\text{N}$ per $1\,\text{m}^2$ acts on the all-negative octahedral face, and the traction is therefore a compression. Subtract N_i from the total traction P_i to find the remainder, the tangential traction T_i:

$$[T_i] = [P_i - N_i] = 2[-1 \quad -1 \quad 2\ ]/3\sqrt{3}\ \text{Pa}\ . \tag{A.91.13}$$

Its magnitude is:

$$T = \sqrt{\frac{6 \times 4}{3 \times 9}} = \sqrt{\frac{2 \times 4}{9}} = 2\sqrt{2}/3\ \text{Pa}\ , \tag{A.91.14}$$

and it acts up the dip of the face. [As can be verified by solving for the traction and its normal and tangential components without the constraint of eq. (A.91.7), the tangential traction is identical throughout the stress field on all planes with the outward normal of eq. (A.91.4); the normal traction component, however, does vary with position due to the existence of a body force.]

Answer 92. Let x_1 be along the wire axis. Then the stress is:

$$[\sigma_{ij}] = \begin{bmatrix} \sigma & 0 & 0 \\ 0 & 0 & 0 \\ 0 & 0 & 0 \end{bmatrix}, \qquad (A.92.1)$$

in arbitrary units. To find the tangential stress acting in a particular glide direction on a particular glide plane, it is convenient to rotate coordinates so that the new axes x_1' and x_2' come to lie in the glide direction and in the direction of the outward normal of the glide plane, respectively (Fig.A.92.1). Such a rotation makes the component σ_{21}' identical with the shear stress τ (the first subscript identifies the outward normal of the plane on which a stress component acts, the second the direction of the component itself). Since only σ_{11} in eq. (A.92.1) is nonzero, only the two matrix elements of the rotation matrix are shown that are needed to rotate σ_{11}:

$$\left(a_{ij} \right) = \begin{pmatrix} \cos\lambda & \ldots & \ldots \\ \cos\varphi & \ldots & \ldots \\ \ldots & \ldots & \ldots \end{pmatrix}, \qquad (A.92.2)$$

and after the rotation:

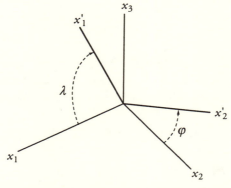

Figure A.92.1. Normal to glide plane and glide direction. After rotation the glide direction lies on x_1', and the glideplane normal lies on x_2'.

$$\tau \equiv \sigma'_{21} = a_{21}\, a_{11}\, \sigma_{11} + 0 + 0 + \cdots \; ; \tag{A.92.3}$$

hence:

$$\tau = \sigma \cos\varphi \cos\lambda \;\; \therefore \tag{A.92.4}$$

Answer 93. Given the stress:

$$\left[\sigma_{ij}\right] = \begin{bmatrix} 9 & -5 & 5 \\ -5 & 4 & 3 \\ 5 & 3 & 8 \end{bmatrix} \text{ kgwt cm}^{-2}, \tag{A.93.1}$$

we note that it is taken to be caused by forces that vary with the local gravitational constant and to act on an unconventional area. We further note that the question asks for the outward-acting forces inside a body, the opposite of those conventionally taken to constitute a stress (which is the set of forces outside a reference body acting on the material on its inside). We follow the method of Answer 87 to find the unit vector l_i corresponding to the given vector:

$$\left[V_i\right] = \begin{bmatrix} 1 & 2 & 2 \end{bmatrix} \text{ cm}, \tag{A.93.2}$$

or rather to its inverse, so as to take into account that we are asked for the forces opposite to those constituting the stress in eq. (A.93.1):

$$\left[l_i\right] = \begin{bmatrix} -1/3 & -2/3 & -2/3 \end{bmatrix}. \tag{A.93.3}$$

Continuing to use the methods of Answer 87, the total traction on the plane with the outward normal l_i is:

$$\left[P_i\right] = \begin{bmatrix} -3 & -3 & -9 \end{bmatrix}. \tag{A.93.4}$$

Being asked to calculate the force exerted on an unconventional $2\,\text{cm}^2$ introduces an unusual factor of two into the calculation of the unusual (asterisk for unconventionality) normal and tangential forces $[{}^*N_i]$ and $[{}^*T_i]$:

$$\left[{}^*N_i\right] = 2\left[P_j\, l_j\, l_i\right] = \begin{bmatrix} -6 & -12 & -12 \end{bmatrix} \text{ kgwt}, \tag{A.93.5}$$

and:

$$\left[{}^*T_i\right] = \left[2P_i - {}^*N_i\right] = \begin{bmatrix} 0 & 6 & -6 \end{bmatrix} \text{ kgwt}. \tag{A.93.6}$$

Answer 94. The problem is suitable for rotation by the Mohr circle construction. The rotation axis has the subscript $R = 2$, hence $a = 3$, $b = 1$. (a) Given the stress:

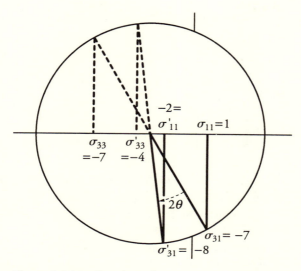

Figure A.94.1. Rotation of a stress tensor about x_2 so as to achieve a specific value for σ'_{31}.

$$[\sigma_{ij}] = \begin{bmatrix} 1 & 0 & -7 \\ 0 & 0 & 0 \\ -7 & 0 & -7 \end{bmatrix} \text{ kPa },$$ (A.94.1)

the original right triangle to the right of the center and below the abscissa (Fig. A.94.1) has as its hypotenuse the radius r, the square of the magnitude of which can be calculated from the known sides:

$$r^2 = 7^2 + 4^2 = 65 .$$ (A.94.2)

The final right triangle in the same quadrant also has r as its hypotenuse, and one of its orthogonal sides has the length 8:

$$r^2 = 65 = 8^2 + 1^2 .$$ (A.94.3)

Thus the rotated tensor is:

$$[\sigma'_{ij}] = \begin{bmatrix} -2 & 0 & -8 \\ 0 & 0 & 0 \\ -8 & 0 & -4 \end{bmatrix} \text{ kPa }.$$ (A.94.4)

(b) The angle of rotation is:

$$\theta = {}^1\!/_2 \left\langle \tan^{-1}(8/1) - \tan^{-1}(7/4) \right\rangle \approx -11.3° ,$$ (A.94.5)

(c) and it moves x_1 toward x_3. (d) The corresponding rotation matrix is:

$$\left(a_{ij}\right) = \begin{pmatrix} \cos 11.3° & 0 & \sin 11.3° \\ 0 & 1 & 0 \\ -\sin 11.3° & 0 & \cos 11.3° \end{pmatrix}, \qquad \text{(A.94.6)}$$

or:

$$\left(a_{ij}\right) \approx \begin{pmatrix} 0.980581 & 0 & 0.196116 \\ 0 & 1 & 0 \\ -0.196116 & 0 & 0.980581 \end{pmatrix}. \qquad \text{(A.94.7)}$$

Answer 95. (a) Figure A.95.1 is the Mohr circle for rotations $0° \le \theta \le 90°$ about an arbitrary axis perpendicular to the unique axis of the given stress, $\sigma_1 = 8$, $\sigma_2 = \sigma_3 = 4$ arbitrary units. (b) For the general case of an axially symmetric stress, inspection of this diagram yields a relationship between θ, the angle from the unique stress axis σ_1 to the pole of the reference plane on which a particular traction acts, the normal traction σ on that reference plane, and the principal-stress difference $\sigma_1 - \sigma_2$. This relationship is:

$$\cos 2\theta = (\sigma - c)/r , \qquad \text{(A.95.1)}$$

where:

$$c = (\sigma_1 + \sigma_2)/2 , \qquad \text{(A.95.2)}$$

and:

$$r = (\sigma_1 - \sigma_2)/2 . \qquad \text{(A.95.3)}$$

Specifically for the given stress for which $c = 6$ and $r = 2$, eq. (A.95.1) becomes:

$$\cos 2\theta = (\sigma - 6)/2 = \sigma/2 - 3 . \qquad \text{(A.95.4)}$$

Inspection of Fig. A.95.1 also yields the general relationship between the angle θ, the principal-stress difference, and the tangential (or shear) component τ of the traction:

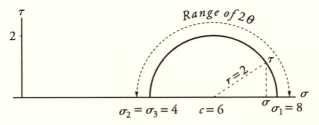

Figure A.95.1. Mohr circle for an axially symmetric stress. Only positive shear stresses shown.

$$\sin 2\theta = \tau/r \ . \tag{A.95.5}$$

For the given stress this is:

$$\sin 2\theta = \tau/2 \ . \tag{A.95.6}$$

(c) To contour the normal stress, solve eq. (A.95.4) for $\sigma = 8, 7.5, 7, 6, 5, 4.5, 4$:

$$
\begin{aligned}
\sigma &= 8 \ , & \cos 2\theta &= \ 1 \ , & \theta &= \ 0° \ ; \\
\sigma &= 7.5, & \cos 2\theta &= \ 3/4, & \theta &\approx 21° \ ; \\
\sigma &= 7 \ , & \cos 2\theta &= \ 1/2, & \theta &= 30° \ ; \\
\sigma &= 6 \ , & \cos 2\theta &= \ 0 \ , & \theta &= 45° \ ; \\
\sigma &= 5 \ , & \cos 2\theta &= -1/2, & \theta &= 60° \ ; \\
\sigma &= 4.5, & \cos 2\theta &= -3/4, & \theta &\approx 69° \ ; \\
\sigma &= 4 \ , & \cos 2\theta &= -1 \ , & \theta &= 90° \ .
\end{aligned}
\tag{A.95.7}
$$

Similarly, to contour the shear stress, solve eq. (A.95.6) for $\tau = 0, 1, 2$:

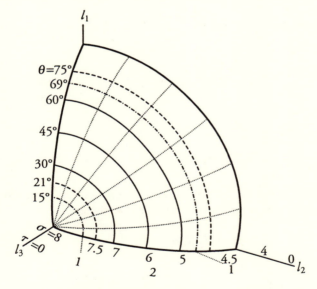

Figure A.95.2. All-positive octant of a unit sphere. Radius vectors represent poles on traction planes due to a specific, axially symmetric stress. Full and dashed contours, normal stress σ at unit and selected half-unit intervals. Dash-dot contours, shear stress τ at unit intervals. (The $\tau = 2$ contour coincides with the $\sigma = 6$ contour.)

$$\tau = 0 \; ; \sin 2\theta = 0 \quad ; \theta = \quad 0°, \; 90° \; ;$$
$$\tau = 1 \; ; \sin 2\theta = 1/2 \; ; \theta = \; 15°, \; 75° \; ;$$
$$\tau = 2 \; ; \sin 2\theta = 1 \quad ; \theta = 45° \; . \tag{A.95.8}$$

The plot of normal stresses (Fig. A.95.2) shows the maximum at the unique axis, decreasing with increasing θ, at first slowly, then more rapidly to $\theta = 45°$, then again more slowly. The shear stresses have their maximum at $\theta = 45°$ and from there decrease to zero, progressively more rapidly as θ approaches either 90° or 0°.

Answer 96. Let the outward normals on the octahedral faces be l_i, then:

$$[l_i] = [\; \pm 1 \quad \pm 1 \quad \pm 1 \;]/\sqrt{3} \; , \tag{A.96.1}$$

with any combination of signs. Then the tractions on the octahedral planes are:

$$[P_i] = [\sigma_{ij} \, l_i] = [\; \pm \sigma_1 \quad \pm \sigma_2 \quad \pm \sigma_3 \;]/\sqrt{3} \; . \tag{A.96.2}$$

Decompose them into normal and tangential components [as in Problem 91(c)]:

$$[N_i] = [l_i \, P_j \, l_j]$$
$$= (\sigma_1 + \sigma_2 + \sigma_3)[\; \pm 1 \quad \pm 1 \quad \pm 1 \;]/3\sqrt{3} \; , \tag{A.96.3}$$

and:

$$[T_i] = [P_i - N_i]$$
$$= [\pm(2\sigma_1 - \sigma_2 - \sigma_3) \pm (-\sigma_1 + 2\sigma_2 - \sigma_3) \pm (-\sigma_1 - \sigma_2 + 2\sigma_3)]/3\sqrt{3} \; . \tag{A.96.4}$$

To find the square of the magnitude of the octahedral shear stress, we dot multiply the traction on the octahedral face with itself:

$$°\tau^2 = T_i \, T_i = T_1^2 + T_2^2 + T_3^2 \; , \tag{A.96.5}$$

or:

$$°\tau^2 = \frac{1}{27}(4\sigma_1^2 + \sigma_2^2 + \sigma_3^2 - 4\sigma_1\sigma_2 + 2\sigma_2\sigma_3 - 4\sigma_3\sigma_1$$
$$+ \sigma_1^2 + 4\sigma_2^2 + \sigma_3^2 - 4\sigma_1\sigma_2 - 4\sigma_2\sigma_3 + 2\sigma_3\sigma_1$$
$$+ \sigma_1^2 + \sigma_2^2 + 4\sigma_3^2 + 2\sigma_1\sigma_2 - 4\sigma_2\sigma_3 - 4\sigma_3\sigma_1)$$
$$= \frac{1}{27}\langle 6(\sigma_1^2 + \sigma_2^2 + \sigma_3^2) - 6(\sigma_2\sigma_3 + \sigma_3\sigma_1 + \sigma_1\sigma_2)\rangle$$
$$= \frac{2}{9}\langle(\sigma_1^2 + \sigma_2^2 + \sigma_3^2) - (\sigma_2\sigma_3 + \sigma_3\sigma_1 + \sigma_1\sigma_2)\rangle \; . \tag{A.96.6}$$

In principal coordinates $^1I = \sigma_1 + \sigma_2 + \sigma_3$ and $^2I = \sigma_2\sigma_3 + \sigma_3\sigma_1 + \sigma_1\sigma_2$ are the first and second invariants; twice the square of the first invariant is:

$$2\,^1I^2 = 2\left(\sigma_1^2 + \sigma_2^2 + \sigma_3^2\right) + 4\left(\sigma_2\,\sigma_3 + \sigma_3\,\sigma_1 + \sigma_1\,\sigma_2\right) , \qquad \text{(A.96.7)}$$

and six times the second:

$$6\,^2I = 6\left(\sigma_2\,\sigma_3 + \sigma_3\,\sigma_1 + \sigma_1\,\sigma_2\right) . \qquad \text{(A.96.8)}$$

Subtract eq. (A.96.8) from eq. (A.96.7):

$$2\,^1I^2 - 6\,^2I = 2\left(\sigma_1^2 + \sigma_2^2 + \sigma_3^2\right) - 2\left(\sigma_2\,\sigma_3 + \sigma_3\,\sigma_1 + \sigma_1\,\sigma_2\right) . \quad \text{(A.96.9)}$$

Comparing eqs. (A.96.6) and (A.96.9) we find:

$$9\,^\circ\tau^2 = 2\,^1I^2 - 6\,^2I \quad \therefore \qquad \text{(A.96.10)}$$

Answer 97. (a) In principal coordinates, the first invariant, $^1I = \Delta_1 + \Delta_2 + \Delta_3$, for the deviator of stress Δ_{ij} is zero (being invariant, it is zero also for all other coordinate orientations). Also in principal coordinates, the second invariant is:

$$^2I = \Delta_2\Delta_3 + \Delta_3\Delta_1 + \Delta_1\Delta_2 , \qquad \text{(A.97.1)}$$

and the scalar product of the deviator with itself, $\Delta_{ij}\Delta_{ij}$ is reduced to the form:

$$\Delta_{ij}\Delta_{ij} = \Delta_1^2 + \Delta_2^2 + \Delta_3^2 . \qquad \text{(A.97.2)}$$

The square of the first invariant is:

$$\begin{aligned}
^1I^2 &= \left(\Delta_1 + \Delta_2 + \Delta_3\right)^2 \\
&= \Delta_1^2 + \Delta_2^2 + \Delta_3^2 + 2\left(\Delta_2\Delta_3 + \Delta_3\Delta_1 + \Delta_1\Delta_2\right) = 0 .
\end{aligned} \qquad \text{(A.97.3)}$$

Substitute eqs. (A.97.1) and (A.97.2) into eq. (A.97.3):

$$^1I^2 = \Delta_{ij}\Delta_{ij} + 2\,^2I = 0 , \qquad \text{(A.97.4)}$$

or solved for ^2I:

$$^2I = -\left(\Delta_{ij}\Delta_{ij}\right)/2 \quad \therefore \qquad \text{(A.97.5)}$$

(b) In principal coordinates the third invariant of the deviator of stress is:

$$^3I = \Delta_1\Delta_2\left(-\Delta_1 - \Delta_2\right) , \qquad \text{(A.97.6)}$$

and in the same coordinates:

$$\Delta_{ij}\Delta_{jk}\Delta_{ki} = \Delta_1^3 + \Delta_2^3 - (\Delta_1 + \Delta_2)^3 , \qquad (A.97.7)$$

which expands to:

$$\Delta_{ij}\Delta_{jk}\Delta_{ki} = \Delta_1^3 + \Delta_2^3 - \Delta_1^3 - 3\Delta_1^2\Delta_2 - 3\Delta_1\Delta_2^2 - \Delta_2^3 , \qquad (A.97.8)$$

and simplifies to:

$$\Delta_{ij}\Delta_{jk}\Delta_{ki}/3 = \Delta_1\Delta_2(-\Delta_1 - \Delta_2) . \qquad (A.97.9)$$

By substitution of eq. (A.97.9) into eq. (A.97.6) we obtain the postulated result:

$$^3I = \Delta_{ij}\Delta_{jk}\Delta_{ki}/3 \;\; \therefore \qquad (A.97.10)$$

Although both calculations (a) and (b) are executed in principal coordinates, their result must hold equally for all other coordinate orientations because only invariant quantities are involved.

Answer 98.   (a) Given the small displacement gradient:

$$[e_{ij}] = \begin{bmatrix} 8 & -1 & -1 \\ 1 & 6 & 0 \\ -5 & 0 & 2 \end{bmatrix} \times 10^{-6} , \qquad (A.98.1)$$

use the separation of e_{ij} into its symmetric and antisymmetric parts, that is, into the small strain $\varepsilon_{ij} = 1/2(e_{ij} + e_{ji})$ and the small rotation $\varpi_{ij} = 1/2(e_{ij} - e_{ji})$:

$$[\varepsilon_{ij}] = \begin{bmatrix} 8 & 0 & -3 \\ 0 & 6 & 0 \\ -3 & 0 & 2 \end{bmatrix} \times 10^{-6} , \qquad (A.98.2)$$

and:

$$[\varpi_{ij}] = \begin{bmatrix} 0 & -1 & 2 \\ 1 & 0 & 0 \\ -2 & 0 & 0 \end{bmatrix} \times 10^{-6} . \qquad (A.98.3)$$

(b) Because ε_{22} is already a principal strain, use the Mohr circle construction to find the other two. With x_2 as the rotation axis, $R = 2$, $a = 3$, $b = 1$. By inspection of the isoceles right triangle in Fig. A.98.1, we find $2\theta = 45°$ and:

$$r = \sqrt{9+9} = 3\sqrt{2} , \qquad (A.98.4)$$

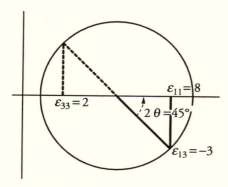

Figure A.98.1. Mohr circle construction. Finding principal values and
directions in the $x_1 - x_3$ plane.

$$r = \sqrt{9+9} = 3\sqrt{2} , \qquad \text{(A.98.4)}$$

and:

$$\left.\begin{array}{l} \varepsilon_1 = 5 + 3\sqrt{2} \\[4pt] \varepsilon_2 = 6 \\[4pt] \varepsilon_3 = 5 - 3\sqrt{2} \end{array}\right\} \times 10^{-6} . \qquad \text{(A.98.5)}$$

To reach the principal axes, the coordinates must be rotated about x_2 by $\theta = +22.5°$, moving x_3 toward x_1. (c) The *physical rotation* ω_i, in radians, is related to the rotation tensor ϖ_{ij} in the following way:

$$\left[\omega_i\right] = \left[\begin{array}{ccc} \varpi_{32} & \varpi_{13} & \varpi_{21} \end{array}\right] . \qquad \text{(A.98.6)}$$

In the present case the rotation vector thus is:

$$\left[\omega_i\right] = \left[\begin{array}{ccc} 0 & 2 & 1 \end{array}\right] \times 10^{-6} \text{ rad} , \qquad \text{(A.98.7)}$$

with the magnitude:

$$\omega = \sqrt{4+1} = \sqrt{5} \times 10^{-6} \text{ rad} . \qquad \text{(A.98.8)}$$

Hence the magnitude of clockwise physical rotation, looking in the positive direction of the vector $\boldsymbol{\omega}$, is:

$$\begin{array}{l} \omega \approx 2.2 \times 10^{-6} \text{rad} , \\[4pt] \approx 1.3° \times 10^{-4} , \\[4pt] \approx 0.46'' \text{ of arc} . \end{array} \qquad \text{(A.98.9)}$$

The direction cosines of the axis of rotation are obtained by normalizing $\boldsymbol{\omega}$:

$$\left[l_i \right] = \left[\omega_i \right]/\omega = \left[\begin{array}{ccc} 0 & 2/\sqrt{5} & 1/\sqrt{5} \end{array} \right] . \qquad \text{(A.98.10)}$$

That the component l_1 is zero indicates that the rotation axis is perpendicular to x_1. The axis points into the all-positive quadrant of the $x_2 - x_3$ plane.

Answer 99. Let the following vectors all lie in a plane orthogonal to the fixed rotation axis $\mathbf{1}$ of a rotating rigid body (Fig. A.99.1), and let that plane contain the origin of the coordinate system of reference. Let vector $\mathbf{A}$ connect $\mathbf{1}$ with the coordinate origin, and let $\mathbf{R}$ be the radius vector from the rotation axis to some point x in this body, further let $\mathbf{x}$ be the position vector (with respect to the coordinate system) of the same point, and finally let $\mathbf{u}$ be the small displacement of the moving point due to a small positive rotation (the following argument is not valid for large rotations). Then the rotation vector $\boldsymbol{\omega}$ that we know from the previous problem is:

$$\boldsymbol{\omega} = \varphi\mathbf{1} , \qquad \text{(A.99.1)}$$

where φ is expressed in radians. Further, with the same restriction to small rotations:

$$\mathbf{u} = \boldsymbol{\omega} \times \mathbf{R} . \qquad \text{(A.99.2)}$$

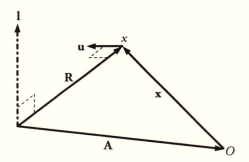

Figure A.99.1. Rigid body rotation about an axis remote from the coordinate origin O. $\mathbf{1}$ – rotation axis perpendicular to the vector triangle $\mathbf{A}, \mathbf{R}, \mathbf{x}$. $\mathbf{A}$ – vector from rotation axis to origin. $\mathbf{R}$ – vector from rotation axis to a point x. $\mathbf{x}$ – position vector for x. $\mathbf{u}$ – small displacement of x perpendicular to $\mathbf{R}$ and $\mathbf{1}$.

Answer 99. Let the following vectors all lie in a plane orthogonal to the fixed rotation axis **l** of a rotating rigid body (Fig. A.99.1), and let that plane contain the origin of the coordinate system of reference. Let vector **A** connect **l** with the coordinate origin, and let **R** be the radius vector from the rotation axis to some point x in this body, further let **x** be the position vector (with respect to the coordinate system) of the same point, and finally let **u** be the small displacement of the moving point due to a small positive rotation (the following argument is not valid for large rotations). Then the rotation vector **ω** that we know from the previous problem is:

$$\boldsymbol{\omega} = \varphi \mathbf{l} , \tag{A.99.1}$$

where φ is expressed in radians. Further, with the same restriction to small rotations:

$$\mathbf{u} = \boldsymbol{\omega} \times \mathbf{R} . \tag{A.99.2}$$

Returning to subscript notation, we find by inspection of Fig. A.99.1:

$$R_i = A_i + x_i . \tag{A.99.3}$$

From eqs. (A.99.1) and (A.99.2), and in the same notation:

$$u_i = \varphi \epsilon_{ijk} l_j R_k . \tag{A.99.4}$$

Substitute eq. (A.99.3) into eq. (A.99.4):

$$u_i = \varphi \epsilon_{ijk} l_j A_k + \varphi \epsilon_{ijk} l_j x_k , \tag{A.99.5}$$

where, considering the fixed position of the rotation axis, the A_i are constants. Differentiate u_i in eq. (A.99.5) to find the displacement gradient $e_{ij} = u_{i,j}$:

$$e_{ij} = \frac{\partial \varphi \epsilon_{imk} l_m x_k}{\partial x_j} . \tag{A.99.6}$$

Note that in eq. (A.99.6) the dummy subscript j of the second term of the right-hand side of eq. (A.99.5) had to be replaced by a different dummy subscript m, j being a free subscript. Because the partial derivative $\partial x_k / \partial x_j$ is unity if $k = j$ and is zero if $k \neq j$ and is thus the equivalent of the Kronecker delta δ_{ij}, we use its substitution property and write:

$$e_{ij} = \varphi \epsilon_{imj} l_m . \tag{A.99.7}$$

It is customary not to leave alphabetic gaps between subscript letters, and we restate eq. (A.99.7) as:

$$e_{ij} = \varphi \epsilon_{ikj} l_k .\tag{A.99.8}$$

Inasmuch as $\epsilon_{jki} = -\epsilon_{ikj}$:

$$e_{ji} = -e_{ij} .\tag{A.99.9}$$

Thus, independently of the position of the rotation axis, the displacement gradient field for a rigid-body rotation is an antisymmetric tensor.

Answer 100. This problem resembles the previous one in most respects. From eq. (A.99.8) the displacement gradient is:

$$u_{i,j} \equiv e_{ij} = \varphi \epsilon_{ikj} l_k .\tag{A.100.1}$$

However, in contrast to the previous problem, the vector from the origin to the rotation axis is a null vector:

$$A_i = 0 .\tag{A.100.2}$$

Therefore, the counterpart of eq. (A.99.5) is simply:

$$u_i = \varphi \epsilon_{ijk} l_j x_k .\tag{A.100.3}$$

Substituting eq. (A.100.1) into eq. (A.100.3) yields the desired answer:

$$u_i = e_{ik} x_k \;\therefore\tag{A.100.4}$$

Note: This result is important and generally useful because the displacement gradient is invariant to coordinate translation, and the equation thus holds for all infinitesimal rigid-body rotations, even if the rotation axis does not pass through the origin.

Answer 101. The physics for this problem is the same as for Problem 99. Thus we can investigate eq. (A.99.8):

$$e_{ij} = \varphi \epsilon_{ikj} l_k .\tag{A.101.1}$$

Because both φ and l_i are now constants, the displacement gradient $[e_{ij}]$ is not a function of position and thus is homogeneous.

Answer 102. This problem, too, is based on the physics of Problem 99. We can thus expand eq. (A.99.8) to:

$$\left[e_{ij} \right] = \left[\varphi \epsilon_{ikj} l_k \right] = \begin{bmatrix} 0 & -\varphi l_3 & \varphi l_2 \\ \varphi l_3 & 0 & -\varphi l_1 \\ -\varphi l_2 & \varphi l_1 & 0 \end{bmatrix} .\tag{A.102.1}$$

Note that x_i does not appear in this equation and that any information about it, such as its mention in the statement of the problem, is irrelevant for the solution. Note also that $\varphi l_i \equiv \omega_i$ in a context like that of Problem 99.

Answer 103. Given the small strain:

$$\left[\varepsilon_{ij}\right] = \begin{bmatrix} 1 & 2 & 1 \\ 2 & 3 & 1 \\ 1 & 1 & 3 \end{bmatrix} \times 10^{-4}, \qquad \text{(A.103.1)}$$

which is homogeneous because none of the components of the strain tensor is a function of position, the displacements are found by means of the general relationship:

$$u_i = {}^{\circ}u_i + \varpi_{ij} x_j + \varepsilon_{ij} x_j \ . \qquad \text{(A.103.2)}$$

Considering that all components of both the displacement at the coordinate origin ${}^{\circ}u_i$ and of the rotation tensor ϖ_{ij} are zero, eq. (A.103.2) simplifies to:

$$u_i = \varepsilon_{ij} x_j \ \text{units of length.} \qquad \text{(A.103.3)}$$

In the specific case of the strain in eq. (A.103.1), this yields:

$$\left[u_i\right] = \left[\, x_1 + 2x_2 + x_3 \quad 2x_1 + 3x_2 + x_3 \quad x_1 + x_2 + 3x_3 \,\right]$$

$$\times 10^{-4} \ \text{units} \ . \qquad \text{(A.103.4)}$$

Note: Even if the units are large, say light years, the strain still remains small.

Answer 104. (a) Given the small displacements:

$$\left[u_i\right] = \left[\, x_2^2 x_3 \quad 2x_1 x_2 x_3 \quad x_1 x_2^2 \,\right] \times 10^{-5}, \qquad \text{(A.104.1)}$$

we differentiate u_i and, upon inspection, find the resulting tensor $e_{ij} = u_{i,j}$ symmetric and thus identical with the corresponding strain tensor ε_{ij}:

$$\left[e_{ij}\right] = \left[\varepsilon_{ij}\right] = \begin{bmatrix} 0 & 2x_2 x_3 & x_2^2 \\ 2x_2 x_3 & 2x_1 x_3 & 2x_1 x_2 \\ x_2^2 & 2x_1 x_2 & 0 \end{bmatrix} \times 10^{-5}. \qquad \text{(A.104.2)}$$

(b) In the strain field of eq. (A.104.2), evaluation at the point $[x_i] = [\,2 \ -3 \ 0\,]$ (conventionally indicated by a downward bar) yields:

$$\left[\varepsilon_{ij}\right]\Big|_{[\ 2\ \ -3\ \ 0\ ]} = \begin{bmatrix} 0 & 0 & 9 \\ 0 & 0 & -12 \\ 9 & -12 & 0 \end{bmatrix} \times 10^{-5}. \qquad \text{(A.104.3)}$$

The specific secular equation $|\varepsilon_{ij} - \lambda\delta_{ij}| = 0$ for the strain of eq. (A.104.3) is:

$$-\lambda^3 + (144 + 81)\lambda = 0 . \qquad \text{(A.104.4)}$$

Because the first and third invariants of the strain tensor are zero, $^1\lambda = 0$ and $^{2,3}\lambda = \pm(^2I)^{1/2} = \pm\sqrt{225}$. Thus in principal coordinates:

$$\varepsilon_1 = 15 \times 10^{-5}, \quad \varepsilon_2 = 0 , \quad \varepsilon_3 = -15 \times 10^{-5}. \qquad \text{(A.104.5)}$$

This is pure shear. (c) We can generalize from eq. (A.104.5) to represent the complete plane with $x_3 = 0$ and only x_1 and x_2 variable:

$$\left[\varepsilon_{ij}\right]\Big|_{[\ x_1\ \ x_2\ \ 0\ ]} = \begin{bmatrix} 0 & 0 & x_2^2 \\ 0 & 0 & 2x_1x_2 \\ x_2^2 & 2x_1x_2 & 0 \end{bmatrix} \times 10^{-5}. \qquad \text{(A.104.6)}$$

Thus in the $(x_3 = 0)$ plane the strain is everywhere pure shear. (d) An analogous evaluation of the $(x_1 = 0)$ plane shows:

$$\left[\varepsilon_{ij}\right]\Big|_{[\ 0\ \ x_2\ \ x_3\ ]} = \begin{bmatrix} 0 & 2x_2x_3 & x_2^2 \\ 2x_2x_3 & 0 & 0 \\ x_2^2 & 0 & 0 \end{bmatrix} \times 10^{-5}. \qquad \text{(A.104.7)}$$

The trace (sum of the diagonal components) of this tensor is everywhere zero; hence there is no dilatation. (e) At the point $[x_i] = [-1\ 0\ 3]$ an evaluation of eq. (A.104.2) yields:

$$\left[\varepsilon_{ij}\right]\Big|_{[-1\ \ 0\ \ 3\ ]} = \begin{bmatrix} 0 & 0 & 0 \\ 0 & -6 & 0 \\ 0 & 0 & 0 \end{bmatrix} \times 10^{-5}. \qquad \text{(A.104.8)}$$

Although the strain at this point has its unique axis parallel to the x_2 coordinate axis, in the conventional order the strain in principal coordinates is:

$$\varepsilon_1 = 0 , \quad \varepsilon_2 = 0 , \quad \varepsilon_3 = -6 \times 10^{-5}. \qquad \text{(A.104.9)}$$

It is thus a uniaxial contraction. (f) The strain everywhere in the $(x_2 = 0)$ plane is:

$$[\varepsilon_{ij}]\Big|_{[\;x_1\;\;0\;\;x_3\;]} = \begin{bmatrix} 0 & 0 & 0 \\ 0 & 2x_1x_3 & 0 \\ 0 & 0 & 0 \end{bmatrix} \times 10^{-5}. \quad \text{(A.104.10)}$$

Because of the signs of the position vector components, extension and contraction are distributed over this plane according to the scheme:

$$\begin{array}{c|c} & x_3 \\ \hline \text{contraction} & \text{extension} \\ \hline \text{extension} & \text{contraction} \end{array} \quad x_1$$

(g) The strain is zero on both the x_1 and x_3 axes. (h) Differentiation of the six essential compatibility equations in the order in which they are listed in eq. (6.14) for the displacement field described by eq. (A.104.1) yields:

$$\begin{aligned} 0 + 0 &= 0 + 0 \; , \\ 2 + 2 &= 2 + 2 \; , \\ 0 + 0 &= 0 + 0 \; , \\ 2 \times 0 &= 0 + 0 \; , \\ 2 \times 0 &= 0 + 0 \; , \\ 2 \times 0 &= 0 + 0 \; , \end{aligned} \quad \text{(A.104.11)}$$

and they thus hold: The displacement field produces a compatible continuum strain field.

Answer 105. Given is the displacement field, in arbitrary units of length:

$$[u_i] = [\,-x_1^2 + 2x_2^2 + 3x_3^2 \quad 3x_1^2 - x_2^2 + 2x_3^2 \quad 2x_1^2 + 3x_2^2 - x_3^2\,] \times 10^{-6}. \quad \text{(A.105.1)}$$

(a) The strain field ε_{ij} is the spatial derivative of a given displacement field, and the dilatation Δ is the trace (sum of diagonal components) of the strain tensor. Hence in general:

$$\Delta = \varepsilon_{ii} = u_{i,\,i} \; . \quad \text{(A.105.2)}$$

Thus for each component u_i only the term containing x_i with the corresponding subscript has a nonzero derivative, and in the specific case:

$$\Delta = \varepsilon_{ii} = u_{i,\,i} = (-2x_1 - 2x_2 - 2x_3) \times 10^{-6}. \quad \text{(A.105.3)}$$

Thus at:

$$[x_i] = \sqrt{3}\,[\,1 \quad 1 \quad 1\,]\ \text{units of length,} \tag{A.105.4}$$

the dilatation is:

$$\Delta = -6\sqrt{3} \times 10^{-6}. \tag{A.105.5}$$

(b) The linear dilatation $'\Delta$, that is, the relative increase in length due to the dilatation in any, not just the x_1 direction, is:

$$'\Delta \equiv \Delta/3 = -2\sqrt{3} \times 10^{-6}. \tag{A.105.6}$$

Note that the dilatation, volumetric or linear, is a strain and is thus dimensionless and independent of the choice of units of length.

Answer 106. The given equation:

$$u_i = \frac{pa^3 x_i}{4\mu(x_k x_k)^{3/2}}, \tag{A.106.1}$$

describes a displacement field as a function of x_i and involving the positive constants p, a, and μ. The quantity in parentheses $(x_k x_k)$ may be taken to represent the square r^2 of the magnitude r of a radius vector from the origin to the position $[x_i]$, and thus inspection of eq. (A.106.1) shows that all displacements $[u_i]$ are radially outward from the origin and inversely proportional to the third power of r. The derivative field of eq. (A.106.1), with x_i in both the numerator and denominator, is:

$$u_{i,j} = \frac{pa^3 \delta_{ij}}{\boxed{4\mu(x_k x_k)^{3/2}}} - \frac{\boxed{pa^3 x_i}\, 12\mu(x_m x_m)^{1/2} x_j}{16\mu^2\,(x_l x_l)^3}, \tag{A.106.2}$$

where the framed expressions are the denominator and numerator of u_i in eq. (A.106.1), the numerator of the first term of eq. (A.106.2) is the space derivative of the numerator in eq. (A.106.1), the denominator of the second term of eq. (A.106.2) the square of the denominator in eq. (A.106.1), and the unframed part of the numerator of the second term of eq. (A.106.2) the space derivative of the denominator of eq. (A.106.1). This last derivative, in turn, results from the application of the chain rule (the sum $x_k x_k$ contains the term x_j^2) as follows:

$$4\mu\left\langle 3/2(x_l x_l)^{1/2}\right\rangle\left\langle 2x_j\right\rangle = 12\mu(x_l x_l)^{1/2} x_j\ .$$

By cancellation, eq. (A.106.2) simplifies to:

$$u_{i,j} = \frac{pa^3 \delta_{ij}}{4\mu(x_k x_k)^{3/2}} - \frac{3pa^3 x_i x_j}{4\mu(x_l x_l)^{5/2}} \cdot \tag{A.106.3}$$

Note that each term has the appropriate two free subscripts of a tensor of the second rank, and also that this tensor is necessarily symmetric because the only nonscalar on the right-hand side is the numerator of the second term, containing the commutative product $x_i x_j = x_j x_i$. Thus:

$$\varpi_{ij} = 0 , \tag{A.106.4}$$

and:

$$\varepsilon_{ij} = u_{i,j} . \tag{A.106.5}$$

Insight into eq. (A.106.2) is improved by factoring scalars:

$$\varepsilon_{ij} = \frac{pa^3}{4\mu(x_k x_k)^{3/2}} \left(\delta_{ij} - \frac{3x_i x_j}{x_l x_l} \right) . \tag{A.106.6}$$

In this form it is obvious that the trace of this tensor:

$$\Delta \equiv \varepsilon_{ii} = \frac{pa^3}{4\mu(x_k x_k)^{3/2}} \left(\delta_{ii} - \frac{3x_i x_i}{x_l x_l} \right) , \tag{A.106.7}$$

vanishes because the quantity in the large parentheses becomes $(3-3)$. Hence the dilatation is zero. Note that this is the elastic solution for an infinite, elastic, isotropic medium enclosing a spherical hole of radius a, centered at the origin, which is filled by a fluid under the pressure p. The bulk modulus of the elastic medium is irrelevant because of the zero dilatation, and only the shear modulus μ enters into the equation.

Answer 107. Infinitesimal strain and rotation are by definition:

$$\varepsilon_{ij} = \frac{1}{2}\left(u_{i,j} + u_{j,i}\right) , \tag{A.107.1}$$

and:

$$\varpi_{ij} = \frac{1}{2}\left(u_{i,j} - u_{j,i}\right) . \tag{A.107.2}$$

Differentiate both with respect to x_i:

$$\varepsilon_{ij,k} = \frac{1}{2}\left(u_{i,jk} + u_{j,ik}\right) , \tag{A.107.3}$$

and:

$$\varpi_{ij,k} = \frac{1}{2}\left(u_{i,jk} - u_{j,ik}\right) . \tag{A.107.4}$$

The choice of subscripts is arbitrary, and we exchange subscripts j and k in eq. (A.107.3):

$$\varepsilon_{ik,j} = \frac{1}{2}\left(u_{i,kj} + u_{k,ij}\right),$$ (A.107.5)

and again subscripts i and j:

$$\varepsilon_{jk,i} = \frac{1}{2}\left(u_{j,ki} + u_{k,ji}\right).$$ (A.107.6)

Subtract eq. (A.107.6) from eq. (A.107.5):

$$\varepsilon_{ik,j} - \varepsilon_{jk,i} = \frac{1}{2}\left(u_{i,kj} - u_{j,ki} + u_{k,ij} - u_{k,ji}\right).$$ (A.107.7)

Because the order of differentiation does not affect results, the last two terms in the parentheses cancel and thus:

$$\varepsilon_{ik,j} - \varepsilon_{jk,i} = \frac{1}{2}\left(u_{i,kj} - u_{j,ki}\right).$$ (A.107.8)

The right-hand side of eq. (A.107.8) differs from that of eq. (A.107.4) only by the order of differentiation. Hence:

$$\varepsilon_{ik,j} - \varepsilon_{jk,i} = \varpi_{ij,k}.$$ (A.107.9)

However, because the strain is homogeneous, that is, constant as a function of x_i, its spatial derivative everywhere is:

$$\varepsilon_{ij,k} = 0.$$ (A.107.10)

Substitute eq. (A.107.10) into eq. (A.107.9) to find that the following holds uniformly:

$$\varpi_{ij,k} = 0.$$ (A.107.11)

Thus if $^{\circ}\varpi_{ij} = 0$, because the rotation does not change with changing x_i, everywhere the following must be true:

$$\varpi_{ij} = 0 \quad \therefore$$ (A.107.12)

Answer 108. Given that for all x_i the following is to hold:

$$b_{ij}x_i x_j = 0,$$ (A.108.1)

let x_i be:

$$[x_i] = [1 \quad 0 \quad 0],$$ (A.108.2)

then the following must be true:

$$b_{11} = 0 \, , \tag{A.108.3}$$

and similarly for:

$$[x_i] = [0 \ 1 \ 0] \, ,$$
$$[x_i] = [0 \ 0 \ 1] \, , \tag{A.108.4}$$

it must be true that:

$$b_{22} = b_{33} = 0 \, . \tag{A.108.5}$$

Let x_i further be:

$$[x_i] = [1 \ 1 \ 0] \, , \tag{A.108.6}$$

then it follows that:

$$b_{12} = b_{21} = 0 \, , \tag{A.108.7}$$

or:

$$b_{12} = -b_{21} \, . \tag{A.108.8}$$

Also setting $[x_i] = [0 \ 1 \ 1]$ and $[x_i] = [1 \ 0 \ 1]$ proves:

$$b_{ji} = -b_{ij} = 0 \ \therefore \tag{A.108.9}$$

Answer 109. (a) The six essential compatibility equations for the given small strain:

$$[\varepsilon_{ij}] = \begin{bmatrix} 6x_1 x_2 & 3x_1^2 & 0 \\ 3x_1^2 & 3x_2^2 & 0 \\ 0 & 0 & 2x_3 \end{bmatrix} \times 10^{-6} \, , \tag{A.109.1}$$

are satisfied as follows:

$$\varepsilon_{11, \, 23} + \varepsilon_{23, \, 11} = \varepsilon_{31, \, 12} + \varepsilon_{12, \, 31} \, ,$$
$$\text{or:} \quad 0 + 0 = 0 + 0 \, ,$$
$$\varepsilon_{22, \, 31} + \varepsilon_{31, \, 22} = \varepsilon_{12, \, 23} + \varepsilon_{23, \, 12} \, ,$$
$$\text{or:} \quad 0 + 0 = 0 + 0 \, ,$$
$$\varepsilon_{33, \, 12} + \varepsilon_{12, \, 33} = \varepsilon_{23, \, 31} + \varepsilon_{31, \, 23} \, ,$$
$$\text{or:} \quad 0 + 0 = 0 + 0 \, ,$$
$$2\varepsilon_{12, \, 12} = \varepsilon_{11, \, 22} + \varepsilon_{22, \, 11} \, ,$$
$$\text{or:} \quad 0 = 0 + 0 \, ,$$
$$2\varepsilon_{23, \, 23} = \varepsilon_{22, \, 33} + \varepsilon_{33, \, 22} \, ,$$
$$\text{or:} \quad 0 = 0 + 0 \, ,$$
$$2\varepsilon_{31, \, 31} = \varepsilon_{33, \, 11} + \varepsilon_{11, \, 33} \, ,$$
$$\text{or:} \quad 0 = 0 + 0 \, . \tag{A.109.2}$$

(b) To find the displacements, eliminate the terms for translation at the origin, $^{\circ}u_i$, and rotation at the origin, $^{\circ}\varpi_{ij}$, from eqs. (6.16) to (6.18) and find u_i, choosing the segments of the integration paths in the same order as in those equations. The component u_1 is:

$$u_1 = \int_{\substack{0 \\ x_3 = x_1 = 0}}^{x'_2} \Big\langle \mathcal{E}_{12} + (x'_2 - x_2)(\mathcal{E}_{12,2} - \mathcal{E}_{22,1}) + x'_3(\mathcal{E}_{12,3} - \mathcal{E}_{32,1}) \Big\rangle dx_2$$

$$+ \int_{\substack{0 \\ x_1 = 0 \\ x_2 = x'_2}}^{x'_3} \Big\langle \mathcal{E}_{13} + (x'_3 - x_3)(\mathcal{E}_{13,3} - \mathcal{E}_{33,1}) \Big\rangle dx_3$$

$$+ \int_{\substack{0 \\ x_2 = x'_2 \\ x_3 = x'_3}}^{x'_1} \mathcal{E}_{11} dx_1 , \tag{A.109.3}$$

or:

$$u_1 = \int_{\substack{0 \\ x_3 = x_1 = 0}}^{x'_2} \Big\langle \mathcal{E}_{12} + (x'_2 - x_2)(0 - 0) + x'_3(0 - 0) \Big\rangle dx_2$$

$$+ \int_{\substack{0 \\ x_1 = 0 \\ x_2 = x'_2}}^{x'_3} \Big\langle \mathcal{E}_{13} + (x'_3 - x_3)(0 - 0) \Big\rangle dx_3$$

$$+ \int_{\substack{0 \\ x_2 = x'_2 \\ x_3 = x'_3}}^{x'_1} \mathcal{E}_{11} dx_1 , \tag{A.109.4}$$

This simplifies to:

$$u_1 = \int_{\substack{0 \\ x_3 = x_1 = 0}}^{x'_2} \mathcal{E}_{12} dx_2 + \int_{\substack{0 \\ x_1 = 0 \\ x_2 = x'_2}}^{x'_3} \mathcal{E}_{13} dx_3 + \int_{\substack{0 \\ x_2 = x'_2 \\ x_3 = x'_3}}^{x'_1} \mathcal{E}_{11} dx_3 . \tag{A.109.5}$$

Note that the first integral is zero because it is evaluated at $x_1 = 0$, the second has a zero integrand, and the third has the value of $3(x'_1)^2 x'_2 \times 10^{-6}$. Thus:

$$u_1 = 3(x'_1)^2 x'_2 \times 10^{-6}. \tag{A.109.6}$$

We integrate similarly to obtain u_2:

$$u_2 = \int\limits_{\substack{0 \\ x_1 = x_2 = 0}}^{x'_3} \left\langle \varepsilon_{23} + (x'_3 - x_3)(\varepsilon_{23,3} - \varepsilon_{33,2}) + x'_1(\varepsilon_{23,1} - \varepsilon_{13,2}) \right\rangle dx_3$$

$$+ \int\limits_{\substack{0 \\ x_2 = 0 \\ x_3 = x'_3}}^{x'_1} \left\langle \varepsilon_{21} + (x'_1 - x_1)(\varepsilon_{21,1} - \varepsilon_{11,2}) \right\rangle dx_1$$

$$+ \int\limits_{\substack{0 \\ x_3 = x'_3 \\ x_1 = x'_1}}^{x'_2} \varepsilon_{22}\, dx_2 \,, \tag{A.109.7}$$

which simplifies to:

$$u_2 = \int\limits_{\substack{0 \\ x_1 = x_2 = 0}}^{x'_3} \varepsilon_{23}\, dx_3 + \int\limits_{\substack{0 \\ x_2 = 0 \\ x_3 = x'_3}}^{x'_1} \varepsilon_{21}\, dx_1 + \int\limits_{\substack{0 \\ x_3 = x'_3 \\ x_1 = x'_1}}^{x'_2} \varepsilon_{22}\, dx_2$$

$$= \left\langle (x'_1)^3 + (x'_2)^3 \right\rangle \times 10^{-6}. \tag{A.109.8}$$

In the same way we obtain and solve:

$$u_3 = \int\limits_{\substack{0 \\ x_2 = x_3 = 0}}^{x'_1} \left\langle \varepsilon_{31} + (x'_1 - x_1)(\varepsilon_{31,1} - \varepsilon_{11,3}) + x'_2(\varepsilon_{31,2} - \varepsilon_{21,3}) \right\rangle dx_1$$

$$+ \int\limits_{\substack{0 \\ x_3 = 0 \\ x_1 = x'_1}}^{x'_2} \left\langle \varepsilon_{32} + (x'_2 - x_2)(\varepsilon_{32,2} - \varepsilon_{22,3}) \right\rangle dx_2$$

$$+ \int\limits_{\substack{0 \\ x_1 = x'_1 \\ x_2 = x'_2}}^{x'_3} \varepsilon_{33}\, dx_3$$

$$= \int\limits_{\substack{0 \\ x_2 = x_3 = 0}}^{x'_1} \varepsilon_{31}\, dx_1 + \int\limits_{\substack{0 \\ x_3 = 0 \\ x_1 = x'_1}}^{x'_2} \varepsilon_{32}\, dx_2 + \int\limits_{\substack{0 \\ x_1 = x'_1 \\ x_2 = x'_2}}^{x'_3} \varepsilon_{33}\, dx_3$$

$$= (x'_3)^2 \times 10^{-6}. \tag{A.109.9}$$

Hence, dropping the primes:

$$[u_i] = [3x_1^2 x_2 \quad x_1^3 + x_2^3 \quad x_3^2] \times 10^{-6} \text{ units of length.} \qquad (A.109.10)$$

Spatially differentiate eq. (A.109.10) to find the displacement gradient, and compare with eq. (A.109.1):

$$\left[\frac{\partial u_i}{\partial x_j}\right] \equiv [e_{ij}] = [\varepsilon_{ij}] = \begin{bmatrix} 6x_1 x_2 & 3x_1^2 & 0 \\ 3x_1^2 & 3x_2^2 & 0 \\ 0 & 0 & 2x_3 \end{bmatrix} \times 10^{-6}, \qquad (A.109.11)$$

Thus:

$$\varpi_{ij} = e_{ij} - \varepsilon_{ij} = 0 . \qquad (A.109.12)$$

The rotation is zero not only at the origin but everywhere.

Answer 110. Inspection of the small plane strain ($\alpha^2 \ll 1$) in the $x_1 - x_2$ plane:

$$[\varepsilon_{ij}] = \begin{bmatrix} 0 & \alpha\sin(2\pi x_1/\lambda) & 0 \\ \alpha\sin(2\pi x_1/\lambda) & 0 & 0 \\ 0 & 0 & 0 \end{bmatrix}, \qquad (A.110.1)$$

shows that it maintains the volume and, therefore, the area in the $x_1 - x_2$ plane constant. (a) A check similar to that performed in the answer to Problem 109 shows that compatibility is satisfied. (b) For simplicity, we define the constant $\beta \equiv 2\pi/\lambda$ and rewrite eq. (A.110.1) as:

$$[\varepsilon_{ij}] = \alpha\sin(\beta x_1) \begin{bmatrix} 0 & 1 & 0 \\ 1 & 0 & 0 \\ 0 & 0 & 0 \end{bmatrix}. \qquad (A.110.2)$$

Inasmuch as translation and rotation at the origin are both zero, as in Problem 109, we use eqs. (A.109.4), (A.109.8), and (A.109.10). The solution for u_1 is:

$$u_1 = \int_0^{x_2'} \varepsilon_{12} \, dx_2 + \int_0^{x_3'} \varepsilon_{13} \, dx_3 + \int_0^{x_1'} \varepsilon_{11} \, dx_1 = 0 . \qquad (A.110.3)$$
$$\underset{x_3 = x_1 = 0}{} \quad \underset{\substack{x_1 = 0 \\ x_2 = x_2'}}{} \quad \underset{\substack{x_2 = x_2' \\ x_3 = x_3'}}{}$$

The first integrand is zero because $x_1 = 0$ and thus $\varepsilon_{12} = \alpha \sin(0) = 0$, and $\varepsilon_{13} = \varepsilon_{11} = 0$. The solution for u_2 is:

$$u_2 = \int_{\substack{0 \\ x_1 = x_2 = 0}}^{x_3'} \varepsilon_{23}\, d x_3$$

$$+ \int_{\substack{0 \\ x_2 = 0 \\ x_3 = x_3'}}^{x_1'} \langle \varepsilon_{21} + (x_1' - x_1)\varepsilon_{21,\,1} \rangle\, d x_1 + \int_{\substack{0 \\ x_3 = x_3' \\ x_1 = x_1'}}^{x_2'} \varepsilon_{22}\, d x_2. \qquad \text{(A.110.4)}$$

The first and last integrands are zero. The first term of the middle integral is:

$$\int_0^{x_1'} \varepsilon_{21}\, d x_1 = \int_0^{x_1'} \alpha \sin(\beta x_1)\, d x_1 \,,$$

and the second:

$$\int_0^{x_1'} x_1' \, \varepsilon_{21,\,1}\, d x_1 = \int_0^{x_1'} x_1' \,\alpha\beta \cos(\beta x_1)\, d x_1 \,,$$

where only the unprimed x_1 is a variable, not x_1'. The third term is:

$$\int_0^{x_1'} x_1 \, \varepsilon_{21,\,1}\, d x_1 = \int_0^{x_1'} x_1 \,\alpha\beta \cos(\beta x_1)\, d x_1 \,.$$

Hence:

$$u_2 = \int_0^{x_1'} \alpha\sin(\beta x_1) + x_1' \,\alpha\beta\cos(\beta x_1) - x_1 \,\alpha\beta\cos(\beta x_1)\, d x_1 \,. \qquad \text{(A.110.5)}$$

Integration (by parts for the third term of the integrand) yields:

$$u_2 = -\frac{\alpha}{\beta}\cos(\beta x_1)\Big|_0^{x_1'} + x_1' \,\alpha\sin(\beta x_1)\Big|_0^{x_1'}$$

$$-\frac{\alpha}{\beta}\langle \cos(\beta x_1) - \beta x_1 \sin(\beta x_1) \rangle \Big|_0^{x_1'} \,, \qquad \text{(A.110.6)}$$

where the evaluation symbols imply that the result of evaluation at the bottom is to be subtracted from that at the top. Such an evaluation of eq. (A.110.6) produces:

$$u_2 = -\,\alpha\cos(\beta x_1')/\beta + \alpha\cos(0)/\beta \boxed{+ x_1' \,\alpha\sin(\beta x_1')}$$

$$-\, x_1' \,\alpha\sin(0)$$

$$-\,\alpha\cos(\beta x_1')/\beta \boxed{-\, x_1' \,\alpha\sin(\beta x_1')} + \alpha\cos(0)/\beta$$

$$= 2\alpha\langle 1 - \cos(\beta x_1')\rangle/\beta \,, \qquad \text{(A.110.7)}$$

since the framed terms cancel, $\sin(0) = 0$, and $\cos(0) = 1$. As is to be expected, u_3 is zero:

$$u_3 = \int_{\substack{0 \\ x_2=x_3=0}}^{x_1'} \varepsilon_{31}\, dx_1 + \int_{\substack{0 \\ x_3=0 \\ x_1=x_1'}}^{x_2'} \varepsilon_{32}\, dx_2 + \int_{\substack{0 \\ x_1=x_1' \\ x_2=x_2'}}^{x_3'} \varepsilon_{33}\, dx_3 = 0\ , \qquad \text{(A.110.8)}$$

because all three integrands are zero. Restoring the full expression for β and dropping the prime, the displacement is:

$$\left[u_i\right] = \left[\begin{matrix} 0 & \alpha\lambda\langle 1-\cos(2\pi x_1/\lambda)\rangle/\pi & 0 \end{matrix}\right]\ , \qquad \text{(A.110.9)}$$

where λ obviously designates the wavelength and $2\alpha\lambda/\pi$ the amplitude, trough to crest, of an indefinitely continuing, all-positive sinusoidal curve (Fig. A.110.1). (c) The displacement gradient is:

$$\left[e_{ij}\right] = \left[\frac{\partial u_i}{\partial x_j}\right] = \begin{bmatrix} 0 & 0 & 0 \\ 2\alpha\sin(2\pi x_1/\lambda) & 0 & 0 \\ 0 & 0 & 0 \end{bmatrix}\ . \qquad \text{(A.110.10)}$$

Hence, by subtraction of the strain tensor eq. (A.110.1) from eq. (A.110.11), the rotation tensor is:

$$\left[\varpi_{ij}\right] = \left[e_{ij}\right] - \left[\varepsilon_{ij}\right] = \alpha\sin\frac{2\pi x_1}{\lambda}\begin{bmatrix} 0 & -1 & 0 \\ 1 & 0 & 0 \\ 0 & 0 & 0 \end{bmatrix}\ . \qquad \text{(A.110.11)}$$

The phenomenon described by eqs. (A.110.1) to (A.110.12) is a simple shear of variable intensity with glide direction parallel to the x_2 axis. (d) Given the same strain, but displacements at the origin:

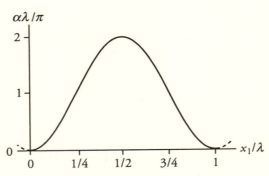

Figure A.110.1. All-positive sinusoidal curve.

$$\left[\overset{\circ *}{u}{}_i\right] = \left[\begin{array}{ccc} \gamma_1 & \gamma_2 & 0 \end{array}\right],$$ (A.110.12)

and clockwise rotation at the origin by a small angle φ about the axis:

$$\left[l_i\right] = \left[\begin{array}{ccc} 0 & 0 & 1 \end{array}\right],$$ (A.110.13)

the rotation tensor at the origin, with φ in radians, is:

$$\left[\overset{\circ *}{\bar\omega}_{ij}\right] = \left[\begin{array}{ccc} 0 & -\varphi & 0 \\ \varphi & 0 & 0 \\ 0 & 0 & 0 \end{array}\right].$$ (A.110.14)

Since:

$$\overset{*}{u}{}_i = \overset{\circ *}{u}{}_i + \overset{*}{\bar\omega}_{ij}\left(x'_j - {}^\circ x_j\right) + u_i,$$ (A.110.15)

and:

$$ {}^\circ x_j = 0,$$ (A.110.16)

and:

$$\overset{*}{\bar\omega}_{ij} = \overset{\circ *}{\bar\omega}_{ij} + \bar\omega_{ij},$$ (A.110.17)

and because both the constant translational displacement and the displacement caused by a constant rotation are simply added to those of eq. (A.110.9), the displacement everywhere is:

$$\left[\overset{*}{u}{}_i\right] = \left[\begin{array}{ccc} \gamma_1 - \varphi x_2 & \gamma_2 + \varphi x_1 + \dfrac{\alpha\lambda}{\pi}\langle 1 - \cos(2\pi x_1/\lambda)\rangle & 0 \end{array}\right].$$ (A.110.18)

Analogously, a constant rotation is added to the variable rotation of eq. (A.110.12):

$$\left[\overset{*}{\bar\omega}_{ij}\right] = \varphi + \alpha\sin\dfrac{2\pi x_1}{\lambda}\left[\begin{array}{ccc} 0 & -1 & 0 \\ 1 & 0 & 0 \\ 0 & 0 & 0 \end{array}\right].$$ (A.110.19)

The strain field is unaffected by the change and is identical with the one given as eq. (A.110.1):

$$\left[\overset{*}{\varepsilon}_{ij}\right] = \left[\varepsilon_{ij}\right].$$ (A.110.20)

Answer 111. A Newtonian liquid of depth h, flowing in steady-state flow (time is not one of the variables in the given equations) over a plane bed inclined by the angle θ, has the strain rate:

$$\left[\dot\varepsilon_{ij}\right] = \left(\dfrac{\rho g x_3 \sin\theta}{2\mu}\right)\left[\begin{array}{ccc} 0 & 0 & -1 \\ 0 & 0 & 0 \\ -1 & 0 & 0 \end{array}\right],$$ (A.111.1)

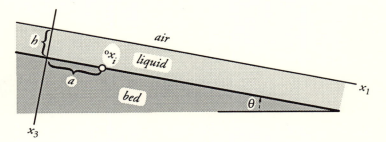

Figure A.111.1. Newtonian flow on an inclined bed. Coordinates
parallel and perpendicular to the bed.

where ρ is the density of the liquid, μ its viscosity, and g the acceleration due
to gravity. At the point:

$$\left[{}^{\circ}x_i \right] = \left[\begin{array}{ccc} a & b & h \end{array} \right] , \tag{A.111.2}$$

at the bottom of the liquid sheet (see Fig. A.111.1) the fluid velocity is zero,
${}^{\circ}v_i = 0$; hence there is no bed slip. The rotation rate at that point is:

$$\left[{}^{\circ}\dot{\omega}_{ij} \right] = \left(\frac{\rho g h \sin\theta}{2\mu} \right) \left[\begin{array}{ccc} 0 & 0 & -1 \\ 0 & 0 & 0 \\ 1 & 0 & 0 \end{array} \right] . \tag{A.111.3}$$

This is a rigid-body rotation about an axis that lies in the plane of the bed
parallel to x_2. Inspection of eq. (A.111.1) shows that $\dot{\varepsilon}_{ij}$ is a plane rate of
strain in the plane perpendicular to x_2; hence:

$$v_2 = 0 , \tag{A.111.4}$$

everywhere in the liquid, and of all components of the rotation rate only two,
$\dot{\omega}_{31} = -\dot{\omega}_{13}$, can be nonzero:

$$\dot{\omega}_{21} = -\dot{\omega}_{12} = \dot{\omega}_{32} = -\dot{\omega}_{23} = 0 . \tag{A.111.5}$$

Rotation and strain rate are, like infinitesimal rotation and strain [eq. (5.19)],
interrelated by:

$$\dot{\omega}_{ij,k} = \dot{\varepsilon}_{ik,j} - \dot{\varepsilon}_{jk,i} , \tag{A.111.6}$$

and rotation rates can be found by integrating eq. (A.111.6). Needed is only
$\dot{\omega}_{31}$:

$$\dot{\varpi}_{31} = {}^{\circ}\dot{\varpi}_{31} + \int_{\substack{a \\ x_2 = b \\ x_3 = h}}^{x_1'} \left(\dot{\varepsilon}_{31,1} - \dot{\varepsilon}_{11,3}\right) dx_1 + \int_{h}^{x_3'} \left(\dot{\varepsilon}_{33,1} - \dot{\varepsilon}_{13,3}\right) dx_3 , \qquad \text{(A.111.7)}$$

Differentiation of the $\dot{\varepsilon}_{ij,k}$ in eq. (A.111.7) yields:

$$\dot{\varpi}_{31} = {}^{\circ}\dot{\varpi}_{31} + \int_{\substack{a \\ x_2 = b \\ x_3 = h}}^{x_1'} \left(0 - 0\right) dx_1 + \int_{h}^{x_3'} \left(0 - \rho g \sin\theta / 2\mu\right) dx_3 , \qquad \text{(A.111.8)}$$

or:

$$\dot{\varpi}_{31} = \left(\frac{\rho g h \sin\theta}{2\mu}\right)\left(\frac{h \, {}^{\circ}\dot{\varpi}_{31} + x_3 - h}{h}\right) , \qquad \text{(A.111.9)}$$

which, considering that ${}^{\circ}\dot{\varpi}_{31} = 1$ and that the h in the numerator of the first and the denominator in the second parentheses cancel, simplifies to:

$$\dot{\varpi}_{31} = \frac{\rho g x_3 \sin\theta}{2\mu} . \qquad \text{(A.111.10)}$$

Hence:

$$\left[\dot{\varpi}_{ij}\right] = \frac{\rho g x_3 \sin\theta}{2\mu} \begin{bmatrix} 0 & 0 & -1 \\ 0 & 0 & 0 \\ 1 & 0 & 0 \end{bmatrix} . \qquad \text{(A.111.11)}$$

Note: This is consistent with the specification of ${}^{\circ}\dot{\varpi}_{ij}$. To find the velocity field, the velocity gradient, which is the sum of the strain and rotation rates, must be integrated:

$$v_1 = \int_{\substack{h \\ x_1 = x_1' \\ x_2 = b}}^{x_3'} \left(\dot{\varepsilon}_{13} + \dot{\varpi}_{13}\right) dx_3 = \int_{\substack{h \\ x_1 = x_1' \\ x_2 = b}}^{x_3'} -\left(\frac{2\rho g x_3 \sin\theta}{2\mu}\right) dx_3 , \qquad \text{(A.111.12)}$$

or:

$$v_1 = \frac{\rho g \left(h^2 - x_3^2\right) \sin\theta}{2\mu} . \qquad \text{(A.111.13)}$$

Note: This velocity component decreases nonlinearly (with the square of the depth h) from the top of the liquid sheet to immediately above its bottom, where it is zero. It is trivially zero in the bed itself. Inasmuch as:

$$\left[\dot{e}_{ij}\right] = \left[\dot{\varpi}_{ij}\right] + \left[\dot{\varepsilon}_{ij}\right] = -\frac{\rho g x_3 \sin\theta}{\mu} \begin{bmatrix} 0 & 0 & 1 \\ 0 & 0 & 0 \\ 0 & 0 & 0 \end{bmatrix} , \qquad \text{(A.111.14)}$$

the remaining velocity components must be zero:

$$v_2 = v_3 = 0 \ . \tag{A.111.15}$$

Summary: There is no slip on the bed and maximum velocity but neither strain nor rotation at the surface, where $x_3 = 0$. Both rates of strain and rotation have their maxima in the liquid in contact with the bed.

Answer 112. (a) Given the flow law:

$$\dot{\varepsilon}_{ij} = A\sigma_{ij} + B\sigma_{kk}\delta_{ij} + C\delta_{ij} \ , \tag{A.112.1}$$

check for invariance by rotating:

$$\dot{\varepsilon'}_{ij} = a_{ik}a_{jl}\dot{\varepsilon}_{kl} = Aa_{ik}a_{jl}\sigma_{kl} + B\sigma_{mm}\delta_{ij} + C\delta_{ij}$$
$$= A\sigma'_{ij} + B\sigma'_{mm}\delta_{ij} + C\delta_{ij} \ . \tag{A.112.2}$$

Thus the flow law is invariant to rotation. (b) If, as assumed, $\dot{\varepsilon}_{ij} = 0$ when $\sigma_{ij} = 0$, then inspection of eq. (A.112.2) shows that necessarily $C = 0$. (c) The stress deviator is:

$$\Delta_{ij} = \sigma_{ij} - \left(\sigma_{kk}\delta_{ij}\right)/3 = \sigma_{ij} + p\delta_{ij} \ , \tag{A.112.3}$$

where p is the pressure. Equation (A.112.1) restated for $C = 0$ and for:

$$A\sigma_{ij} = A\Delta_{ij} + \left(A\sigma_{kk}\delta_{ij}\right)/3 \ , \tag{A.112.4}$$

becomes:

$$\dot{\varepsilon}_{ij} = A\Delta_{ij} + \left(A\sigma_{kk}\delta_{ij}\right)/3 + B\sigma_{kk}\delta_{ij}$$
$$= A\Delta_{ij} + (A/3 + B)\sigma_{kk}\delta_{ij}$$
$$= A\Delta_{ij} - (A + 3B)p\delta_{ij} \ . \tag{A.112.5}$$

In analogy to the stress deviator, we define the distortion rate as:

$$\dot{D}_{ij} = \dot{\varepsilon}_{ij} - \left(\dot{\varepsilon}_{kk}\delta_{ij}\right)/3 \ . \tag{A.112.6}$$

The flow law in terms of the deviator, distortion rate, and pressure is:

$$\dot{D}_{ij} + \left(\dot{\varepsilon}_{kk}\delta_{ij}\right)/3 = A\Delta_{ij} - (A + 3B)p\delta_{ij} \ . \tag{A.112.7}$$

(d) This must be a combination of two separate equations because a strain rate proportional to δ_{ij} is necessarily invariant to coordinate rotation and can, therefore, be only a change of volume. The first separate equation in eq. (A.112.7) is:

$$\dot{D}_{ij} = A\Delta_{ij} ,$$ (A.112.8)

the second:

$$\left(\dot{\varepsilon}_{kk}\delta_{ij}\right)/3 = -\left(A + 3B\right) p\delta_{ij} ,$$ (A.112.9)

or:

$$\dot{\varepsilon}_{kk} = -3p\left(A + 3B\right) .$$ (A.112.10)

(e) This equation is physically not reasonable, considering that it implies an indefinitely continuing rate of volume change at a given pressure. The following must therefore necessarily hold:

$$A = -3B .$$ (A.112.11)

(f) In an incompressible fluid $\dot{\varepsilon}_{kk} = 0$, and the flow law reduces to eq. (A.112.8).

Answer 113. The problem was stated using the form:

$$|V_{ij}| = \begin{vmatrix} u_1 & u_2 & u_3 \\ v_1 & v_2 & v_3 \\ w_1 & w_2 & w_3 \end{vmatrix} .$$ (A.113.1)

This is the equivalent of defining the elements of the matrix V_{ij} as follows:

$$V_{1j} \equiv u_j ,$$
$$V_{2j} \equiv v_j ,$$
$$V_{3j} \equiv w_j .$$ (A.113.2)

Let (a_{ij}) effect an arbitrary rotation of the three vectors **u**, **v**, and **w**; then the determinant for the rotated matrix is:

$$|V'_{ij}| = \begin{vmatrix} a_{11}u_1 & a_{12}u_2 & a_{13}u_3 \\ a_{21}v_1 & a_{22}v_2 & a_{23}v_3 \\ a_{31}w_1 & a_{32}w_2 & a_{33}w_3 \end{vmatrix} .$$ (A.113.3)

Substituting eqs. (A.113.2) into eq. (A.113.3) and using symbolic subscript notation, we obtain:

$$|V'_{ij}| = |a_{ik}V_{jk}| ,$$ (A.113.4)

where the dummy subscript k replaces the repeated numerical subscripts in eq. (A.113.3). According to the result of Problem 56, the right-hand side of eq. (A.113.4) can be decomposed:

$$\left| V'_{ij} \right| = \left| a_{ik} V_{jk} \right| = \left| a_{ik} \right| \left| V_{jk} \right| , \tag{A.113.5}$$

which, because $\left| a_{jj} \right| = 1$, further simplifies to:

$$\left| V'_{ij} \right| = \left| V_{jk} \right| = \left| V_{ij} \right| \therefore \tag{A.113.6}$$

Answer 114. (a) Figure A.114.1 shows the parallelepiped defined by three noncoplanar vectors $^{k}A_{i}$. The two vectors defining the parallelogram at its base $^{1}A_{i}$ and $^{2}A_{i}$ enclose the angle θ, and their cross product C_{k} is:

$$C_{k} = \epsilon_{ijk} \, ^{1}A_{i} \, ^{2}A_{j} = C \, l_{k} , \tag{A.114.1}$$

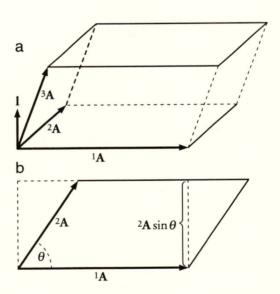

Figure A.114.1. The parallelepiped defined by three vectors. (a) Perspective view. The unit vector **1** is normal to the plane of the base parallelogram defined by the first two vectors. (b) Downward view of the base parallelogram.

where l_k is the upward-normal unit vector on the base parallelogram and C the magnitude of the cross product. The top view (Fig. A.114.1b) of the parallelogram at the base of the parallelepiped illustrates that its area B:

$$B = {}^1\!A\,{}^2\!A \sin\theta \equiv C \; , \tag{A.114.2}$$

equals the magnitude of the cross product ${}^1\mathbf{A} \times {}^2\mathbf{A}$. Substitute this result, expressed in subscript notation, into eq. (A.114.1):

$$C_k = \epsilon_{ijk}\,{}^1\!A_i\,{}^2\!A_j = B\,l_k \; , \tag{A.114.3}$$

and find the upward-normal unit vector:

$$l_k = \frac{C_k}{B} = \frac{\epsilon_{ijk}\,{}^1\!A_i\,{}^2\!A_j}{B} \; . \tag{A.114.4}$$

Because the top and bottom faces of a parallelepiped are parallel, its volume V is the product of the area of its base B with its height H:

$$V = BH \; . \tag{A.114.5}$$

The height H is the projection of ${}^3\mathbf{A}$ onto a line normal to the parallelepiped base (Fig. A.114.1a) and thus the dot product of ${}^3\mathbf{A}$ with the unit vector $\mathbf{l}$, the $[l_i]$ of eq. (A.114.1):

$$H = {}^3\!A_i\,l_i \; . \tag{A.114.6}$$

Substitute eq. (A.114.4) into eq. (A.114.6):

$$H = \frac{C_k\,{}^3\!A_k}{B} = \frac{\epsilon_{ijk}\,{}^1\!A_i\,{}^2\!A_j\,{}^3\!A_k}{B} \; , \tag{A.114.7}$$

and substitute eq. (A.114.7) into eq. (A.114.5) to obtain the volume (the preceding identity follows from the result of Problem 55):

$$V = BH = C_k\,{}^3\!A_k = \epsilon_{ijk}\,{}^1\!A_i\,{}^2\!A_j\,{}^3\!A_k \equiv \left| {}^k\!A_i \right| \quad \therefore \tag{A.114.8}$$

(b) Assume that a strain moved material points from original coordinates a_i to final coordinates x_i. Let the final coordinates x_i be functions of original positions a_i:

$$x_i = x_i(a_1, a_2, a_3) \; . \tag{A.114.9}$$

Let an elemental cube in the original coordinates be characterized by three equal, orthogonal, infinitesimal vectors $d^k a_i$; then an elemental final parallelepiped should be similarly defined by the generally different, nonorthogonal infinitesimal vectors $d^k x_i$:

$$d^k x_i = \frac{\partial x_i}{\partial a_j} d^k a_j \ . \tag{A.114.10}$$

Calling the initial elemental volume dv and the final dV, and using the results of (a), we obtain:

$$dV = \left| d^k x_i \right| = \left| \frac{\partial x_i}{\partial a_j} d^k a_j \right| = \left| \frac{\partial x_i}{\partial a_j} \right| \left| d^k a_j \right| = \left| \frac{\partial x_i}{\partial a_j} \right| dv \ . \tag{A.114.11}$$

Hence, the ratio of new to old volume is:

$$\frac{dV}{dv} = \left| \frac{\partial x_i}{\partial a_j} \right| \ . \tag{A.114.12}$$

Then the volume strain Δ is:

$$\Delta \equiv \frac{dV - dv}{dv} = \frac{dV}{dv} - 1 = \left| \frac{\partial x_i}{\partial a_j} \right| - 1 \ \therefore \tag{A.114.13}$$

Note that this is the Lagrangian (function of the original positions) volume strain. (c) To calculate the volume strain as a function of the final positions of material points, we first determine the original coordinates a_i as a function of their final counterparts:

$$a_i = a_i (x_1, x_2, x_3) \ . \tag{A.114.14}$$

Now let an elemental cube in the final coordinates be characterized by three equal, orthogonal, infinitesimal vectors $d^k x_i$, then a materially identical, elemental, original parallelepiped should be defined by the generally different, nonorthogonal infinitesimal vectors $d^k a_i$:

$$d^k a_i = \frac{\partial a_i}{\partial x_j} d^k x_j \ . \tag{A.114.15}$$

The corresponding relationship between volume elements is:

$$dv = \left| d^k a_i \right| = \left| \frac{\partial a_i}{\partial x_j} \right| dV \ , \tag{A.114.16}$$

the volume ratio:

$$\frac{dv}{dV} = \left|\frac{\partial a_i}{\partial x_j}\right| , \tag{A.114.17}$$

and the volume strain:

$$\Delta = \left|\frac{\partial a_i}{\partial x_j}\right|^{-1} - 1 \quad \therefore \tag{A.114.18}$$

Note that this is the Eulerian (function of the final positions) volume strain.

Answer 115. Displacements are the differences between new and old positions of the same material point:

$$u_i = x_i - a_i . \tag{A.115.1}$$

The Lagrangian displacement gradient (differentiated with respect to original coordinates a_i) is:

$$\frac{\partial u_i}{\partial a_j} = \frac{\partial x_i}{\partial a_j} - \frac{\partial a_i}{\partial a_j} , \tag{A.115.2}$$

or, because $\partial a_i / \partial a_j$ is unity when $i = j$ and zero when $i \neq j$:

$$\frac{\partial u_i}{\partial a_j} = \frac{\partial x_i}{\partial a_j} - \delta_{ij} , \tag{A.115.3}$$

According to the results of Problem 114, the dilatation is:

$$\Delta = \left|\frac{\partial x_i}{\partial a_j}\right| - 1 . \tag{A.115.4}$$

Substitute eq. (A.115.3) into eq. (A.115.4):

$$\Delta = \left|\frac{\partial u_i}{\partial a_j} + \delta_{ij}\right| - 1 , \tag{A.115.5}$$

or explicitly:

$$\Delta = \begin{vmatrix} 1 + \dfrac{\partial u_1}{\partial a_1} & \dfrac{\partial u_1}{\partial a_2} & \dfrac{\partial u_1}{\partial a_3} \\[2mm] \dfrac{\partial u_2}{\partial a_1} & 1 + \dfrac{\partial u_2}{\partial a_2} & \dfrac{\partial u_2}{\partial a_3} \\[2mm] \dfrac{\partial u_3}{\partial a_1} & \dfrac{\partial u_3}{\partial a_2} & 1 + \dfrac{\partial u_3}{\partial a_3} \end{vmatrix} - 1 , \tag{A.115.6}$$

or:

$$\Delta = 1 + \frac{\partial u_1}{\partial a_1} + \frac{\partial u_2}{\partial a_2} + \frac{\partial u_3}{\partial a_3} + (\text{products of derivatives}) - 1 . \tag{A.115.7}$$

Hence:

$$\frac{\partial u_i}{\partial a_i} \rightarrow \varepsilon_{ii} \quad \text{as} \quad \frac{\partial u_i}{\partial a_j} \rightarrow 0 \ , \tag{A.115.8}$$

and thus for infinitesimal strains:

$$\Delta = \varepsilon_{ii} \ \therefore \tag{A.115.9}$$

Answer 116. To show that the finite dilatation Δ is given by:

$$\Delta = \left| \delta_{ij} + 2\mathfrak{E}_{ij} \right|^{1/2} - 1 = \left| \delta_{ij} - 2e_{ij} \right|^{-1/2} - 1 \ , \tag{A.116.1}$$

where the square roots are taken positive, first use the definition of Green's tensor:

$$2\mathfrak{E}_{ij} = \frac{\partial x_k}{\partial a_i} \frac{\partial x_k}{\partial a_j} - \delta_{ij} \ . \tag{A.116.2}$$

Substitute eq. (A.116.2) into the first part of eq. (A.116.1) and obtain:

$$\left| \delta_{ij} + 2\mathfrak{E}_{ij} \right|^{1/2} - 1 = \left| \frac{\partial x_k}{\partial a_i} \frac{\partial x_k}{\partial a_j} \right|^{1/2} - 1 \ . \tag{A.116.3}$$

Inasmuch as $|c_{ij}| = |a_{ij}||b_{ij}|$ if either $c_{ij} = a_{ik}b_{jk}$ or $c_{ij} = a_{ki}b_{kj}$ and inasmuch as in the present case $a_{ij} = b_{ij}$, the right-hand side can be rewritten:

$$\left| \delta_{ij} + 2\mathfrak{E}_{ij} \right|^{1/2} - 1 = \left| \frac{\partial x_i}{\partial a_j} \right| - 1 \ , \tag{A.116.4}$$

so that the determinant on the right-hand side becomes the Jacobian, which must be positive and finite. Considering the results of Problem 114, we conclude:

$$\left| \delta_{ij} + 2\mathfrak{E}_{ij} \right|^{1/2} - 1 = \Delta \ \therefore \tag{A.116.5}$$

Similarly, use the definition of Almansi's tensor:

$$2e_{ij} = \delta_{ij} - \frac{\partial a_k}{\partial x_i} \frac{\partial a_k}{\partial x_j} \ , \tag{A.116.6}$$

substitute into the second part of eq. (A.116.1), and obtain:

$$\left| \delta_{ij} - 2e_{ij} \right|^{-1/2} - 1 = \left| \frac{\partial a_k}{\partial x_i} \frac{\partial a_k}{\partial x_j} \right|^{-1/2} - 1 = \left| \frac{\partial a_i}{\partial x_j} \right|^{-1} = \Delta \ \therefore \tag{A.116.7}$$

Answer 117. (a) Given is the linear transformation rule from initial to final positions:

$$x_i(a_1, a_2, a_3) = A_{ij}a_j , \qquad (A.117.1)$$

where the finite, linear transformation matrix (A_{ij}) is:

$$\left(A_{ij}\right) = \begin{pmatrix} 1 & 1 & -4 \\ -3 & -2 & 2 \\ 5 & 3 & 1 \end{pmatrix}. \qquad (A.117.2)$$

Because the Lagrangian spatial derivative $\partial x_i/\partial a_j$ for eq. (A.117.1) is:

$$\frac{\partial x_i}{\partial a_j} = A_{ij} , \qquad (A.117.3)$$

Green's tensor becomes:

$$\mathfrak{E}_{ij} = \frac{1}{2}\left(\frac{\partial x_k}{\partial a_i}\frac{\partial x_k}{\partial a_j} - \delta_{ij}\right) = \frac{1}{2}\left(A_{ki}A_{kj} - \delta_{ij}\right). \qquad (A.117.4)$$

The explicit Green's tensor for the transformation matrix in eq. (A.117.2) is:

$$\left[\mathfrak{E}_{ij}\right] = \frac{1}{2}\begin{bmatrix} 1+9+25-1 & 1+6+15 & -4-6+5 \\ 1+6+15 & 1+4+9-1 & -4-4+3 \\ -4-6+5 & -4-4+3 & 16+4+1-1 \end{bmatrix}, \qquad (A.117.5)$$

or:

$$\left[\mathfrak{E}_{ij}\right] = \frac{1}{2}\begin{bmatrix} 34 & 22 & -5 \\ 22 & 13 & -5 \\ -5 & -5 & 20 \end{bmatrix}. \qquad (A.117.6)$$

(b) To find the Eulerian Almansi's tensor, consider eq. (A.117.1) as a set of simultaneous linear equations in the a_i. That set has the solutions:

$$a_i = A_{ij}^{-1}x_j = \frac{\overset{*}{A}_{ij}x_j}{\left|A_{ij}\right|} , \qquad (A.117.7)$$

where $\overset{*}{A}_{ij}$ are the cofactors of A_{ij} and $|A_{ij}| = 1$; hence $A_{ij}^{-1} = \overset{*}{A}_{ji}$. The sign factors for the cofactors are:

$$\begin{pmatrix} +1 & -1 & +1 \\ -1 & +1 & -1 \\ +1 & -1 & +1 \end{pmatrix}.$$

Thus the inverse transformation matrix and transposed cofactors become:

$$\left(A_{ij}^{-1}\right) = \left(\overset{*}{A}_{ij}\right) = \begin{pmatrix} -8 & -13 & -6 \\ 13 & 21 & 10 \\ 1 & 2 & 1 \end{pmatrix} . \tag{A.117.8}$$

(Here it is useful to perform the check $A_{ik} A_{kj}^{-1} = \delta_{ij}$.) Almansi's tensor is:

$$e_{ij} = \frac{1}{2}\left(\delta_{ij} - \frac{\partial a_k}{\partial x_i}\frac{\partial a_k}{\partial x_j}\right) = \frac{1}{2}\left(\delta_{ij} - A_{ki}^{-1} A_{kj}^{-1}\right) , \tag{A.117.9}$$

or:

$$\left[e_{ij}\right] = \frac{1}{2}\begin{bmatrix} 1-(64+169+1) & -(104+273+2) & -(48+130+1) \\ -(104+273+2) & 1-(169+441+4) & -(78+210+2) \\ -(48+130+1) & -(78+210+2) & 1-(36+100+1) \end{bmatrix}$$

$$= \frac{1}{2}\begin{bmatrix} -233 & -379 & -179 \\ -379 & -613 & -290 \\ -179 & -290 & -136 \end{bmatrix} . \tag{A.117.10}$$

(c) The dilatation for this strain is:

$$\Delta = \frac{V-v}{v} = \left|A_{ij}\right| - 1 = 0 . \tag{A.117.11}$$

The strain is a finite, homogeneous strain at constant volume.

Answer 118. In subscript notation the given equation is:

$$T_{ij} = a_{ik}\, a_{jl}\, U_{mk}\, V_{ml} - W_{ji} . \tag{A.118.1}$$

For the conversion to matrix notation it is necessary to adjoin matrices with the same dummy subscript and to move these subscripts to proximal positions, if necessary by transposition. We begin with the matrix product that has the dummy subscript m and find it necessary to form the transpose of $[U_{ij}]$:

$$U_{km}^{\mathsf{T}} = U_{mk} . \tag{A.118.2}$$

To fit into the proper subscript order, a_{ij} and W_{ji} must also be similarly transposed, and the equation becomes:

$$T_{ij} = a_{ik}\, U_{km}^{\mathsf{T}}\, V_{ml}\, a_{lj}^{\mathsf{T}} - W_{ij}^{\mathsf{T}} . \tag{A.118.3}$$

The dummy subscripts $k, m,$ and l are now in proximal position, and the free subscripts i and j have the same order in all three terms; we can, therefore, state eq. (A.118.3) in matrix form:

$$\mathbb{T} = \mathbb{R} \, \mathbb{U}^\mathsf{T} \, \mathbb{V} \, \mathbb{R}^\mathsf{T} - \mathbb{W}^\mathsf{T} \ . \tag{A.118.4}$$

Note that the rotation matrix, usually shown as (a_{ij}) in subscript notation, becomes $\mathbb{R}$ in matrix notation.

Answer 119. (a) To convert the given equation:

$$\mathbb{D} = \mathbb{A}^\mathsf{T} \mathbb{B}^\mathsf{T} \mathbb{B} \mathbb{A} + \mathbb{C} \ , \tag{A.119.1}$$

to subscript notation, we give all three terms the customary subscripts i and j. Thus the left-hand side becomes D_{ij} and the last term C_{ij}. To the first matrix in the first right-hand term, we give the free subscript i in the first position and start supplying dummy subscripts until we arrive at the last matrix, which must carry the free subscript in the second position:

$$D_{ij} = A^\mathsf{T}_{ik} B^\mathsf{T}_{kl} B_{lm} A_{mj} + C_{ij} \ . \tag{A.119.2}$$

Because transposition can be expressed by subscript order, this becomes:

$$D_{ij} = A_{ki} B_{lk} B_{lm} A_{mj} + C_{ij} \ , \tag{A.119.3}$$

or perhaps, because of the arbitrary order of factors:

$$D_{ij} = A_{ki} A_{mj} B_{lk} B_{lm} + C_{ij} \ . \tag{A.119.4}$$

(b) By the same procedure the matrix equation:

$$\mathbb{E} = \mathbb{B} \mathbb{A} \mathbb{A}^\mathsf{T} \mathbb{B}^\mathsf{T} + \mathbb{C} \tag{A.119.5}$$

becomes:

$$E_{ij} = A_{kl} A_{ml} B_{ik} B_{jm} + C_{ij} \ . \tag{A.119.6}$$

Note that the right-hand sides of eqs. (A.119.1) and (A.119.5) differ only by the order of the factors in the first term. If $\mathbb{A}$ and $\mathbb{B}$ were transformation matrices, this would indicate that each transformation was preceded by the one denoted to its right in the term. The same difference is expressed by the contrasting arrangements of the subscripts in eqs. (A.119.4) and (A.119.6); it is difficult, however, to recognize any sequence of transformations in the subscripted versions.

Answer 120. (a) Given the transformation matrix:

$$\left(A_{ij} \right) = \begin{pmatrix} 3.103\,716\,5 & 1.998\,225\,3 & -0.611\,587\,9 \\ -0.324\,590\,4 & -0.611\,587\,9 & 4.529\,967\,7 \\ 5.026\,090\,4 & 3.103\,716\,5 & -0.324\,590\,4 \end{pmatrix}, \quad \text{(A.120.1)}$$

its symmetric square (exact to six places) is:

$$\left[A_{ki}\,A_{kj} \right] = \begin{bmatrix} 35 & 22 & -5 \\ 22 & 14 & -5 \\ -5 & -5 & 21 \end{bmatrix}. \quad \text{(A.120.2)}$$

Thus Green's tensor is:

$$\left[\mathfrak{E}_{ij} \right] = \tfrac{1}{2}\left[A_{ki}\,A_{kj} - \delta_{ij} \right] = \tfrac{1}{2}\begin{bmatrix} 34 & 22 & -5 \\ 22 & 13 & -5 \\ -5 & -5 & 20 \end{bmatrix}, \quad \text{(A.120.3)}$$

the same as in eq. (A.117.6). (b) By standard methods, the principal values of this tensor are:

$$\mathfrak{E}_1 = 24.746\,286\,3, \quad \mathfrak{E}_2 = 9.253\,206\,1, \quad \mathfrak{E}_3 = -0.499\,492\,4, \quad \text{(A.120.4)}$$

and the direction cosines from principal to reference coordinates are:

$$\left(b_{ij} \right) = \begin{pmatrix} 0.819\,712\,9 & 0.234\,458\,9 & 0.522\,589\,5 \\ 0.525\,419\,6 & 0.055\,459\,3 & -0.849\,033\,9 \\ -0.228\,046\,0 & 0.970\,542\,8 & -0.077\,728\,6 \end{pmatrix}, \quad \text{(A.120.5)}$$

indicating the following approximate direction angles:

$$\left(\beta_{ij} \right) \approx \begin{pmatrix} 35° & 76° & 58° \\ 58° & 87° & 148° \\ 103° & 14° & 94° \end{pmatrix}. \quad \text{(A.120.6)}$$

(c) The principal elongations indicated by Green's tensor are:

$$^i\lambda = \left(1 + 2\,{}^i\mathfrak{E} \right)^{1/2} = 7.105\,812\,6, \quad 4.416\,606\,4, \quad 0.031\,863\,8. \quad \text{(A.120.7)}$$

Their product is unity to seven places, and hence the dilatation is $\Delta = 0$.
(d) These elongations form the principal values of the stretch tensor, $^i\lambda \equiv {}^iS$, and hence:

$$\left[{}^P S_{ij} \right] = \begin{bmatrix} 7.105\,812\,6 & 0 & 0 \\ 0 & 4.416\,606\,4 & 0 \\ 0 & 0 & 0.031\,863\,8 \end{bmatrix}, \qquad \text{(A.120.8)}$$

and the inverse:

$$\left[{}^P S_{ij}^{-1} \right] = \begin{bmatrix} 0.140\,729\,9 & 0 & 0 \\ 0 & 0.226\,418\,2 & 0 \\ 0 & 0 & 31.383\,603\,4 \end{bmatrix}. \qquad \text{(A.120.9)}$$

To rotate the stretch and inverse stretch tensors from principal to reference coordinates, we use the matrix of eq. (A.120.5):

$$S_{ij} = b_{ik}\,b_{jl}\,{}^P S_{kl}, \qquad S_{ij}^{-1} = b_{ik}\,b_{jl}\,{}^P S_{kl}^{-1}, \qquad \text{(A.120.10)}$$

which yields:

$$\left[S_{ij} \right] = \begin{bmatrix} 5.026\,090\,4 & 3.103\,716\,5 & -0.324\,590\,4 \\ 3.103\,716\,5 & 1.998\,225\,3 & -0.611\,587\,9 \\ -0.324\,590\,4 & -0.611\,587\,9 & 4.529\,967\,0 \end{bmatrix}, \qquad \text{(A.120.11)}$$

and:

$$\left[S_{ij}^{-1} \right] = \begin{bmatrix} 8.677\,855\,0 & -13.861\,217\,7 & -1.249\,590\,7 \\ -13.861\,217\,7 & 22.662\,665\,2 & 2.066\,459\,5 \\ -1.249\,590\,7 & 2.066\,459\,5 & 0.410\,205\,3 \end{bmatrix}. \qquad \text{(A.120.12)}$$

(e) After squaring the principal stretches of eq. (A.120.9) to obtain:

$${}^P S_1^2 = 50.492\,572, \quad {}^P S_2^2 = 19.506\,412, \quad {}^P S_3^2 = 0.001\,015\,300,$$

the following checks are found to hold within six places:

$$\left[S_{ki}\,S_{kj} \right] = \left[A_{ki}\,A_{kj} \right] = \left[b_{ik}\,b_{jl}\,{}^P S_{ij}^2 \right], \quad \left[S_{ki}^{-1}\,S_{kj} \right] = \left[\delta_{ij} \right].$$

(f) To find the matrix for the rotation that could have produced the transformation S_{ij} if it followed the strain, it is required to postmultiply A_{ij} by S_{ij}^{-1}. A schematic diagram (Fig. A.120.1) illustrates the logic of this procedure. The calculation yields the following to four places:

$$\left(a_{ij} \right) = \left(A_{ik}\,S_{kj}^{-1} \right) = \begin{pmatrix} 0 & 1 & 0 \\ 0 & 0 & 1 \\ 1 & 0 & 0 \end{pmatrix}. \qquad \text{(A.120.13)}$$

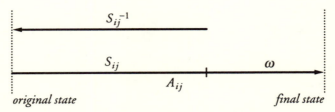

Figure A.120.1. Finding the rotation ω implicit in a transformation A_{ij} in which the rotation follows a strain S_{ij}.

$$\left(a_{ij}\right) = \left(A_{ik}\, S_{kj}^{-1}\right) = \begin{pmatrix} 0 & 1 & 0 \\ 0 & 0 & 1 \\ 1 & 0 & 0 \end{pmatrix}. \qquad (A.120.13)$$

(g) This is the rotation encountered in Problem 49. It rotates coordinate axes through 120° counterclockwise about the axis with the direction:

$$[l_i] = [\begin{array}{ccc} 1 & 1 & 1 \end{array}]/\sqrt{3} , \qquad (A.120.14)$$

carrying axes $a_1 \rightarrow a'_3$, $a_2 \rightarrow a'_1$, and $a_3 \rightarrow a'_2$. Counterclockwise coordinate rotation implies a clockwise sense of the material rotation, seen in the direction of l_i.

Answer 121. (a) Given a transformation matrix:

$$\left(B_{ij}\right) = \begin{pmatrix} -0.324\,590\,4 & 5.026\,090\,4 & 3.103\,716\,5 \\ -0.611\,587\,9 & 3.103\,716\,5 & 1.998\,225\,3 \\ 4.529\,967\,7 & -0.324\,590\,4 & -0.611\,587\,9 \end{pmatrix}, \quad (A.121.1)$$

Almansi's strain tensor e_{ij} is found by first inverting this matrix:

$$\left(B_{ij}^{-1}\right) = \begin{pmatrix} -1.249\,591\,4 & 2.066\,460\,7 & 0.410\,205\,3 \\ 8.677\,861\,2 & -13.861\,227\,6 & -1.249\,591\,4 \\ -13.861\,227\,6 & 22.662\,681\,3 & 2.066\,460\,7 \end{pmatrix}. \quad (A.121.2)$$

Next, find the tensor that is the symmetric square of this matrix. To three places, it is:

$$\left[B_{ki}^{-1}\, B_{kj}^{-1}\right] = \begin{bmatrix} 269 & -437 & -40 \\ -437 & 710 & 65 \\ -40 & 65 & 6 \end{bmatrix}. \qquad (A.121.3)$$

From this, Almansi's tensor, to the same precision, is:

$$\left[e_{ij}\right] = \tfrac{1}{2}\left[\delta_{ij} - B_{ki}^{-1} B_{kj}^{-1}\right] = \tfrac{1}{2}\begin{bmatrix} -268 & 437 & 40 \\ 437 & -709 & -65 \\ 40 & -65 & -5 \end{bmatrix}, \quad (A.121.4)$$

or:

$$\left[e_{ij}\right] = \begin{bmatrix} -134.0 & 218.5 & 20.0 \\ 218.5 & -354.5 & -32.5 \\ 20.0 & -32.5 & -2.5 \end{bmatrix}. \quad (A.121.5)$$

To four significant figures, the principal values of Almansi's tensor are:

$$e_1 = 0.490\,1 \; , \quad e_2 = 0.474\,4 \; , \quad e_3 = -492.0 \; . \quad (A.121.6)$$

(b) The principal elongations $^i\lambda$, or $^iS = \left(1 - 2\,^ie\right)^{-1/2}$, are:

$$S_1 = 7.112 \; , \quad S_2 = 4.415 \; , \quad S_3 = 0.031\,86 \; . \quad (A.121.7)$$

Equation (A.121.7) agrees with eq. (A.120.8) to three significant figures. Because of this identity, the dilatation is $\Delta = 0$, as in the previous problem. A rotation matrix (b_{ij}) rotates e_{ij} from principal [eq. (A.121.6)] to reference coordinates [eq. (A.121.5)]; it is:

$$\left(b_{ij}\right) = \begin{pmatrix} 0.818\,7 & 0.234\,3 & 0.522\,6 \\ 0.525\,2 & 0.055\,3 & -0.849\,0 \\ -0.232\,3 & 0.970\,6 & -0.077\,7 \end{pmatrix}, \quad (A.121.8)$$

indicating approximate direction angles:

$$\left(\beta_{ij}\right) \approx \begin{pmatrix} 35° & 76° & 58° \\ 58° & 87° & 148° \\ 103° & 14° & 94° \end{pmatrix}. \quad (A.121.9)$$

The same matrix rotates the symmetric tensor S_{ij} from principal to reference coordinates. Considering that both the principal values and the orientation of S_{ij} are identical with those of Problem 120, the sought tensor is that of eq. (A.120.11). (c) Assuming that the transformation B_{ij} of eq. (A.121.1) results from a rotation characterized by a matrix (c_{ij}) followed by the strain described by Almansi's tensor e_{ij} or the stretch tensor S_{ij}, find the matrix (c_{ij}^{-1}) for the inverse of the rotation by postmultiplying B_{ij}^{-1} with S_{ij} according to the logical diagram Fig. A.121.1:

$$\left(c_{ij}^{-1}\right) = \left(B_{ik}^{-1} S_{kj}\right) = \begin{pmatrix} 0 & 0 & 1 \\ 1 & 0 & 0 \\ 0 & 1 & 0 \end{pmatrix}, \quad (A.121.10)$$

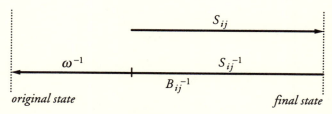

Figure A.121.1. Finding the inverse of the rotation ω implicit in a transformation B_{ij} in which the strain S_{ij} follows the rotation ω.

and hence:

$$\left(c_{ij}\right) = \left(c_{ji}^{-1}\right) = \begin{pmatrix} 0 & 1 & 0 \\ 0 & 0 & 1 \\ 1 & 0 & 0 \end{pmatrix} . \qquad (A.121.11)$$

As in the previous problem, this matrix rotates coordinate axes through 120° counterclockwise about the axis with the direction:

$$[l_i] = [\ 1 \quad 1 \quad 1\ ]/\sqrt{3}\ , \qquad (A.121.12)$$

carrying axes $x_1 \rightarrow x_3'$, $x_2 \rightarrow x_1'$, and $x_3 \rightarrow x_2'$. In a fixed reference frame and seen in the direction of l_i, this indicates clockwise rotation.

Answer 122. (a) Assume that a strain followed a rotation to produce the transformation matrix A_{ij} of eq. (A.120.1). To calculate the matrix representing the hypothetical rotation according to the same logical diagram as Fig. A.121.1, except that A replaces B, it is necessary to invert A_{ij}:

$$\left(A_{ij}^{-1}\right) = \begin{pmatrix} -13.861\,242\,0 & -1.249\,592\,7 & 8.677\,870\,1 \\ 22.662\,704\,4 & 2.066\,462\,8 & -13.861\,242\,0 \\ 2.066\,462\,8 & 0.410\,205\,5 & -1.249\,592\,7 \end{pmatrix} . \qquad (A.122.1)$$

The hypothetical "inverse rotation" is $\overset{*}{a}{}_{ij}^{-1} = A_{ik}^{-1} S_{kj}$, and the "rotation" $\overset{*}{a}{}_{ji} = \overset{*}{a}{}_{ij}^{-1}$. Hence:

$$\left(\overset{*}{a}{}_{ij}\right) = \begin{pmatrix} -76.4 & 124.8 & 12.1 \\ -50.8 & 82.9 & 8.0 \\ 44.6 & -71.4 & -6.6 \end{pmatrix} . \qquad (A.122.2)$$

Although its determinant is $|\overset{*}{a}{}_{ij}| = 1$, this matrix cannot be the description of a rotation because its components violate the necessary condition for cosines that $-1 \leq a_{ij} \leq 1$.   (b) Assume that a rotation followed a strain to produce the transformation matrix B_{ij} of eq. (A.121.1).   To calculate the matrix representing the hypothetical rotation according to the same logical diagram as Fig. A.120.1, except that B replaces A, we postmultiply B_{ij} with S_{ij}^{-1}, already known from eq. (A.120.12), as follows:

$$\overset{*}{b}_{ij} = B_{ik} S_{kj}^{-1} . \tag{A.122.3}$$

The numerical result is, within the numerical error, the same as in eq. (A.122.2), hence cannot possibly represent a rotation.   (c) To resolve which of the two possible orders of strain and rotation could have produced the transformation matrix of eq. (A.117.2), assume first that strain followed rotation.   The appropriate logical diagram in that case is Fig. A.121.1, replacing B by A, and the operations for finding the hypothetical rotation matrix are:

$$\overset{**}{a}_{ij}^{-1} = A_{ik}^{-1} S_{kj} , \quad \overset{**}{a}_{ij} = \overset{**}{a}_{ji}^{-1} . \tag{A.122.4}$$

The numerical result is:

$$\left(\overset{**}{a}_{ij} \right) = \begin{pmatrix} -78.6 & 127.3 & 10.9 \\ -47.1 & 76.2 & 6.5 \\ -16.6 & 28.2 & 3.0 \end{pmatrix} . \tag{A.122.5}$$

This matrix, although its determinant is unity, cannot represent a rotation. Assuming, however, that rotation followed strain, the logical diagram is that of Fig. A.120.1 and the operation is:

$$\overset{**}{b}_{ij} = A_{ik} S_{kj}^{-1} , \tag{A.122.6}$$

with the numerical result:

$$\left(\overset{**}{b}_{ij} \right) = \begin{pmatrix} -0.1850 & 0.5356 & -0.8240 \\ -0.8103 & 0.3912 & 0.4363 \\ 0.5560 & 0.7484 & 0.3616 \end{pmatrix} . \tag{A.122.7}$$

This is a suitable rotation matrix, the determinant of which is unity within numerical error limits.   The rotation axis is:

$$[l_i] = [\, 0.1600 \quad -0.7067 \quad -0.6892 \,] , \tag{A.122.8}$$

and the angle of coordinate rotation is $102.5°$ clockwise, implying a $102.5°$ counterclockwise physical rotation in a fixed coordinate system.

Answer 123. (a) Choose, in original coordinates a_i, a plane:

$$\boxed{A_i}\, a_i - \boxed{1} = \boxed{0}\ , \tag{A.123.1}$$

where the frames contain constants, three A_i in the first, and numerical constants in the other two. The three a_i are variable. The numerical constants are arbitrary and were chosen for convenience only; replacing them produces planes parallel to the one chosen here. A_i/A are the direction cosines of the normal on the plane. Given a transformation matrix α_{ij} from the state before to that after the deformation, use its inverse α_{ij}^{-1} to find a_i in terms of the final coordinates x_i (see *Summary of Formulæ*):

$$a_i = \left(\alpha_{ij}^{-1} + \delta_{ij}\right) x_j - \beta_i\ . \tag{A.123.2}$$

Substitute eq. (A.123.2) into eq. (A.123.1):

$$\boxed{A_i\left(\alpha_{ij}^{-1} + \delta_{ij}\right)}\, x_j - \boxed{\left(A_i\beta_i + 1\right)} = \boxed{0}\ . \tag{A.123.3}$$

The frames again contain constants only, three in the first and one each in the other two; the three x_j are variable. Equation (A.123.3) is thus the equation of a plane that generally differs from that of eq. (A.123.1), even if the two coordinate systems a_i and x_i coincide in origin and directions of axes. Set the three constants of the first frame equal to B_i:

$$A_i\left(\alpha_{ij}^{-1} + \delta_{ij}\right) \equiv B_i\ , \tag{A.123.4}$$

then B_i/B are the direction cosines of the normal on the plane *after* deformation. Hence, originally parallel planes remain parallel and plane after a homogeneous deformation. (b) Parallel lines being definable as the intersections of two sets of parallel planes, they, too, remain straight and parallel after a homogeneous deformation. (c) Consider, to start with, a special ellipsoid, the sphere, and for the transformation eq. (A.123.2) choose for simplicity $\beta_i = 0$; the equation of a sphere in the original coordinate may then be taken to be:

$$a_k a_k = 1\ . \tag{A.123.5}$$

Set:

$$C_{ij}^{-1} \equiv \alpha_{ij}^{-1} + \delta_{ij}\ . \tag{A.123.6}$$

Then:

$$a_i = C_{ij}^{-1} x_j\ . \tag{A.123.7}$$

Substitute eq. (A.123.7) into eq. (A.123.5):

$$\left(C_{kj}^{-1} x_j\right)\left(C_{ki}^{-1} x_i\right) = 1 \ , \tag{A.123.8}$$

or:

$$x_i\, x_j\, C_{ki}^{-1}\, C_{kj}^{-1} = 1 \ . \tag{A.123.9}$$

Even if C_{ij} and C_{ij}^{-1} were asymmetric (they are symmetric), their products with their own transposes, $C_{ki}\, C_{kj}$ and $C_{ki}^{-1}\, C_{kj}^{-1}$, would necessarily have to be symmetric. We can thus write:

$$C_{ki}^{-1}\, C_{kj}^{-1} \equiv S_{ij} \ . \tag{A.123.10}$$

Substitute eq. (A.123.10) into eq. (A.123.9), and obtain:

$$S_{ij}\, x_i\, x_j = 1 \ , \tag{A.123.11}$$

which is the general equation of a quadratic centered on the coordinate origin. If we put restrictions on C_{ij} and C_{ij}^{-1}, namely, that their Jacobians $|C_{ij}|$ and $|C_{ij}^{-1}|$ should be positive and finite [see eq. (6.4)], we ensure that the quadratic of eq. (A.123.11) is in fact the equation of an ellipsoid. We can now restate α_{ij}^{-1}, as defined by eq. (A.123.6), in terms of the half-axes of that ellipsoid in its principal coordinates:

$$\alpha_1^{-1} = 1/A - 1 \ , \quad \alpha_2^{-1} = 1/B - 1 \ , \quad \alpha_3^{-1} = 1/C - 1 \ . \tag{A.123.12}$$

Hence, in original coordinates of the same orientation:

$$\left[a_i\right] = \left[\ x_1/A \quad x_2/B \quad x_3/C \ \right] , \tag{A.123.13}$$

and, by substitution of these components of a_i into eq. (A.123.5):

$$x_1^2/A^2 + x_2^2/B^2 + x_3^2/C^2 = 1 \ , \tag{A.123.14}$$

which is the standard equation for a centered ellipsoid with half-axes A, B, C in principal coordinates. Because what was shown to be valid for principal coordinates must be equally true in differently oriented coordinate systems, we have shown that a sphere is transformed into an ellipsoid. (d) Further, because any linear transformation may be regarded as the result of two successive linear transformations, the sphere with which we started in eq. (A.123.5) may be taken as the result of that *special* transformation of an earlier, general ellipsoid, that reduces it to a sphere. Hence we have demonstrated that any homogeneous linear transformation with a nonzero, finite, and positive Jacobian transforms an ellipsoid into another ellipsoid.

Answer 124. (a) Let original coordinates a_i be related to the final coordinates x_i by:

$$a_i = A_{ij} x_j , \qquad (A.124.1)$$

replace A_{ij} by $(\gamma_{ij} + \delta_{ij})$:

$$a_i = (\gamma_{ij} + \delta_{ij}) x_j , \qquad (A.124.2)$$

and use the substitution property of the Kronecker delta:

$$a_i = \gamma_{ij} x_j + x_i . \qquad (A.124.3)$$

Further, because $u_i = x_i - a_i$:

$$u_i = x_i - x_i - \gamma_{ij} x_j = -\gamma_{ij} x_j , \qquad (A.124.4)$$

and as a consequence:

$$e_{ij} = \frac{1}{2} \left(\frac{\partial u_i}{\partial x_j} + \frac{\partial u_j}{\partial x_i} - \frac{\partial u_k}{\partial x_i} \frac{\partial u_k}{\partial x_j} \right) = -\frac{1}{2} \left(\gamma_{ij} + \gamma_{ji} + \gamma_{ki} \gamma_{kj} \right) . \qquad (A.124.5)$$

The sphere inscribed into the original coordinates is given as:

$$c \, a_k a_k = 1 . \qquad (A.124.6)$$

As we are interested only in directions, it is permissible to set $c = 1$; thus:

$$a_k a_k = 1 . \qquad (A.124.7)$$

Substitute eq. (A.124.1) into eq. (A.124.7):

$$\left(A_{ki} x_i \right) \left(A_{kj} x_j \right) = 1 , \qquad (A.124.8)$$

or:

$$x_i x_j A_{ki} A_{kj} = 1 . \qquad (A.124.9)$$

Because the order of multiplication is arbitrary, $A_{ki} A_{kj} = A_{kj} A_{ki}$; and we can set:

$$A_{ki} A_{kj} \equiv S_{ij} , \qquad (A.124.10)$$

and restate eq. (A.124.9) as:

$$S_{ij} x_i x_j = 1 . \qquad (A.124.11)$$

Because the transformation of eq. (A.124.1) must be continuous, eq. (A.124.11) represents an ellipsoid rather than some other quadric. Thus:

$$S_{ij} = A_{ki} A_{kj}$$
$$= (\gamma_{ki} + \delta_{ki})(\gamma_{kj} + \delta_{kj})$$
$$= \gamma_{ij} + \gamma_{ji} + \gamma_{ki} \gamma_{kj} + \delta_{ij}$$
$$= -2e_{ij} + \delta_{ij} \; . \qquad (A.124.12)$$

It follows that:

$$e_{ij} = \tfrac{1}{2}(\delta_{ij} - S_{ij}) \; , \qquad (A.124.13)$$

and that S_{ij} and e_{ij} are related to each other by multiplication with a constant factor and the addition of another constant. Neither operation affects the principal directions of the two tensors, and these directions are therefore the same in both, Q. E. D. (b) This demonstration for Almansi's tensor has no counterpart for Green's tensor $\mathfrak{G}_{ij}$ because γ_{ij} could well be asymmetric, which would imply a rotation in addition to a strain. Thus the principal directions of $\mathfrak{G}_{ij}$ are unrelated to those of S_{ij} unless γ_{ij} is symmetrical.

Answer 125. (a) Nye's theoretical flow law is:

$$\dot{\varepsilon} = (\sigma/B)^n \; , \qquad (A.125.1)$$

where the parameter $\dot{\varepsilon}$ is related to the second invariant of a constant-volume strain rate:

$$^1I \equiv \dot{\varepsilon}_{ii} = 0 \; , \quad ^2I = -\dot{\varepsilon}_{ij} \dot{\varepsilon}_{ij}/2 \; , \qquad (A.125.2)$$

by changing its sign and taking its square root:

$$\dot{\varepsilon}^2 = \dot{\varepsilon}_{ij} \dot{\varepsilon}_{ij}/2 \; . \qquad (A.125.3)$$

Similarly, σ is the square root of the second invariant (with changed sign) of the deviator of stress Δ_{ij} in arbitrary units [see eq. (A.97.5)]:

$$\sigma^2 = \Delta_{ij} \Delta_{ij}/2 \; . \qquad (A.125.4)$$

Figure A.125.1. shows that the viscosity of a Newtonian liquid is independent of the shear stress, whereas ice following Nye's law "weakens" (viscosity decreases) with increasing shear stress. (b) Glen established the following experimental flow law for ice :

$$\dot{L} = G L S^q \; , \qquad (A.125.5)$$

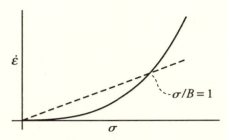

Figure A.125.1. Relationship of strain rate and stress. Arbitrary units. Continuous line, Nye's ice. Dashed, Newtonian liquid.

where $\dot{L}$ is the rate of shortening, L the axial length of the specimen, S the axial compressive stress, and G and q are experimental constants, and assuming that ice is incompressible, the strain rate is:

$$\left[\dot{\varepsilon}_{ij}\right] = \begin{bmatrix} -\dot{L}/L & 0 & 0 \\ 0 & \dot{L}/2L & 0 \\ 0 & 0 & \dot{L}/2L \end{bmatrix}$$

$$= \frac{\dot{L}}{L} \begin{bmatrix} -1 & 0 & 0 \\ 0 & 1/2 & 0 \\ 0 & 0 & 1/2 \end{bmatrix}, \tag{A.125.6}$$

and the stress in arbitrary units:

$$\left[\sigma_{ij}\right] = \begin{bmatrix} -S & 0 & 0 \\ 0 & 0 & 0 \\ 0 & 0 & 0 \end{bmatrix}. \tag{A.125.7}$$

Note that the confining pressure $\sigma_{22} = \sigma_{33} = 0$ differs from pressure, the mean normal compressive stress $p = S/3$. The negative second invariant of Glen's strain rate, needed for comparison with Nye's postulate, is:

$$\frac{\dot{\varepsilon}_{ij}\,\dot{\varepsilon}_{ij}}{2} = \dot{\varepsilon}^2 = \frac{\dot{L}^2}{2L^2}\left(1 + \frac{1}{4} + \frac{1}{4}\right) = \frac{3}{4}\frac{\dot{L}^2}{2L^2}, \tag{A.125.8}$$

hence:

$$\dot{\varepsilon} = \sqrt{3}\,\dot{L}/2L . \tag{A.125.9}$$

The deviator of Glen's stress is:

$$\left[\Delta_{ij}\right] = \left[\sigma_{ij} - \delta_{ij}\,\sigma_{kk}/3\right]$$

$$= \frac{S}{3}\begin{bmatrix} -3+1 & 0 & 0 \\ 0 & 0+1 & 0 \\ 0 & 0 & 0+1 \end{bmatrix} = \frac{S}{3}\begin{bmatrix} -2 & 0 & 0 \\ 0 & 1 & 0 \\ 0 & 0 & 1 \end{bmatrix}, \qquad \text{(A.125.10)}$$

and its negative second invariant:

$$\Delta_{ij}\,\Delta_{ij}/2 = \sigma^2 = (4+1+1)\,S^2/18 = S^2/3 \; ; \qquad \text{(A.125.11)}$$

hence:

$$\sigma = S/\sqrt{3} \; . \qquad \text{(A.125.12)}$$

(c) For the experimental equivalents of the variables in the theory, substitute eqs. (A.125.9) and (A.125.12) into eq. (A.125.1):

$$\sqrt{3}\,\dot{L}/2L = S^n/(3^{n/2}\,B^n) \; , \qquad \text{(A.125.13)}$$

and solve for $\dot{L}$:

$$\dot{L} = \frac{2L\,S^n}{3^{(n+1)/2}\,B^n} \; . \qquad \text{(A.125.14)}$$

By comparison with eq. (A.125.5) we find Glen's equivalents to Nye's constants n and B:

$$n \equiv q \; , \quad B \equiv \left\langle 2/\!\left(3^{(q+1)/2}\,G\right)\right\rangle^{1/q}. \qquad \text{(A.125.15)}$$

(d) Steinemann found the engineering shear rate $\dot{\gamma}$ to be related to the shear stress τ by the equation:

$$\dot{\gamma} = K\,\tau^s \; , \qquad \text{(A.125.16)}$$

where K and s are constants. The strain rate tensor (note the factor of $1/2$) is:

$$\left[\dot{\varepsilon}_{ij}\right] = \tfrac{1}{2}\begin{bmatrix} 0 & \dot{\gamma} & 0 \\ \dot{\gamma} & 0 & 0 \\ 0 & 0 & 0 \end{bmatrix}. \qquad \text{(A.125.17)}$$

This is a deviatoric strain rate at constant volume. The stress, also deviatoric, is:

$$\left[\sigma_{ij}\right] = \left[\Delta_{ij}\right] = \begin{bmatrix} 0 & \tau & 0 \\ \tau & 0 & 0 \\ 0 & 0 & 0 \end{bmatrix}. \qquad \text{(A.125.18)}$$

(e) Hence the negative second invariants of strain rate and stress for this experiment are:

$$\dot{\varepsilon}^2 = \dot{\gamma}^2/4 \ , \tag{A.125.19}$$

and:

$$\sigma^2 = \tau^2, \tag{A.125.20}$$

and the theoretical variables themselves, in terms of the experimental ones:

$$\dot{\varepsilon} = \dot{\gamma}/2 \ , \tag{A.125.21}$$

and:

$$\sigma = \tau \ . \tag{A.125.22}$$

Substitute eq. (A.125.21) into eq. (A.125.16) and obtain:

$$\dot{\gamma} = 2\dot{\varepsilon} = 2\tau^n/B^n \ . \tag{A.125.23}$$

(No need to replace τ by its equivalent σ.) Comparison with eq. (A.125.1) yields theoretical in terms of experimental constants as follows:

$$n \equiv s \ , \quad B \equiv (2/K)^{1/s}. \tag{A.125.24}$$

Thus Nye's equation matches both experiments. Experimental results commonly need this type of transformation to demonstrate their equivalence to theoretical or other experimental results. (f) The relations between the two sets of experimental constants are:

$$s \equiv q \ , \quad K \equiv 3^{(q+1)/2}G \ . \tag{A.125.25}$$

Answer 126. (a) The strain rate in the ablating region of an ice sheet glacier is given as:

$$\left[\varepsilon_{ij}\right] = \frac{A}{h} \begin{bmatrix} -1 & -x_2/(h^2 - x_2^2)^{1/2} & 0 \\ -x_2/(h^2 - x_2^2)^{1/2} & 1 & 0 \\ 0 & 0 & 0 \end{bmatrix}, \tag{A.126.1}$$

where A, the rate of ablation, and h, the depth of the ice sheet, are constants; this implies that the glacier maintains a steady state. Figure A.126.1 shows the orientation of the coordinate system. At the origin, at some particular point on the glacier surface, the velocity is given as:

$$\left[{}^{\circ}u_i\right] = \begin{bmatrix} U & -A & 0 \end{bmatrix}, \tag{A.126.2}$$

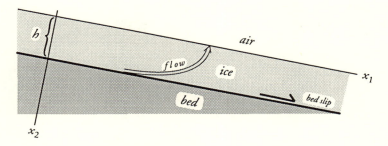

Figure A.126.1. Ablating sheet glacier. The coordinate origin is at the surface, at a distance h from the bed. Bed and surface are parallel, and the x_1 direction lies in the steepest direction of the glacier surface.

where U would be different for a differently chosen coordinate origin. The rotation rate at the origin is zero:

$$\overset{\circ}{\varpi}_{ij} = 0 \ . \tag{A.126.3}$$

Because $\overset{\circ}{u}_3 = 0$, $\dot{\varepsilon}_{31} = \dot{\varepsilon}_{32} = \dot{\varepsilon}_{33} = 0$, $\overset{\circ}{\varpi}_{ij} = 0$, and none of the $\dot{\varepsilon}_{ij}$ is a function of x_3:

$$u_3 = 0 \ . \tag{A.126.4}$$

By integration as in eq. (5.33) the surface-normal velocity component is (omitting terms with zero factors):

$$u_2 = -A + \frac{A}{h} \int_{0}^{x_2'} \dot{\varepsilon}_{22} \, dx_2 \Big|_{x_1 = x_1'} = -A + \frac{A}{h} \int_{0}^{x_2'} (1) \, dx_2 \Big|_{x_1 = x_1'}$$

$$= -A + A\, x_2 / h \ . \tag{A.126.5}$$

We find an upward (negative) flow, which decreases linearly with depth, from the ablation rate $-A$ at the surface, to zero at the bed. The bed-parallel flow rate [eq. (5.32)] is:

$$u_1 = U + \frac{A}{h} \int_{0}^{x_2'} \left\langle \dot{\varepsilon}_{12} + (x_2' - x_2) \, \dot{\varepsilon}_{12,\,2} \right\rangle dx_2 \Big|_{x_1 = 0} + \frac{A}{h} \int_{0}^{x_1'} \dot{\varepsilon}_{11} \, dx_1 \Big|_{x_2 = x_2'} \ . \tag{A.126.6}$$

The second integrand has the value -1. For evaluation of the first we differentiate:

$$\frac{h}{A}\dot{\varepsilon}_{12,2} = \left(-h^2+\boxed{x_2^2}\right)\Big/\left(h^2-x_2^2\right)^{3/2}-\boxed{x_2^2}\Big/\left(h^2-x_2^2\right)^{3/2}, \qquad (A.126.7)$$

and cancel the framed terms:

$$\frac{h}{A}\dot{\varepsilon}_{12,2} = -h^2\Big/\left(h^2-x_2^2\right)^{3/2}. \qquad (A.126.8)$$

Thus:

$$u_1 = U + \frac{A}{h}\int_0^{x_2'} \left\langle -x_2\Big/\left(h^2-x_2^2\right)^{1/2}+\left(x_2'-x_2\right)-h^2\Big/\left(h^2-x_2^2\right)^{3/2}\right\rangle dx_2$$

$$+ \frac{A}{h}\int_0^{x_1'} (-1)\,dx_1 \ . \qquad (A.126.9)$$

Evaluate the second integral and raise the first term in the first to the common denominator:

$$u_1 = U - \frac{A\,x_1}{h}$$

$$+ \frac{A}{h}\int_0^{x_2'} \left\langle \left(\boxed{-h^2x_2}+x_2^3 - h^2 x_2'\boxed{+h^2x_2}\right)\Big/\left(h^2-x_2\right)^{3/2}\right\rangle dx_2 \ . \qquad (A.126.10)$$

We evaluate the remaining integral (framed terms cancel) with the help of the two-item table of integrals (the hint in the question):

$$u_1 = U - \frac{A\,x_1}{h} + \frac{2A}{h}\left\langle \left(h^2-x_2^2\right)^{1/2}-h\right\rangle \ . \qquad (A.126.11)$$

The first term is constant, the second decreases linearly down-glacier, and the last decreases nonlinearly, at first slightly from zero at the surface (the bracket becomes $h - h = 0$), and then faster and faster with depth to a residual of $-2A$ at the bottom (where $x_2 = h$, and the parentheses become $h^2 - x_2^2 = 0$): the glacier slips on the bed. The velocity field is:

$$\left[u_i\right] = \left[\left\{U - \frac{A\,x_1}{h} + \frac{2A}{h}\left\langle \left(h^2-x_2^2\right)^{1/2}-h\right\rangle\right\}\right.$$

$$\left.\left\{-\frac{A}{h}\left(h-x_2\right)\right\}\ \left\{0\right\}\right] \ . \qquad (A.126.12)$$

(b) In order to find the rotation rate, it is necessary to obtain the velocity gradient by differentiating eq. (A.126.12):

$$\left[\dot{e}_{ij}\right] \equiv \left[u_{i,j}\right] = \frac{A}{h}\begin{bmatrix} -1 & -2x_2\left(h^2-x_2^2\right)^{-1/2} & 0 \\ 0 & 1 & 0 \\ 0 & 0 & 0 \end{bmatrix} . \quad (A.126.13)$$

By subtraction of eq. (A.126.1) from eq. (A.126.13) the rotation rate is:

$$\left[\dot{\omega}_{ij}\right] = \left[\dot{e}_{ij}\right] - \left[\dot{\varepsilon}_{ij}\right] = \frac{A\,x_2^2}{h}\left(h^2-x_2^2\right)^{-1/2}\begin{bmatrix} 0 & -1 & 0 \\ 1 & 0 & 0 \\ 0 & 0 & 0 \end{bmatrix} . \quad (A.126.14)$$

(c) *At the surface* the coordinates are the principal coordinates of the strain; the ice is shortened by normal stress acting along x_1; it is lengthened along x_2, but the lengthening is exactly compensated by ablation from above and by the ice rising from below. Using the prescript t to identify points x_i at the top, eq. (A.126.13) simplifies to:

$$\left[{}^t u_i\right] = \left[\begin{array}{ccc} U - A\,x_1/h & -A & 0 \end{array}\right] . \quad (A.126.15)$$

At some sufficiently large x_1, $U = A\,x_1/h$ and the glacier stops its flow and ends. To make the physics match the mathematical formulation, some support on the downside is needed, say, in the form of a terminal moraine. The velocity gradient and strain rate are identically:

$$\left[{}^t\dot{e}_{ij}\right] = \left[{}^t\dot{\varepsilon}_{ij}\right] = \frac{A}{h}\begin{bmatrix} -1 & 0 & 0 \\ 0 & 1 & 0 \\ 0 & 0 & 0 \end{bmatrix} , \quad (A.126.16)$$

and therefore the ice at the surface does not rotate:

$$ {}^t\dot{\omega}_{ij} = 0 . \quad (A.126.17)$$

(d) *At the bed* (prescript b) the velocity is bed slip only:

$$\left[{}^b u_i\right] = \left[\begin{array}{ccc} U - A(2h + x_1)/h & 0 & 0 \end{array}\right] . \quad (A.126.18)$$

The velocity gradient, strain rate, and rotation rate tensors are all degenerate, their nonzero off-diagonal components going to $\pm\infty$. These terms become large but "reasonable" and an upward velocity component u_2 starts appearing at very small distances from the bottom. (e) *Between bed and surface* the "rising" of the ice expressed by the velocity component u_2 increases linearly upward; furthermore, the high rate of simple shearing near the bottom decreases asymptotically toward the surface as:

$$\dot{\gamma} = \dot{e}_{12} = -2A\,x_2/h\left(h^2 - x_2^2\right)^{1/2}. \qquad \text{(A.126.19)}$$

This equation can be normalized by setting $h = 1$ and varying x_2 from 0 to 1. The negatives of the shear rate throughout the glacier interior, divided by A, are, from top to bottom:

x_2	0	0.1	0.2	0.3	0.4	0.5	0.6
$-\dot{\gamma}/A$	0	0.20	0.41	0.63	0.87	1.15	1.50

x_2	0.7	0.8	0.9	0.99	0.999	1.0
$-\dot{\gamma}/A$	1.96	2.67	4.13	14.0	44.7	∞

This mathematical description of sheet-glacier flow down an inclined bed differs from that of the flow of a Newtonian liquid (Problem 111) by bed slip, the effect of ablation, and a nonlinear flow law.

Answer 127. (a) In a deep and narrow valley glacier, the sides of which can be treated as essentially vertical, the stress is taken to be:

$$\left[\sigma_{ij}\right] = \begin{bmatrix} L & -A\,x_2 & 0 \\ -A\,x_2 & 0 & 0 \\ 0 & 0 & -B\,x_3 \end{bmatrix}, \qquad \text{(A.127.1)}$$

where L is the longitudinal stress, $A \equiv \rho g \sin\alpha$, $B \equiv \rho g \cos\alpha$, g is the acceleration due to gravity, and ρ and α are the density and the surface slope of the ice. Both A and B are positive constants (the glacier slope angle α is constant and cannot exceed $90°$). The coordinate origin is at the glacier surface midway between the walls, which are planes $x_2 = \pm w$, and the coordinate x_3 is measured downward into the ice perpendicularly to the glacier surface at the plane $x_3 = 0$. By inspection of eq. (A.127.1):

$$\sigma_{33} = -B\,x_3 = \sigma_3 . \qquad \text{(A.127.2)}$$

This is a principal stress, and it is always negative (compressive) because B is positive. The normal stress across the glacier σ_{22}, generally not a principal stress, is zero because the parallel walls bounding the glacier do not move in- or outward. The secular equation for the stress thus is:

$$\left\langle -\lambda\,(L-\lambda) - A^2\,x_2^2 \right\rangle \left\langle -B\,x_3 - \lambda \right\rangle = 0 \;, \tag{A.127.3}$$

with two solutions that are independent of eq. (A.127.2):

$$^1\lambda,\,^2\lambda \;=\; \sigma_1,\,\sigma_2 \;=\; \tfrac{1}{2}\left\langle L \mp \left(L^2 + 4A^2\,x_2^2\right)^{1/2}\right\rangle \;. \tag{A.127.4}$$

The quantity in parentheses in eq. (A.127.4) is necessarily positive, and its positive square root is always equal to or larger than the absolute value $|L|$ of L (vertical bars around a scalar symbol for absolute value). Using the negative square root in solving eq. (A.127.4) for σ_1, the three possible cases for the sign of L are:

1. $L < 0$ (compressive): $\sigma_1 < 0$ (compressive) and $\sigma_2 \geq 0$ (tensile),
2. $L = 0$: $\sigma_1 = -2A\,x_2 \leq 0$ (compressive) and $\sigma_2 = 2A\,x_2 \geq 0$ (tensile),
3. $L > 0$ (tensile): $\sigma_1 \leq 0$ (compressive) and $\sigma_2 > 0$ (tensile).

Therefore, considering that invariably $\sigma_3 < 0$ (compressive), of the principal stresses one and only one, σ_2, is tensile. (b) Because x_3 is a principal axis, σ_2 lies in the x_1–x_2 plane and the crevasses, normal to σ_2, are thus invariably perpendicular to the glacier surface. The three contrasting cases may be illustrated by schematic Mohr diagrams, for all of which $R = 3$, $a = 1$, $b = 2$, representing the two-dimensional stress:

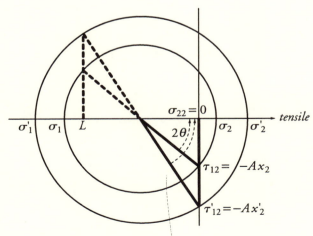

Figure A.127.1. Stress in a narrow valley glacier with compressive longitudinal stress, for positive x_2.

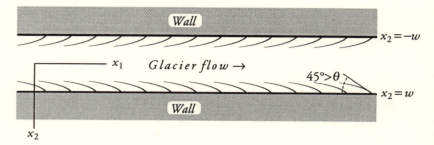

Figure A.127.2. Crevasses in a narrow valley glacier with compresssive longitudinal stress.

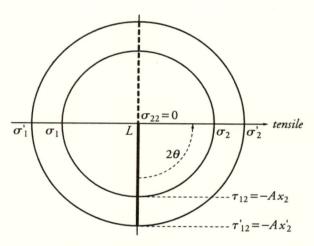

Figure A.127.3. Stress in a narrow valley glacier with zero longitudinal stress, for positive x_2.

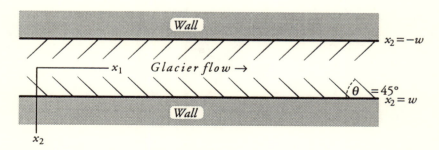

Figure A.127.4. Crevasses in a narrow valley glacier with zero longitudinal stress.

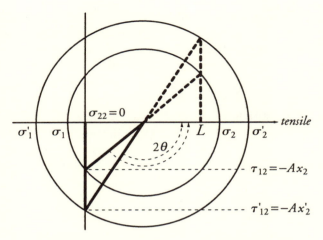

Figure A.127.5.   Stress in a narrow valley glacier with
tensile longitudinal stress, for positive x_2.

$$[\sigma_{\alpha\beta}] = \begin{bmatrix} L & -A\,x_2 \\ -A\,x_2 & 0 \end{bmatrix}. \qquad\qquad (A.127.5)$$

(c) $L < 0$ (compressive, Fig. A.127.1).   The absolute value of the shear stress $\tau_{12} = A\,x_2$ increases with distance from the median line of the glacier, and so does the contrast between the principal stresses σ_1 and σ_2. The angle θ of the principal direction of the compressive principal stress σ_1 with the glacier axis x_1 is:

$$\theta = \tan^{-1}\!\left(2A\,x_2/|L|\right)/2 = \tan^{-1}\!\left(C\,x_2\right)/2 , \qquad (A.127.6)$$

because in a glacier $2A/|L|$ is locally a positive constant C. The angle between the glacier axis, or glacier wall, and the trace of the crevasse that

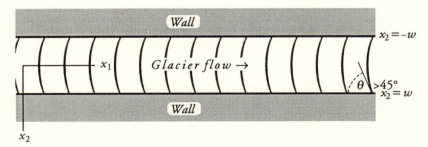

Figure A.127.6.   Crevasses in a narrow valley glacier with tensile longitudinal stress.

follows the direction of the compressive principal stress increases with the distance x_2 from the median line but remains less than 45° at its maximum at the wall. A threshold tensile stress is required to overcome the tensile strength of the ice and to open a crevasse, and no crevasses form near the median of the glacier where the tensile stress is weak. Thus the crevasse pattern for compressive longitudinal stress must resemble the scheme shown in Fig. A.127.2. (d) $L = 0$ (zero longitudinal stress, Fig. A.127.3). The absolute value of the shear stress $\tau_{12} = -Ax_2$ increases with distance from the median line of the glacier, and so does the tensile principal stresses σ_2. The angle θ of the principal direction of the compressive principal stress σ_1 with the glacier axis x_1 is constant at 45°. Hence the crevasse pattern for zero longitudinal stress must resemble the scheme shown in Fig. A.127.4. (e) $L > 0$ (longitudinal stress tensile, Fig. A.127.5). The absolute value of the shear stress $\tau_{12} = -Ax_2$ increases with distance from the median line of the glacier, and the tensile principal stress σ_2 also increases outward, from L at the median line. The angle θ of the principal direction of the compressive principal stress σ_1 with the glacier axis x_1 decreases from 90° at the median line to an angle greater than 45° at the wall. Provided the tensile stress L exceeds the tensile strength of ice, there is no median gap in the crevasse pattern, which is shown schematically in Fig. A.127.6.

Answer 128. (a) In a body subject to a homogeneous stress consisting of the compressive traction S acting along x_1 and a confining pressure P along x_2 and x_3 (Fig. A.128.1), the stress tensor is:

$$[\sigma_{ij}] = \begin{bmatrix} S & 0 & 0 \\ 0 & P & 0 \\ 0 & 0 & P \end{bmatrix}, \tag{A.128.1}$$

where, according to a convention usual in geology, compressive stress is taken to be positive. Note that the "confining pressure" P is distinct from the "hydrostatic pressure" p, the mean normal stress (without change of sign because of the change in convention). (b) Figure A.128.2a is the Mohr diagram for several states of stress with increasing S at constant confining pressure P up to that magnitude ^fS at which the Mohr circle touches the failure envelope. That envelope is a straight line with slope $\mu \equiv \tan\varphi$ and with an ordinate intercept $^o\tau$. (c) By inspection of Fig. A.128.2a the radius of the Mohr circle at failure $(S = {}^fS)$ is:

$$R = \left({}^fS - P\right)/2 , \tag{A.128.2}$$

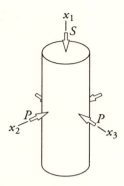

Figure A.128.1. Compression
of a cylinder by the traction S
and a confining pressure P.

and the intercept b of the failure envelope with the normal-stress axis
(abscissa) is:

$$b = {}^{\circ}\tau/\tan\varphi ,$$
(A.128.3)

where the cohesion ${}^{\circ}\tau$ is the intercept of the failure envelope with the shear-
stress axis (ordinate) and φ its angle with the normal-stress axis (abscissa).
Further:

$$f_S - P = 2R ,$$
(A.128.4)

and thus:

$$f_S + P = 2R + 2P .$$
(A.128.5)

Inspection of Figs. A.128.2b and A.128.2c yields two different definitions of
$\sin\varphi$, one in terms of standard elements of a Mohr diagram, the other in terms
of the coefficient of internal friction:

$$\sin\varphi = R/(P + R + b) = \mu/(1 + \mu^2)^{1/2}.$$
(A.128.6)

Figure A.128.2c also shows that $\tan\varphi = \mu$, which, substituted into eq. (A.128.3),
yields:

$$b = {}^{\circ}\tau/\mu ,$$
(A.128.7)

Expand eq. (A.128.6) to:

$$2R/\langle(2R + 2P) + 2b\rangle = \mu/(1 + \mu^2)^{1/2},$$
(A.128.8)

substitute eqs. (A.128.4), (A.128.5), and (A.128.7) into eq. (A.128.8):

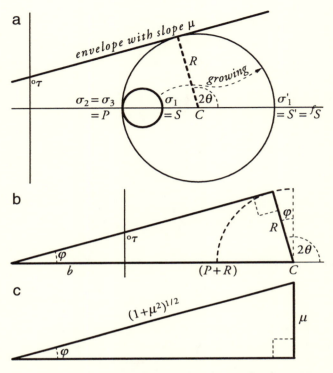

Figure A.128.2. Mohr diagram for the Coulomb criterion of shear failure. (a) Superposed Mohr diagrams for compressive stresses $S < {}^fS$ and $S' = {}^fS$ at constant confining pressure P. (b) Representation in Mohr space of 2θ, where θ is the angle in ordinary space between the pole to the failure plane and the x_1 axis. (c) Angle φ and coefficient μ of "internal friction." Base of the triangle has unit length.

$$\left({}^fS - P\right) \Big/ \left({}^fS + P + 2{}^\circ\tau/\mu\right) = \mu \Big/ \left(1 + \mu^2\right)^{1/2}, \qquad (A.128.9)$$

to solve for ${}^fS$, rearrange:

$$ {}^fS + P + 2{}^\circ\tau/\mu = \left\langle\left({}^fS - P\right)\left(1 + \mu^2\right)^{1/2}\right\rangle \Big/ \mu , \qquad (A.128.10)$$

expand:

$$ {}^fS = \left\langle {}^fS\left(1 + \mu^2\right)^{1/2} - P\left(1 + \mu^2\right)^{1/2} - P\mu - 2{}^\circ\tau \right\rangle \Big/ \mu , \qquad (A.128.11)$$

collect terms:

$$f_S\langle\mu-\left(1+\mu^2\right)^{1/2}\rangle/\mu = -\langle P\left(1+\mu^2\right)^{1/2}+P\mu+2^\circ\tau\rangle/\mu , \quad \text{(A.128.12)}$$

cancel $1/\mu$ and divide by the term in the left-hand angle brackets:

$$f_S = \langle P\left(1+\mu^2\right)^{1/2}+\mu\left(2P-P\right)+2^\circ\tau\rangle/\langle\left(1+\mu^2\right)^{1/2}-\mu\rangle , \quad \text{(A.128.13)}$$

and simplify to:

$$f_S = P+2\langle P\mu+^\circ\tau\rangle/\langle\left(1+\mu^2\right)^{1/2}-\mu\rangle . \quad \text{(A.128.14)}$$

(d) Inspect Fig. A.128.2b to find 2θ:

$$2\theta-\varphi = \pi/2 , \quad \text{(A.128.15)}$$

and thus:

$$\theta = \pi/4 + \varphi/2 . \quad \text{(A.128.16)}$$

(e) Considering that -cotan $\alpha = \tan\left(\pi/2+\alpha\right)$, inspection of Figure A.128.1c shows that:

$$\theta = \pi/4 + \tan^{-1}\mu/2 ; \quad \text{(A.128.17)}$$

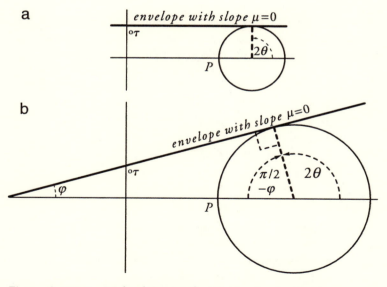

Figure A.128.3. Mohr diagrams for Coulomb criterion at constant confining pressure and coefficient of "internal friction" – μ.
(a) Frictionless case $\mu = 0$. (b) Finite friction $\mu > 0$.

hence:

$$\theta = \left\langle \tan^{-1}(-1/\mu) \right\rangle / 2 \ . \tag{A.128.18}$$

For a given $^{\circ}\tau$ and P, as μ goes from zero to infinity, 2θ increases from $\pi/2$ to π (Fig.A.128.2); thus:

$$\pi/4 \le \theta < \pi/2 \ , \tag{A.128.19}$$

or, put differently, $\theta \ge 45°$ for all $\mu \ge 0$.

Answer 129. *The strain rate approach to the problem of rotated garnets.* (a) Figure A.129.1 is a sketch of the trace of a materially marked fiducial surface inside and out of a garnet that grew while it was rotated with respect to the fiducial plane, parallel to the x_1–x_2 plane at a distance from the crystal, about an axis parallel to x_1. The shape of the fiducial surface in the vicinity of the crystal is left undetermined and will not be considered in the following development. Far from the garnet inclusion, the strain is plane, and the rotation axis is also the principal direction of zero instantaneous and zero cumulative strain. Let garnet-rolling time start at time $t = 0$. Then, if the coordinate origin moves with a material point, $v_1 = v_3 = u_1 = u_3 = 0$, and:

$$u_2\big|_t = v_2 t = k x_3 t \ , \tag{A.129.1}$$

where k is a constant. This represents a simple shear rate with displacements in the x_2 direction and an undeformed shear plane in x_1–x_2, which coincides with the fiducial plane. (b) By inspection of Fig. A.129.2:

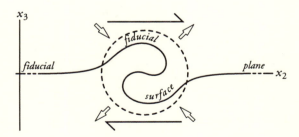

Figure A.129.1.   Snowball garnet.   Dashed, approximately circular crystal outline (garnet grows commonly as a nearly regular dodecahedron). Heavy line, fiducial surface, say, a discrete foliation plane in a schist. Half-arrows, sense of simple shear. Hollow arrows, principal directions of instantaneous strain rate.

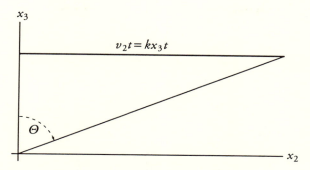

Figure A.129.2. Cumulative strain due to a uniform simple shear rate with x_1-x_2 as the shear plane. Θ is a negative angle because the positive x_1 direction points toward the viewer.

$$\tan \Theta = -kt . \qquad (A.129.2)$$

The velocity is:

$$\left[v_i \right] = \left[\begin{array}{ccc} 0 & kx_3 & 0 \end{array} \right] , \qquad (A.129.3)$$

and the velocity gradient:

$$\left[\dot{e}_{ij} \right] = \left[\frac{\partial v_i}{\partial x_j} \right] = \left[\begin{array}{ccc} 0 & 0 & 0 \\ 0 & 0 & k \\ 0 & 0 & 0 \end{array} \right] . \qquad (A.129.4)$$

The strain rate is the symmetric part of the velocity gradient [eq. (A.129.4)]:

$$\left[\dot{\varepsilon}_{ij} \right] = \frac{k}{2} \left[\begin{array}{ccc} 0 & 0 & 0 \\ 0 & 0 & 1 \\ 0 & 1 & 0 \end{array} \right] , \qquad (A.129.5)$$

and the rotation rate is its antisymmetric part:

$$\left[\dot{\omega}_{ij} \right] = \frac{k}{2} \left[\begin{array}{ccc} 0 & 0 & 0 \\ 0 & 0 & 1 \\ 0 & -1 & 0 \end{array} \right] . \qquad (A.129.6)$$

Rotation affects both the inside of the garnet inclusion and the material in the far field at a distance outside, but the garnet interior is protected from strain. Distinguishing the two by prescripts G and F (for far field), we find for the garnet interior the strain rate:

$$^{G}\dot{\varepsilon}_{ij} = 0 , \qquad (A.129.7)$$

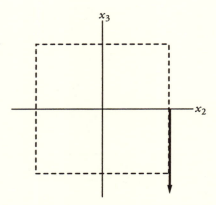

Figure A.129.3. Orientation of displacement and velocity components
u_{32} and v_{32} in the x_2-x_3 plane.

with the component (see Fig. A.129.3):

$$^G\dot{\varepsilon}_{32} = 0 \ , \tag{A.129.8}$$

and the rigid-body rotation rate:

$$^G\dot{\varpi}_{ij} = \dot{\varpi}_{ij} \ , \tag{A.129.9}$$

with the component:

$$^G\dot{\varpi}_{32} = -k/2 \ , \tag{A.129.10}$$

and a velocity gradient that is the sum of eqs. (A.129.7) and (A.129.9), and of (A.129.8) and (A.129.10), respectively:

$$^G\dot{e}_{ij} = \dot{\varpi}_{ij} \ , \tag{A.129.11}$$

with the component:

$$^G\dot{e}_{32} = -k/2 \ . \tag{A.129.12}$$

In contrast to the garnet interior, the strain rate in the far field is:

$$^F\dot{\varepsilon}_{ij} = \dot{\varepsilon}_{ij} \ , \tag{A.129.13}$$

with the component:

$$^F\dot{\varepsilon}_{32} = k/2 \ , \tag{A.129.14}$$

and the rotation is:

$$^F\dot{\varpi}_{ij} = \dot{\varpi}_{ij} \ , \tag{A.129.15}$$

with the component:

$$^F\dot{\varpi}_{32} = -k/2 \ , \tag{A.129.16}$$

and a velocity gradient that is the sum of eqs. (A.129.13) and (A.129.15), and of (A.129.14) and (A.129.16), respectively:

$$^F\dot{e}_{ij} = \dot{e}_{ij} \; , \qquad\qquad (A.129.17)$$

with the component:

$$^F\dot{e}_{32} = 0 \; . \qquad\qquad (A.129.18)$$

Hence, relative to the fiducial plane in the far field, the garnet inclusion, and with it the oldest segment of the fiducial surface, rotates at the rate:

$$\dot{\Phi} = -k/2 \; . \qquad\qquad (A.129.19)$$

Note that this is a counterclockwise physical rotation, as seen in the positive x_1 direction. The cumulative rotation thus is:

$$\Phi = -k\,t/2 \; . \qquad\qquad (A.129.20)$$

Since from eq. (A.129.2) $-k\,t = \tan\Theta$, we obtain a solution for the angle Φ of cumulative physical rotation in terms of the observed angle Θ made by the innermost segment of the fiducial surface within the inclusion with the far-field fiducial plane:

$$\Phi = \tan\Theta/2 \; . \qquad\qquad (A.129.21)$$

Thus, as the absolute magnitude of Θ approaches $\pi/2$, the absolute magnitude of Φ can become very large and comprise many complete revolutions, each of 2π.

Answer 130. *The finite-strain approach to the problem of rotated garnets.* (a) With reference to original coordinates a_i coinciding with the coordinates x_i used in Problem 129, and from inspection of Fig. A.130.1, the finite displacements are:

$$\left[u_i \right] = \left[\; 0 \quad k\,t\,a_3 \quad 0 \; \right] . \qquad\qquad (A.130.1)$$

Using the definition of Green's tensor in terms of displacements:

$$\mathfrak{E}_{ij} = \frac{1}{2}\left(\frac{\partial u_i}{\partial a_j} + \frac{\partial u_j}{\partial a_i} + \frac{\partial u_k}{\partial a_i}\frac{\partial u_k}{\partial a_j} \right) , \qquad\qquad (A.130.2)$$

that tensor is here:

$$\left[\mathfrak{E}_{ij} \right] = \begin{bmatrix} 0 & 0 & 0 \\ 0 & 0 & k\,t/2 \\ 0 & k\,t/2 & k^2 t^2/2 \end{bmatrix} . \qquad\qquad (A.130.3)$$

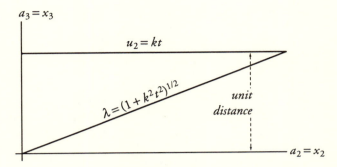

Figure A.130.1. Finite displacements at $a_3 = 1$ for simple shear with a_1–a_2 as the shear plane. Elongation λ shown for the direction $[l_i] = [0\ 0\ 1]$.

Because the elongation λ is in general:

$$\lambda = \left(1 + 2\mathfrak{E}_{ij}\, l_i\, l_j\right)^{1/2}, \qquad\qquad (A.130.4)$$

for the unit vector in the a_3 direction $[l_i] = [0\ 0\ 1]$, it is:

$$\lambda = \left(1 + k^2 t^2\right)^{1/2}. \qquad\qquad (A.130.5)$$

(b) Find the principal directions of Green's tensor by the Mohr circle construction, with $R = 1$, $a = 2$, and $b = 3$ (Fig. A.130.2). The angle θ from direction of maximum strain to the coordinate axis a_3 can be found by inspection:

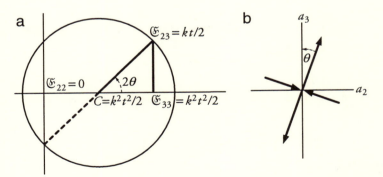

Figure A.130.2. (a) Mohr circle construction for the principal directions of Green's tensor, eq. (A.130.3), and (b) sketch of the corresponding physical orientations. The long and short heavy arrows represent orientations of the strain axes.

$$\tan 2\theta = \frac{4kt}{2k^2t^2} = \frac{2}{kt} . \qquad \text{(A.130.6)}$$

Solve for kt and substitute the trigonometric identity:

$$\tan 2\theta = \frac{2\tan\theta}{1 - \tan^2\theta} ,$$

to obtain:

$$kt = \frac{2}{\tan 2\theta} = \frac{1 - \tan^2\theta}{\tan\theta} , \qquad \text{(A.130.7)}$$

or:

$$kt = 1/\tan\theta - \tan\theta . \qquad \text{(A.130.8)}$$

(c) Figure A.130.3 shows the traces, in the coinciding $a_2\text{–}a_3$ and $x_2\text{–}x_3$ coordinate planes, of the unique two sets of material planes that are orthogonal both before and after the strain, which makes them principal planes; inspection of the triangles to the left of the vertical line shows [from here to eq. (A.130.13) the sign of angles will be neglected]:

$$\tan(\theta + \varphi) = \frac{1}{(1/\tan\theta) - kt} = 1/\tan\theta , \qquad \text{(A.130.9)}$$

where φ is the angle of post-strain rigid-body rotation that restores material planes originally parallel to the a_1 and a_2 coordinate axes to parallelism with the x_1 and x_2 axes. Inspection of the triangles to the right of the vertical line shows:

$$\tan(\theta + \varphi) = \tan\theta + kt = 1/\tan\theta , \qquad \text{(A.130.10)}$$

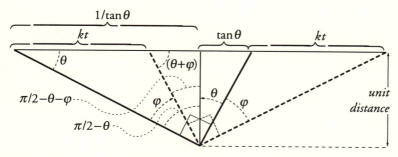

Figure A.130.3. Material planes that are orthogonal both before and after a finite simple shear. The thin horizontal line is parallel to $a_2 = x_2$ at unit distance from the origin along the vertical $a_2 = x_2$ axis (thin line). Heavy lines, traces of original material planes; dashed heavy lines, traces of the corresponding material lines after the strain.

and, applying to the equation of the first and last expression of eq. (A.130.10) the trigonometric identity:

$$\tan(\alpha + \beta) = \frac{\tan \alpha + \tan \beta}{1 - \tan \alpha \tan \beta} \, ,$$

yields:

$$\tan(\theta + \varphi) \equiv \frac{\tan \theta + \tan \varphi}{1 - \tan \theta \tan \varphi} = 1/\tan \theta \, . \qquad \text{(A.130.11)}$$

Multiply the right-hand equation in eq. (A.130.11) by the denominator of the middle expression, and obtain:

$$\tan \theta + \tan \varphi = \frac{1}{\tan \theta} - \frac{\tan \theta \tan \varphi}{\tan \theta} \, . \qquad \text{(A.130.12)}$$

Cancel $(\tan \theta)$ in the last term, collect the $(\tan \varphi)$ on the left-hand side, and interpret the resulting right-hand side in terms of eq. (A.130.8):

$$2 \tan \varphi = 1/\tan \theta - \tan \theta \equiv kt \, . \qquad \text{(A.130.13)}$$

Hence, restoring the negative sign to φ and using eq. (A.129.2) together with the equivalence of Figs. A.130.1 and A.129.2, obtain:

$$\tan \varphi = -kt/2 = \tan \Theta/2 \, . \qquad \text{(A.130.14)}$$

(d) This result, which implies that $|\varphi| < 90°$ in all cases, differs from the result obtained in the strain-rate approach, $\Phi = (1/2) \tan \Theta$. The discrepancy arises from having chosen a different set of assumptions: in the present problem we have implicitly assumed *first* an irrotational finite strain, *then* a rotation φ about the x_1 axis. (We could also have assumed, but did not, that a rotation preceded a strain.) The strain-rate approach of Problem 129, on the other hand, assumed, in a manner of speaking, that infinitesimal strain and rotation increments alternated. The two solutions approach each other for small φ or Φ, where $\varphi \approx \sin \varphi \approx \tan \varphi$. Thus, for example:

$$5° = 0.0873 \text{ radians,}$$
$$\sin 5° = 0.0872,$$
$$\tan 5° = 0.0875.$$

For large angles of apparent rotation, the strain-rate approach is the only realistic model of a possible strain history. It keeps the orientation of the shear planes (far-field fiducial surfaces) constant; the finite-strain approach does not. Although the finite-strain approach describes the deformed state of the far field correctly, it does so in terms that are not applicable to features observed inside the garnet inclusion.   (e) A specific solution for the case $\Theta = -60°$ is shown in Fig. A.130.4. In that case, according to eqs. (A.129.2) and (A.130.6), $\theta = 24.55°$ and $kt = -\tan \Theta = 1.732$. By Mohr circle construction, the principal values of Green's tensor are:

$$\left.\begin{array}{c} \mathfrak{E}_3 \\ \mathfrak{E}_2 \end{array}\right\} = \frac{k^2 t^2}{4} \pm \left(\frac{k^4 t^4}{16} + \frac{k^2 t^2}{4}\right)^{1/2} = \frac{3 \pm \sqrt{21}}{4} \,, \qquad (A.130.15)$$

and the corresponding principal elongations are:

$$^3\lambda = \left(1 + 2\mathfrak{E}_3\right)^{1/2} = \left\langle \left(5 + \sqrt{21}\right)/2 \right\rangle^{1/2} \approx 2.189 \,, \qquad (A.130.16)$$

and:

$$^2\lambda = \left(1 + 2\mathfrak{E}_2\right)^{1/2} = \left\langle \left(5 - \sqrt{21}\right)/2 \right\rangle^{1/2} \approx 0.457 \,, \qquad (A.130.17)$$

and, since $\tan \varphi = (1/2) \tan \Theta$, $\varphi \approx 40.89°$.

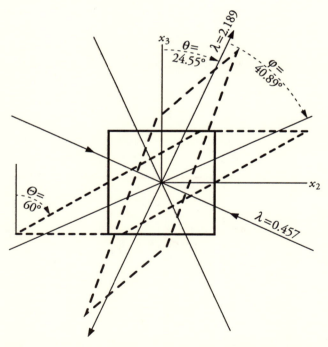

Figure A.130.4.   Finite simple shear through 60°. Solid heavy line, original shape. Long-dashed heavy line, shape and intermediate orientation before rotation.   Short-dashed heavy line, final shape and orientation.

Answer 131. *The analytic-geometry approach to a specific problem of rotated garnets.* We solve for an observed complete revolution of the fiducial surface inside the garnet crystal with respect to the far-field fiducial plane. With such a rotation of 360°:

$$\Phi = 2\pi = \tan\Theta/2 \ , \tag{A.131.1}$$

or according to eq. (A.129.21):

$$\tan\Theta = 4\pi = 2\Phi \ , \tag{A.131.2}$$

In an Eulerian reference frame with coinciding coordinates a_i and x_i (see Fig. A.131.1), original positions of material points in terms of their final positions are:

$$\left[a_i\right] = \left[\begin{array}{ccc} x_1 & x_2 - 4\pi x_3 & x_3 \end{array}\right] \ . \tag{A.131.3}$$

Let the analytical equation for an original sphere, say a spherical pebble, be:

$$a_i\, a_i = 1 \ , \tag{A.131.4}$$

or:

$$a_1^2 + a_2^2 + a_3^2 = 1 \ , \tag{A.131.5}$$

Then, by inspection of Fig. A.131.1 and use of eq. (A.131.2), the corresponding final ellipsoid is:

$$x_1^2 + \left(x_2 - 4\pi x_3\right)^2 + x_3^2 = 1 \ , \tag{A.131.6}$$

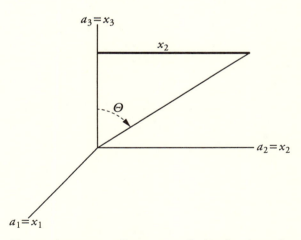

Figure A.131.1. Geometry of simple shear in coinciding original and final coordinate systems a_i and x_i.

or, explicitly:

$$x_1^2 + x_2^2 - 8\pi x_2 x_3 + \left(16\pi^2 + 1\right) x_3^2 = 1 \ . \qquad (A.131.7)$$

We informally call the third term, the one containing the product $x_2 x_3$, the "two-variable term," and we eliminate it by an appropriate coordinate rotation about the x_1 axis:

$$\left(b_{ij}\right) = \begin{pmatrix} 1 & 0 & 0 \\ 0 & \cos\theta & -\sin\theta \\ 0 & \sin\theta & \cos\theta \end{pmatrix}, \qquad (A.131.8)$$

then:

$$x_1' = x_1 \ ,$$
$$x_2' = x_2 \cos\theta - x_3 \sin\theta \ ,$$
$$x_3' = x_2 \sin\theta + x_3 \cos\theta \ . \qquad (A.131.9)$$

Note that this is the equivalent of finding the appropriate (b_{ij}) by Mohr circle construction for the symmetric tensor:

$$\left[S_{ij}\right] = \begin{bmatrix} 0 & 0 & 0 \\ 0 & 0 & 8\pi \\ 0 & 8\pi & 16\pi^2 \end{bmatrix},$$

which represents the ellipsoid of eq. (A.131.6). Execute the rotation by substituting eqs. (A.131.9) into eq. (A.131.6). For convenience, drop the primes. After collecting terms, this results in:

$$x_1^2 + x_2^2 \left\langle \boxed{\cos^2\theta} - 8\pi \sin\theta \cos\theta + \left(16\pi^2 + \boxed{1}\right)\sin^2\theta \right\rangle$$
$$+ x_3^2 \left\langle \boxed{\sin^2\theta} + 8\pi \sin\theta \cos\theta + \left(16\pi^2 + \boxed{1}\right)\cos^2\theta \right\rangle$$
$$+ x_2 x_3 \left\langle \begin{array}{c} \boxed{-\sin\theta \cos\theta} - 4\pi\left(\cos^2\theta - \sin^2\theta\right) \\ + \left(16\pi^2 + \boxed{1}\right)\sin\theta \cos\theta \end{array} \right\rangle = 1 \ . \qquad (A.131.10)$$

In the first two lines, the framed terms combine to unity, and in the third they cancel; however, all the terms on the left-hand side of the third line may be dropped, considering that the rotation has reduced the "two-variable term" to zero. Hence:

$$x_1^2 + x_2^2 \left\langle 1 - 8\pi \sin\theta \cos\theta + 16\pi^2 \sin^2\theta \right\rangle$$
$$+ x_3^2 \left\langle 1 + 8\pi \sin\theta \cos\theta + 16\pi^2 \cos^2\theta \right\rangle = 1 \ . \qquad (A.131.11)$$

This is the sought equation, in principal coordinates, of the ellipsoid representing the deformed pebble. Independently, to cause the "two-variable term" to vanish, the expression in angle brackets from the third line in eq. (A.131.10) must be zero, and the resulting equation, divided by 4π, is:

$$\sin^2\theta - \cos^2\theta + 4\pi\sin\theta\cos\theta = 0 \ . \tag{A.131.12}$$

Dividing eq. (A.131.12) by $\cos^2\theta$ yields a quadratic equation in $\tan\theta$:

$$\tan^2\theta + 4\pi\tan\theta - 1 = 0 \ . \tag{A.131.13}$$

Hence:

$$\tan\theta = -2\pi \pm \left(4\pi^2 + 1\right)^{1/2}$$
$$\approx -6.2832 \pm 6.3623$$
$$\approx \begin{cases} 0.0791 \\ -12.6455 \end{cases} \ . \tag{A.131.14}$$

Thus:

$$\theta \approx \begin{cases} 4.52° \\ -85.48° \end{cases} , \tag{A.131.15}$$

are the orientations of the two principal axes that are inclined to the coordinate system. Using the first of the solutions (A.131.15), we obtain numerical values:

$$\cos\theta \approx 0.9969$$
$$\sin\theta \approx 0.0788$$
$$\sin\theta\cos\theta \approx 0.07855$$
$$\cos^2\theta \approx 0.9938$$
$$\sin^2\theta \approx 0.006209$$

Substitute these numerical values into eq. (A.131.11):

$$x_1^2 + 0.0063\,x_2^2 + 159.91\,x_3^2 = 1 \ . \tag{A.131.16}$$

To give this equation its standard form, express the coefficients as squared denominators (take the square root of their reciprocals):

$$\frac{x_1^2}{1^2} + \frac{x_2^2}{12.65^2} + \frac{x_3^2}{0.0791^2} = 1 \ . \tag{A.131.17}$$

Thus, to three significant figures, the axial ratios of an originally spherical, now ellipsoidal, pebble are $0.0791/1/12.65$.

Answer 132. *The tensor approach to a specific problem of snowball garnets.* As in Problem 131, the original positions of material points are taken to be:

$$\left[a_i\right] = \left[\begin{array}{ccc} x_1 & x_2 - 4\pi x_3 & x_3 \end{array}\right] . \tag{A.132.1}$$

Because $u_i = x_i - a_i$, the displacements are:

$$\left[u_i\right] = \left[\begin{array}{ccc} 0 & 4\pi x_3 & 0 \end{array}\right] . \tag{A.132.2}$$

Almansi's tensor measures strain as a function of final coordinates:

$$\left[e_{ij}\right] = \left|\begin{array}{ccc} 0 & 0 & 0 \\ 0 & 0 & 2\pi \\ 0 & 2\pi & -8\pi^2 \end{array}\right| . \tag{A.132.3}$$

The secular equation for this tensor is:

$$e^3 + 8\pi^2 e^2 - 4\pi^2 e = 0 . \tag{A.132.4}$$

With the third invariant $^3I = 0$, one of the solutions is zero, and by division with e, the remainder constitutes a quadratic with the solutions:

$$e = -4\pi^2 \pm \left(16\pi^4 + 4\pi^2\right)^{1/2}. \tag{A.132.5}$$

Factor by 2π and obtain the principal values of Almansi's tensor:

$$e = \begin{cases} 0 \\ 2\pi\langle -2\pi \pm (4\pi^2 + 1)^{1/2}\rangle \end{cases} . \tag{A.132.6}$$

With the order established by the context and indicated by prescripts (to avoid confusion with vector components), this yields numerically:

$$^1e = 0 , \quad ^2e \approx -79.4542 , \quad ^3e \approx 0.4970 , \tag{A.132.7}$$

and principal elongations $^i\lambda = (1 - 2\,^ie)^{-1/2}$:

$$^1\lambda = 1 , \quad ^2\lambda \approx 12.65 , \quad ^3\lambda \approx 0.0791 . \tag{A.132.8}$$

The numerical results are those of eq. (A.131.17). To determine the orientation of the axes, let the unit vector along the long axis be $[l_i] = [0\ l_2\ l_3]$, so that $e_{ij}l_j - \,^2e\,l_i = 0$ [eq. (3.50)]. Then:

$$e_{23}l_3 - \,^2e\,l_2 = 0 , \tag{A.132.9}$$

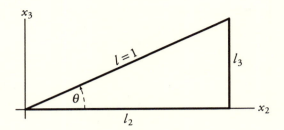

Figure A.132.1. Decomposition of the unit vector l_i parallel to a principal tensor axis.

or in the present case:

$$2\pi\, l_3 - 0.4970\, l_2 \approx 0\ . \tag{A.132.10}$$

By inspection of Fig. A.132.1, the tangent of θ is:

$$\tan\theta = l_3/l_2\ , \tag{A.132.11}$$

or in the present case:

$$\tan\theta \approx 0.4970/2\pi \approx 0.0791\ , \tag{A.132.12}$$

a value that happens to equal the minimum principal elongation $^{min}\lambda$, which is the inverse of the maximum elongation $^{max}\lambda$. The angle θ is:

$$\theta \approx 4.52°\ . \tag{A.132.13}$$

Equation (A.132.12) suggests that, at least in simple shear, the following may hold:

$$\tan\theta = {}^{min}\lambda = 1/{}^{max}\lambda\ . \tag{A.132.14}$$

To verify this guess, use eq. (A.129.21), according to which $\tan\Theta = 2\Phi$, to rewrite eq. (A.132.1) as:

$$\left[a_i\right] = \left[\ x_1 \quad x_2 - 2\Phi x_3 \quad x_3\ \right]\ , \tag{A.132.15}$$

eq. (A.132.2) as:

$$\left[u_i\right] = \left[\ 0 \quad 2\Phi x_3 \quad 0\ \right]\ , \tag{A.132.16}$$

and eq. (A.132.3) as:

$$\left[e_{ij}\right] = \begin{bmatrix} 0 & 0 & 0 \\ 0 & 0 & \Phi \\ 0 & \Phi & -2\Phi^2 \end{bmatrix}\ . \tag{A.132.17}$$

The secular equation for this tensor is:

$$e^3 + 2\Phi^2 e^2 - \Phi^2 e = 0 \ , \tag{A.132.18}$$

which, with the third invariant $^3I = 0$, reduces to a quadratic equation having the solutions:

$$e = \begin{cases} 0 \\ -\Phi^2 \pm \Phi(\Phi^2 + 1)^{1/2} \end{cases}, \tag{A.132.19}$$

where the positive and negative square roots are for the short and long axes, respectively. The principal elongations, calculated as for eq. (A.132.8), are:

$$^1\lambda = 1 \ ,$$
$$^2\lambda = \langle 1 + 2\Phi^2 - 2\Phi(\Phi^2 + 1)^{1/2} \rangle^{-1/2},$$
$$^3\lambda = \langle 1 + 2\Phi^2 + 2\Phi(\Phi^2 + 1)^{1/2} \rangle^{-1/2}. \tag{A.132.20}$$

The angular brackets in the last two principal components may be written as perfect squares, as, for example:

$$\langle (\Phi^2 + 1) - 2\Phi(\Phi^2 + 1)^{1/2} + \Phi^2 \rangle \ ,$$

and the roots in eqs. (A.132.20) can be extracted:

$$^1\lambda = 1 \ , \quad ^2\lambda = \langle (\Phi^2 + 1)^{1/2} - \Phi \rangle^{-1}, \quad ^3\lambda = \langle (\Phi^2 + 1)^{1/2} + \Phi \rangle^{-1}. \tag{A.132.21}$$

With the same unit vector along the long axis as in eq. (A.132.9) and Fig. A.132.1, and solved by the same method, the principal value of Almansi's tensor in the direction of the long strain axis is:

$$^2e = -\Phi^2 + \Phi(\Phi^2 + 1)^{1/2}. \tag{A.132.22}$$

Because according to Fig. A.132.1, $\tan\theta = l_3/l_2$ and because $^2e\, l_2 = e_{23}\, l_3$ [eq. (A.132.9)]:

$$\tan\theta = l_3/l_2 = \,^2e/e_{23} \ . \tag{A.132.23}$$

Thus:

$$\tan\theta = \langle -\Phi^2 + \Phi(\Phi^2 + 1)^{1/2} \rangle / \Phi = -\Phi + (\Phi^2 + 1)^{1/2}. \tag{A.132.24}$$

Hence:

$$1/\,^{max}\lambda = (\Phi^2 + 1)^{1/2} - \Phi = \tan\theta = \,^{min}\lambda \ \therefore \tag{A.132.25}$$

The conjecture is correct, but note that this was shown to be so *only for simple shear*. It happens not to be true for any other combination of strain and rotation.

Answer 133. (a) Let $S_{ij} = S_{ji}$ be a transformation matrix and tensor that represents a finite, homogeneous strain without rotation, translation, or change of volume:

$$x_i = S_{ij}\, a_j \,. \tag{A.133.1}$$

To find Green's tensor, use:

$$u_i = x_i - a_i = S_{ij}\, a_j - a_i \,, \tag{A.133.2}$$

and the definition of Green's tensor:

$$\mathfrak{E}_{ij} = \frac{1}{2}\left(\frac{\partial u_i}{\partial a_j} + \frac{\partial u_j}{\partial a_i} + \frac{\partial u_k}{\partial a_i}\frac{\partial u_k}{\partial a_j}\right), \tag{A.133.3}$$

to write:

$$\mathfrak{E}_{ij} = S_{ij} - \delta_{ij} + \left(S_{ki} - \delta_{ki}\right)\left(S_{kj} - \delta_{kj}\right)/2$$
$$= S_{ij} - \delta_{ij} + \left(S_{ki}\,S_{kj}\right)/2 - S_{ij} + \delta_{ij}/2 \,, \tag{A.133.4}$$

which simplifies further to:

$$\mathfrak{E}_{ij} = \left(S_{ki}\,S_{kj} - \delta_{ij}\right)/2 \,. \tag{A.133.5}$$

Note that this procedure is independent of the symmetry of S_{ij} and could thus also have been performed on an asymmetric transformation matrix. Because $\lambda^2 - 1 = 2\mathfrak{E}_{ij}\,l_i\,l_j$, where l_i are the direction cosines of an original material line subject to the elongation λ, we write:

$$\lambda^2 - 1 = S_{ki}\,S_{kj}\,l_i\,l_j - \delta_{ij}\,l_i\,l_j \,, \tag{A.133.6}$$

which simplifies to:

$$\lambda^2 - 1 = S_{ki}\,S_{kj}\,l_i\,l_j - 1 \,, \tag{A.133.7}$$

and further to:

$$\lambda^2 = S_{ki}\,S_{kj}\,l_i\,l_j \,. \tag{A.133.8}$$

(b) Let a unit sphere be centered at the coordinate origin. Consider a cone from the center of this sphere to its surface, ending there in an arbitrarily shaped but compact element of surface with the area db (Fig. A.133.1). Let the radius **r** from the origin to the center of the surface element have the

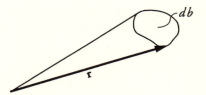

Figure A.133.1.   Solid-angle
element from the center to the
surface of a unit sphere ($r = 1$).

orientation r_i. A unit sphere has the surface area 4π and the volume $4\pi/3$. Thus the surface element occupies a fraction $db/4\pi$ of the total surface area of the sphere, and the complete cone has a volume dv that occupies an equal proportional fraction of the total volume of the sphere:

$$dv = \frac{4\pi}{3}\frac{db}{4\pi} = \frac{db}{3} \ . \tag{A.133.9}$$

(c) Apply the strain to unit sphere and cone and, using r_i as direction cosines, determine the length R of the strained line R_i, which is the material equivalent of r_i:

$$R^2 = S_{ki} S_{kj} r_i r_j \ . \tag{A.133.10}$$

After a constant-volume strain, the new cone volume dV is unchanged from the original dv:

$$dV = dv = db/3 \ . \tag{A.133.11}$$

It is of no concern that the surface element on an ellipsoid produced by the strain is no longer orthogonal to the radius vector R_i, instead, we need to know the area of the intercept of the deformed cone with a sphere of radius R. That sphere has the volume $4\pi R^3/3$. Dividing it by the volume of the cone $db/3$, we obtain the volume fraction ${}^vF$ of that sphere that is occupied by the deformed cone:

$$ {}^vF = \frac{db}{3}\frac{3}{4\pi R^3} = \frac{db}{4\pi R^3} \ . \tag{A.133.12}$$

Since the complete surface of a sphere with radius R is $4\pi R^2$ and the area fraction dB subtended by the cone on the sphere is proportional to its volume fraction, the area fraction is:

$$dB = \frac{db\, 4\pi R^2}{4\pi R^3} = \frac{db}{R} \ , \tag{A.133.13}$$

(d) To normalize this area fraction, we find the area subtended by the same cone on a unit sphere. This is achieved by applying the scale factor $1/R^2$ to dB:

$$d b' = \frac{d B}{R^2} = \frac{d b}{R^3} . \tag{A.133.14}$$

Substitute eq. (A.133.10) into eq. (A.133.14):

$$d b' = \left(S_{ki} S_{kj} r_i r_j \right)^{-3/2} d b , \tag{A.133.15}$$

or by substitution of eq. (A.133.8) into eq. (A.133.15):

$$d b' = \lambda^{-3} d b , \tag{A.133.16}$$

where λ is the elongation in the r_i direction.

Answer 134. (a) The equation:

$$r_i l_i = 0 , \tag{A.134.1}$$

characterizes a plane containing all vectors r_i perpendicular to the unit vector l_i, the pole of that plane. We apply to this plane the strain of Problem 133 by means of the transformation matrix S_{ij} in eq. (A.133.1), except that it is here unnecessary to insist on conservation of volume. The transformed vectors of eq. (A.134.1) are:

$$R_i = S_{ij} r_j , \tag{A.134.2}$$

and thus:

$$r_i = S_{ij}^{-1} R_j . \tag{A.134.3}$$

(b) Substitute eq. (A.134.3) into eq. (A.134.1):

$$\left(S_{ij}^{-1} R_j \right) l_i = 0 , \tag{A.134.4}$$

which may also be written as:

$$R_j \left(S_{ij}^{-1} l_i \right) = 0 , \tag{A.134.5}$$

or, by using the subscripts in alphabetical order and because $S_{ij}^{-1} = S_{ji}^{-1}$, as:

$$R_i \left(S_{ij}^{-1} l_j \right) = 0 , \tag{A.134.6}$$

Whereas eq. (A.134.4) states that the transformed vectors R_i are no longer orthogonal to l_i, eq. (A.134.6) is of the form:

$$R_i L_i = 0 , \tag{A.134.7}$$

which indicates that a new pole L_i exists for the plane defined by these transformed vectors. For both eqs. (A.134.6) and (A.134.7) to hold in the same context, the following must be true:

$$L_i = K \, S_{ij}^{-1} l_j , \tag{A.134.8}$$

where K is an arbitrary scalar constant. Because a scale factor like K is invariant to rotation, this equation states that the orientation of the pole of a material plane changes with a strain, as would the orientation of a material line undergoing the inverse of the physical strain.

Answer 135. (a) March's formula for tabular markers [eq. (7.7)] is:

$${}^i\varepsilon = {}^{iP}\rho^{-1/3} - 1 , \tag{A.135.1}$$

where ${}^{iP}\rho$ are the normalized principal pole densities $({}^{1P}\rho\,{}^{2P}\rho\,{}^{3P}\rho = 1)$ and strain is conventionally defined as $\varepsilon = \Delta l / {}^o l$. With this definition, the principal elongations are:

$${}^i\lambda = {}^i\varepsilon + 1 . \tag{A.135.2}$$

March's formula can therefore be restated as:

$${}^i\lambda = {}^{iP}\rho^{-1/3}, \tag{A.135.3}$$

and solved for ${}^{iP}\rho$:

$${}^{iP}\rho = {}^i\lambda^{-3}. \tag{A.135.4}$$

Let N be the number of pole intercepts in an original surface element on the unit sphere with the area db, analogously to the axis intercepts in Answer 133, then the pole density N/db may be normalized by means of a coefficient C such that the measure of the original pole density ${}^{Po}\rho$ is unity (every direction is a principal direction for a uniform density distribution):

$$1 = {}^{Po}\rho = C N / db . \tag{A.135.5}$$

Remembering from Answer 134 that pole orientations of transformed material planes behave as if they were the orientations of material lines subject to the inverse of the actual strain, we find the area db' of the inversely transformed surface element as in eq. (A.133.16), except that we substitute for the elongation λ its inverse λ^{-1}:

$$db' = \lambda^3 db \ , \tag{A.135.6}$$

which implies also:

$$db' = \left(S_{ki}^{-1} S_{kj}^{-1} l_i l_j\right)^{-3/2} db \ , \tag{A.135.7}$$

and:

$$\frac{db'}{db} = \lambda^3 \ . \tag{A.135.8}$$

This transformation conserves the number N of poles in the surface element, and the normalized pole density in principal coordinates is therefore:

$${}^{iP}\rho = {}^i\lambda^{-3} \frac{C\,N}{db} \ . \tag{A.135.9}$$

However, by eq. (A.135.5) $CN/db = 1$, and we have derived March's formula for pole densities of tabular markers, modified to solve for principal normalized pole densities ${}^{iP}\rho$ as functions of principal elongations ${}^i\lambda$, as in eq. (A.135.3):

$${}^{iP}\rho = {}^i\lambda^{-3} \ \therefore \tag{A.135.10}$$

(b) Using once again the fact that rod axes change orientation under a strain the way plate poles do under its inverse, it follows from eq. (A.135.10) that March's formula for rod-shaped markers solved for the normalized axis density ${}^{iA}\rho$ is:

$${}^{iA}\rho = {}^i\lambda^3 \ , \tag{A.135.11}$$

solved for ${}^i\lambda$:

$${}^i\lambda = {}^{iA}\rho^{1/3}, \tag{A.135.12}$$

and for ${}^i\varepsilon$:

$${}^i\varepsilon = {}^{iA}\rho^{1/3} - 1 \ \therefore \tag{A.135.13}$$

For complete derivations of March's formulas, by both geometric and tensor-mathematical methods, see Lipshie (1984), pp. 299-320.

Answer 136. (a) The geographic coordinates, N, E, and down, are principal coordinates for the compaction strain, a strain that implies a 60% volume loss:

$${}^{1c}\varepsilon = 0 \ , \quad {}^{2c}\varepsilon = 0 \ , \quad {}^{3c}\varepsilon = -0.6 \ . \tag{A.136.1}$$

Tilt the original, horizontal bedding in two steps to achieve the required strike and dip; to produce a 30° W dip, rotate coordinates about an angle $\theta = 30°$ clockwise about x_1:

$$
(a_{ij}) = \begin{pmatrix} 1 & 0 & 0 \\ 0 & \cos\theta & \sin\theta \\ 0 & -\sin\theta & \cos\theta \end{pmatrix}
$$

$$
\approx \begin{pmatrix} 1 & 0 & 0 \\ 0 & 0.866 & 0.500 \\ 0 & -0.500 & 0.866 \end{pmatrix}. \qquad \text{(A.136.2)}
$$

Note that for the numerical calculation at least three more decimal places must be used than here shown; this is necessary to avoid excessive cumulative errors in the final results. The resulting strike is 0°. Modify the strike by rotating coordinates next about an angle $\theta' = 45°$ clockwise about x_3:

$$
(b_{ij}) = \begin{pmatrix} \cos\theta' & \sin\theta' & 0 \\ -\sin\theta' & \cos\theta' & 0 \\ 0 & 0 & 1 \end{pmatrix}
$$

$$
\approx \begin{pmatrix} 0.707 & 0.707 & 0 \\ -0.707 & 0.707 & 0 \\ 0 & 0 & 1 \end{pmatrix}. \qquad \text{(A.136.3)}
$$

Conveniently, combine the two rotations before letting them act on the tensor of eq. (A.136.1):

$$
(c_{ij}) \approx \begin{pmatrix} 0.707 & 0.612 & 0.354 \\ -0.707 & 0.612 & 0.354 \\ 0 & -0.500 & 0.866 \end{pmatrix}. \qquad \text{(A.136.4)}
$$

Refer the pretectonic compaction strain to geographic coordinates after the tilt by performing the rotation ${}^{c}\varepsilon_{ij} = c_{ik} c_{jl} {}^{Pc}\varepsilon_{kl}$ (prescript P for a tensor in principal coordinates):

$$
[{}^{c}\varepsilon_{ij}] \approx \begin{bmatrix} -0.075 & -0.075 & -0.184 \\ -0.075 & -0.075 & -0.184 \\ -0.184 & -0.184 & -0.450 \end{bmatrix}, \qquad \text{(A.136.5)}
$$

and find the corresponding stretch tensor $C_{ij} = c_{ik} c_{jl} {}^{Pc}\varepsilon_{kl} + \delta_{ij}$:

$$
[C_{ij}] \approx \begin{bmatrix} 0.925 & -0.075 & -0.184 \\ -0.075 & 0.925 & -0.184 \\ -0.184 & -0.184 & 0.550 \end{bmatrix}. \qquad \text{(A.136.6)}
$$

The next event to be mathematically modeled is the tectonic strain, for which the geographic coordinates are again principal coordinates:

$$^{11}\varepsilon = -0.5 \ , \quad ^{22}\varepsilon = 0 \ , \quad ^{33}\varepsilon = 1.0 \ . \tag{A.136.7}$$

The corresponding stretch tensor $T_{ij} = {}^{t}\varepsilon_{ij} + \delta_{ij}$ is:

$$\left[T_{ij} \right] = \begin{bmatrix} 0.5 & 0 & 0 \\ 0 & 1 & 0 \\ 0 & 0 & 2 \end{bmatrix} . \tag{A.136.8}$$

The combined transformation matrix for compaction followed by tectonic strain is $\mathbb{F} = \mathbb{T}\mathbb{C}$:

$$\left[F_{ij} \right] \approx \begin{bmatrix} 0.4625 & -0.0375 & -0.0919 \\ -0.0750 & 0.9250 & -0.1837 \\ -0.3674 & -0.3674 & 1.1000 \end{bmatrix} . \tag{A.136.9}$$

(b) Thus the transformation matrix $\mathbb{F}$ is not symmetric although it is the product of two symmetric matrices. (c) To interpret this transformation in alternative ways, we perform both the right and the left polar decompositions. For right decomposition we need the symmetric square matrix $\mathbb{F}^2 = \mathbb{F}^T \mathbb{F}$:

$$\left[F_{ij}^2 \right] \approx \begin{bmatrix} 0.3545 & 0.0483 & -0.4329 \\ 0.0483 & 0.9920 & -0.5707 \\ -0.4329 & -0.5707 & 1.2522 \end{bmatrix} . \tag{A.136.10}$$

For left decomposition we need the symmetric square of the transpose of $\mathbb{F}$, $(\mathbb{F}^T)^2 = \mathbb{F}\mathbb{F}^T$:

$$\left[\left(F^T \right)_{ij}^2 \right] \approx \begin{bmatrix} 0.2238 & 0.0525 & -0.2572 \\ 0.0525 & 0.8950 & -0.5144 \\ -0.2572 & -0.5144 & 1.4800 \end{bmatrix} . \tag{A.136.11}$$

The principal values of both $\mathbb{F}^2$ and $(\mathbb{F}^T)^2$, calculated by standard methods, are:

$$^{1}F^2 \approx 0.1343 \ , \quad ^{2}F^2 \approx 0.6606 \ , \quad ^{3}F^2 \approx 1.8039 \ , \tag{A.136.12}$$

and the two symmetric tensors thus differ only by their orientations. The principal directions of $\mathbb{F}^2$ with respect to geographic coordinates are:

$$
(d_{ij}) \approx \begin{pmatrix} 0.849 & 0.259 & 0.461 \\ -0.464 & 0.782 & 0.415 \\ -0.253 & -0.566 & 0.784 \end{pmatrix} , \tag{A.136.13}
$$

which implies the following trends and plunges for the three principal axes:

$$
\begin{aligned}
\mathsf{T} &\approx 17°, \ \mathsf{P} \approx 27°, \\
\mathsf{T} &\approx 121°, \ \mathsf{P} \approx 24°, \\
\mathsf{T} &\approx 246°, \ \mathsf{P} \approx 52°.
\end{aligned}
$$

The principal directions of $(\mathbb{F}^{\mathsf{T}})^2$ are:

$$
(f_{ij}) \approx \begin{pmatrix} 0.929 & 0.248 & 0.273 \\ -0.347 & 0.840 & 0.418 \\ -0.125 & -0.483 & 0.867 \end{pmatrix} , \tag{A.136.14}
$$

implying the trends and plunges:

$$
\begin{aligned}
\mathsf{T} &\approx 15°, \ \mathsf{P} \approx 16°, \\
\mathsf{T} &\approx 112°, \ \mathsf{P} \approx 25°, \\
\mathsf{T} &\approx 256°, \ \mathsf{P} \approx 60°.
\end{aligned}
$$

The square roots of the principal values of $\mathbb{F}^2$ in eq. (A.136.12) are the principal values of both stretch tensors $\mathbb{U}$ and $\mathbb{V}$:

$$
{}^1U = {}^1V \approx 0.366 , \ {}^2U = {}^2V \approx 0.813 , \ {}^3U = {}^3V \approx 1.343 , \tag{A.136.15}
$$

and the corresponding strain in both the respective principal coordinates is:

$$
{}^1\varepsilon \approx -0.634 , \ {}^2\varepsilon \approx -0.187 , \ {}^3\varepsilon \approx 0.343 . \tag{A.136.16}
$$

Refer $\mathbb{U}$ to geographic coordinates by rotating it by means of the inverse of d_{ij}, $d_{ij}^{-1} = d_{ji}$; the resulting right-decomposed stretch tensor is:

$$
[U_{ij}] \approx \begin{bmatrix} 0.525 & -0.022 & -0.280 \\ -0.022 & 0.953 & -0.289 \\ -0.280 & -0.289 & 1.044 \end{bmatrix} , \tag{A.136.17}
$$

and similarly rotated by f_{ji}, the left-decomposed stretch tensor is:

$$
[V_{ij}] \approx \begin{bmatrix} 0.435 & -0.071 & -0.171 \\ -0.071 & 0.909 & -0.252 \\ -0.171 & -0.252 & 1.178 \end{bmatrix} . \tag{A.136.18}
$$

To calculate the rigid-body rotation $\mathbb{R}$, the inverse decomposed stretches are needed. They are most conveniently inverted in their principal form, say, as $^iU^{-1} \equiv 1/^iU$:

$$^1U^{-1} = {}^1V^{-1} \qquad {}^2U^{-1} = {}^2V^{-1} \qquad {}^3U^{-1} = {}^3V^{-1}$$
$$\approx 2.729 \, , \qquad\qquad \approx 1.230 \, , \qquad\qquad \approx 0.745 \, . \qquad \text{(A.136.19)}$$

Rotated to reference coordinates by d_{ji} and f_{ji}, they are:

$$\left[U_{ij}^{-1} \right] \approx \begin{bmatrix} 2.279 & 0.257 & 0.683 \\ 0.257 & 1.175 & 0.395 \\ 0.683 & 0.395 & 1.250 \end{bmatrix}, \qquad \text{(A.136.20)}$$

and:

$$\left[V_{ij}^{-1} \right] \approx \begin{bmatrix} 2.517 & 0.317 & 0.432 \\ 0.317 & 1.210 & 0.305 \\ 0.432 & 0.305 & 0.977 \end{bmatrix}. \qquad \text{(A.136.21)}$$

Calculated either by means of $\mathbb{R} = \mathbb{F}\mathbb{U}^{-1}$ or of $\mathbb{R} = \mathbb{V}^{-1}\mathbb{F}$, $\mathbb{R}$ is:

$$\left(R_{ij} \right) \approx \begin{pmatrix} 0.982 & 0.040 & 0.186 \\ -0.056 & 0.995 & 0.084 \\ -0.182 & -0.093 & 0.979 \end{pmatrix}, \qquad \text{(A.136.22)}$$

where $|R_{ij}| = 1$. This matrix represents [eqs. (3.66) and (3.67)] a coordinate rotation of $21°$ clockwise or a physical rotation of $21°$ counterclockwise about an axis with the direction cosines:

$$\left[l_i \right] \approx \begin{bmatrix} 0.422 & -0.877 & 0.229 \end{bmatrix}. \qquad \text{(A.136.23)}$$

The negative end of this unit vector trends $296°$ and plunges $13°$. Looking in the down-plunge direction of this axis, the coordinate rotation is counterclockwise and the rotation with respect to fixed coordinates clockwise. (d) Before the tectonic strain the bedding had a strike of $135°$ and a SW dip of $30°$. Its pole thus trended $45°$ and plunged $60°$. Thus the unit vector along the downward pole was:

$$\left[u_i \right] \approx \begin{bmatrix} 0.354 & 0.354 & 0.866 \end{bmatrix}. \qquad \text{(A.136.24)}$$

According to eq. (A.132.8), a vector U_i parallel to the pole of a material plane deformed by the strain with the stretch tensor T_{ij} is related to its predeformation counterpart u_i by:

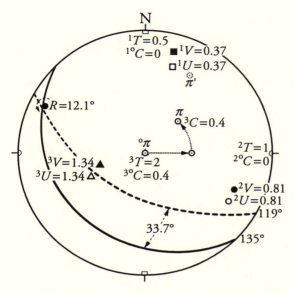

N

¹T=0.5
¹°C=0 ■¹V=0.37
□¹U=0.37
π̈'

*⁻R=12.1°

π
○³C=0.4

°π
○···○

³V=1.34▲
³U=1.34△   ³T=2
³°C=0.4

²T=1
²°C=0

●²V=0.81
○²U=0.81
119°

33.7°

135°

Figure A.136.1. Orientations of bedding planes and poles original, intermediate, and final, and of the rotation axis. Equal-area projection, lower hemisphere.

$$U_i = K\,T_{ij}^{-1}\,u_j \;,$$ (A.136.25)

and we find the pole of the deformed bed by normalizing $u'_i = U_i/U$:

$$[u'_i] \approx [\; 0.784 \quad 0.392 \quad 0.480 \;]\;,$$ (A.136.26)

which trends 27° and plunges 29°. Thus bedding in the final state strikes 119° and dips 61° to the SSW. The angle between pretectonic and final bedding is 34°; see Fig. A.136.1.

☙

Summary of Formulæ

Trigonometry

Double angle:
$$\cos 2\theta = 2\cos^2\theta - 1$$
$$\tan 2\theta = \frac{2\tan\theta}{1 - \tan^2\theta}$$

Two angles:
$$\cos(\alpha + \beta) = \cos\alpha\cos\beta - \sin\alpha\sin\beta$$
$$\tan(\alpha + \beta) = \frac{\tan\alpha + \tan\beta}{1 - \tan\alpha\tan\beta}$$

Coordinates

Let x_1, x_2, and x_3 stand for cartesian coordinates x, y, and z, and x_i for any of the three.

Subscripted variables

Einstein summation convention:
An alphabetic subscript repeated in the same term (dummy subscript) implies summation over the range of that subscript.
$$a_i b_i \equiv a_1 b_1 + a_2 b_2 + a_3 b_3$$

Dummy subscripts may only occur in pairs, and the same pair may only be used once per term. Alphabetic subscripts that are not repeated are called *free subscripts*.

Dummy subscripts
are so called because they may be replaced, term by term, by other alphabetic symbols:
$$a_i b_i \equiv a_j b_j$$
as long as symbols already used for free subscripts or for another pair of dummy subscripts in the same term are avoided.

Free subscripts:
Any number of free alphabetic subscripts may occur in one term, but each of these free subscripts must also occur in every other term of the same equation.

Implied spatial differentiation:
A comma in the subscript implies differentiation with respect to x_i etc., if the subscripts i, etc., follow the comma. For example:
$$\varphi_{,i} \equiv \frac{\partial\varphi}{\partial x_i}$$
$$V_{i,j} \equiv \frac{\partial V_i}{\partial x_j}$$
$$T_{i,jk} \equiv \frac{\partial^2 T_i}{\partial x_j \partial x_k}$$

Equations

Quadratic:
$$x^2 + ax + b = 0$$
$$x = -a/2 \pm \sqrt{(a/2)^2 - b}$$

Cubic:
$$x^3 + ax^2 + bx + c = 0$$
substitute:
$$x = y - a/3$$
then:
$$y^3 - py + q = 0$$
where:
$$p = \frac{a^2}{3} - b, \quad q = c + \frac{2a^3}{27} - \frac{ab}{3}$$

The *discriminant* is:
$$D = \frac{q^2}{4} - \frac{p^3}{27}$$

If $D > 0$, then one root is real, two are complex conjugates. If $D = 0$, then all roots are real, and two of them are identical:
$$^1y = 2u, \quad ^2y = {}^3y = -u$$

where:
$$u = \sqrt[3]{-q/2}$$

If $D < 0$, then:

$$^1y = 2\sqrt{p/3}\,\cos(\varphi/3)$$

$$^2y = 2\sqrt{p/3}\,\cos\left(\frac{\varphi + 2\pi}{3}\right)$$

$$^3y = 2\sqrt{p/3}\,\cos\left(\frac{\varphi + 4\pi}{3}\right)$$

where:

$$\cos\varphi = -q/2\sqrt{p^3/27}$$

To recover ^ix, substitute a and ^iy into:

$$^ix = {}^iy - a/3$$

FUNCTIONS

Constraints, Lagrange multiplier:
To maximize a function:

$$u = f(x_i)$$

subject to the constraint:

$$\varphi(x_i) = 0$$

the constraint implying that the x_i in these two equations are *not* independent, multiply the constraint equation by the arbitrary Lagrange multiplier λ. Then solve:

$$\frac{\partial f}{\partial x_i} + \lambda\frac{\partial\varphi}{\partial x_i} = 0$$

for λ together with:

$$\varphi = 0$$

Tangent planes and normal lines
Let the equation of a surface be:

$$u(x_i) = 0$$

then the equation of the plane tangent at the point $^\circ x_i$ is:

$$(x_i - {}^\circ x_i)\,u_{,i}(^\circ x_i) = 0$$

and the normal n_i to the tangent plane is:

$$n_i = u_{,i}(^\circ x_i)$$

Note that in expressions such as:

$$u_{,i}(^\circ x_i)$$

the subscript in the "function of" parentheses merely indicates an integer value identical with that of the subscript outside, and its repetition invokes neither multiplication nor summation.

ALTERNATING MATRIX

$\epsilon_{ijk} = 1$, if $i, j, k = 1, 2, 3$ or
  cyclic permutations $2, 3, 1$ or $3, 1, 2$
$\epsilon_{ijk} = -1$, if $i, j, k = 3, 2, 1$ or
  cyclic permutations $1, 3, 2$ or $2, 1, 3$
$\epsilon_{ijk} = 0$, if neither of these

DETERMINANTS

The determinant of a 3×3 matrix $\mathbb{A}$ is:

$$|\mathbb{A}| \equiv |A_{ij}| \equiv \epsilon_{ijk}A_{1i}A_{2j}A_{3k}$$
$$= \epsilon_{ijk}A_{i1}A_{j2}A_{k3}$$

and thus also of its transpose:

$$|\mathbb{A}| = |\mathbb{A}^\mathsf{T}|, \quad |A_{ij}| = |A_{ji}|$$

Multiplying one row (or column) of a determinant by a scalar factor multiplies the determinant by that factor; thus, for example:

$$\begin{vmatrix} xa & xb & xc \\ d & e & f \\ g & h & i \end{vmatrix} = x\begin{vmatrix} a & b & c \\ d & e & f \\ g & h & i \end{vmatrix}$$

A determinant is unaltered if a multiple of one row (or column) is added to another row (or column), as, for example, in:

$$\begin{vmatrix} a & b & c \\ d & e & f \\ g & h & i \end{vmatrix} = \begin{vmatrix} a & b & c \\ d+xa & e+xb & f+xc \\ g & h & i \end{vmatrix}$$

The determinant derived from

$|A_{ij}|$ by interchanging two of its columns (or rows) equals $-|A_{ij}|$, and if two columns (or rows) of $|A_{ij}|$ are identical, then $|A_{ij}| = 0$. Thus:

$$\begin{vmatrix} a & b & c \\ d & e & f \\ g & h & i \end{vmatrix} = - \begin{vmatrix} d & e & f \\ a & b & c \\ g & h & i \end{vmatrix}$$

and:

$$\begin{vmatrix} a & b & c \\ a & b & c \\ d & e & f \end{vmatrix} = 0$$

If each element of a row (or column) of a determinant is the sum of two terms, then this determinant equals the sum of two determinants, each having a row (or column) consisting of the separate summands and remaining rows (or columns) that are identical with those of the original determinant; for example:

$$\begin{vmatrix} a+d & b+e & c+f \\ g & h & i \\ j & k & l \end{vmatrix}$$

$$= \begin{vmatrix} a & b & c \\ g & h & i \\ j & k & l \end{vmatrix} + \begin{vmatrix} d & e & f \\ g & h & i \\ j & k & l \end{vmatrix}$$

If $|A_{ij}|$ and $|B_{ij}|$ are determinants and $C_{ij} = A_{ik} B_{jk}$, then:

$$\left| C_{ij} \right| = \left| A_{pq} \right| \left| B_{rs} \right|$$

VECTORS

Magnitude V of a vector $\mathbf{V}$, its length, is the positive square root of V^2:

$$V^2 = V_i V_i \equiv V_1^2 + V_2^2 + V_3^2$$
$$V = (V_i V_i)^{1/2}$$

Sum $\mathbf{S}$ of vectors $\mathbf{A}$ and $\mathbf{B}$, a vector:

$$S_i = A_i + B_i$$

Dot product D of vectors $\mathbf{A}$ and $\mathbf{B}$ forming an angle θ, a scalar:

$$D = \mathbf{A} \bullet \mathbf{B} \equiv AB \cos \theta$$
$$D = A_i B_i$$
$$\equiv A_1 B_1 + A_2 B_2 + A_3 B_3$$

The dot product is A times the projection of $\mathbf{B}$ on $\mathbf{A}$ or B times the projection of $\mathbf{A}$ on $\mathbf{B}$. Two vectors are perpendicular if and only if their dot product is zero.

$$(S\mathbf{A}) \bullet \mathbf{B} = S(\mathbf{A} \bullet \mathbf{B})$$

unless one of the vectors is a differential operator and S is not a constant.

Unit vectors along coordinate axes
Let unit vectors $\mathbf{i}$, $\mathbf{j}$, and $\mathbf{k}$ be parallel to the cartesian coordinates x_1, x_2, and x_3.

Cross product $\mathbf{C}$ of vectors $\mathbf{A}$ with $\mathbf{B}$ forming an angle θ ($\mathbf{u}$ is the unit vector perpendicular to the $\mathbf{AB}$ plane and positive in the screw rule direction for θ positive from $\mathbf{A}$ to $\mathbf{B}$), a vector:

$$\mathbf{C} = \pm\mathbf{A} \times \mathbf{B} \equiv \pm\mathbf{u} AB \sin\theta$$

or:

$$\mathbf{C} = \pm \begin{vmatrix} \mathbf{i} & \mathbf{j} & \mathbf{k} \\ A_1 & A_2 & A_3 \\ B_1 & B_2 & B_3 \end{vmatrix}$$

or:

$$\mathbf{C} = \pm \left\langle \begin{array}{c} \mathbf{i}\,(A_2 B_3 - A_3 B_2) \\ -\mathbf{j}\,(A_1 B_3 - A_3 B_1) \\ +\mathbf{k}(A_1 B_2 - A_2 B_1) \end{array} \right\rangle$$

or:

$$C_i = \pm \epsilon_{ijk} A_j B_k$$

The negative of $\pm$ applies if the coordinate system is left handed.

$$\mathbf{A} \times \mathbf{B} = -\mathbf{B} \times \mathbf{A}$$
$$\mathbf{A} \times \mathbf{B} \neq \mathbf{B} \times \mathbf{A}$$
$$(S\mathbf{A} \times \mathbf{B}) = S(\mathbf{A} \times \mathbf{B})$$
$$\mathbf{i} \times \mathbf{i} = \mathbf{j} \times \mathbf{j} = \mathbf{k} \times \mathbf{k} = 0$$
$$\mathbf{i} \times \mathbf{j} = -\mathbf{j} \times \mathbf{i} = \mathbf{k} , \text{ etc.}$$

Two vectors are parallel if and only if their cross product is a zero vector.

Differentiation of vectors with respect to time

$$\mathbf{V} = \mathbf{V}(t)$$

$$\frac{d\mathbf{V}}{dt} \equiv \mathbf{i}\frac{dV_1}{dt} + \mathbf{j}\frac{dV_2}{dt} + \mathbf{k}\frac{dV_3}{dt}$$

and

$$\frac{d}{dt}(\mathbf{A} \bullet \mathbf{B}) = \frac{d\mathbf{A}}{dt} \bullet \mathbf{B} + \mathbf{A} \bullet \frac{d\mathbf{B}}{dt}$$
$$\frac{d}{dt}(\mathbf{A} \times \mathbf{B}) = \frac{d\mathbf{A}}{dt} \times \mathbf{B} + \mathbf{A} \times \frac{d\mathbf{B}}{dt}$$

FIELDS

Scalar fields

$$\varphi(x_i, t) = 0$$

Vector fields

$$\mathbf{v}(x_i, t) = \mathbf{0}$$

consist of three scalar fields.

Differential operators

Del (or nabla), a vector:

$$\nabla \equiv \mathbf{i}\frac{\partial}{\partial x_1} + \mathbf{j}\frac{\partial}{\partial x_2} + \mathbf{k}\frac{\partial}{\partial x_3}$$

or:

$$(\nabla)_i \equiv \frac{\partial}{\partial x_i}$$

Gradient, a vector:

$$\operatorname{grad}\varphi \equiv \nabla\varphi \equiv [\varphi_{,i}]$$

$$\nabla\varphi = \mathbf{i}\frac{\partial\varphi}{\partial x_1} + \mathbf{j}\frac{\partial\varphi}{\partial x_2} + \mathbf{k}\frac{\partial\varphi}{\partial x_3}$$

in the direction of *most steeply increasing* φ.

Divergence, a scalar:

$$\operatorname{div}\mathbf{V} \equiv \nabla\bullet\mathbf{V} \equiv \frac{\partial V_1}{\partial x_1} + \frac{\partial V_2}{\partial x_2} + \frac{\partial V_3}{\partial x_3}$$

and

$$\operatorname{div}[V_i] \equiv V_{i,i} \equiv \frac{\partial V_i}{\partial x_i}$$

Curl, an axial vector:

$$\operatorname{curl}\mathbf{V} \equiv \nabla\times\mathbf{V}$$

$$\equiv \begin{vmatrix} \mathbf{i} & \mathbf{j} & \mathbf{k} \\ \dfrac{\partial}{\partial x_1} & \dfrac{\partial}{\partial x_2} & \dfrac{\partial}{\partial x_3} \\ V_1 & V_2 & V_3 \end{vmatrix}$$

$$\equiv \mathbf{i}\left(\frac{\partial V_3}{\partial x_2} - \frac{\partial V_2}{\partial x_3}\right)$$
$$- \mathbf{j}\left(\frac{\partial V_3}{\partial x_1} - \frac{\partial V_1}{\partial x_3}\right)$$
$$+ \mathbf{k}\left(\frac{\partial V_2}{\partial x_1} - \frac{\partial V_1}{\partial x_2}\right)$$

$$\equiv [\epsilon_{ijk} V_{k,j}]$$

Characteristics of vector fields

Solenoidal vector fields satisfy either:

$$\nabla\bullet\mathbf{V} = 0$$

or:

$$\mathbf{V} = \nabla\times\mathbf{A}$$

where $\mathbf{A}$ is a vector function.

Irrotational vector fields satisfy either:

$$\nabla\times\mathbf{V} = \mathbf{0}$$

or:

$$\mathbf{V} = \nabla\varphi$$

where φ is a scalar function.

Well-behaved vector fields in a simply connected region satisfy one of the following four conditions:

$$\mathbf{V} = \nabla\times\mathbf{A} + \nabla\varphi$$
$$\nabla\times\mathbf{V} = \nabla\times(\nabla\times\mathbf{A})$$
$$\nabla\bullet(\nabla\times\mathbf{A}) = 0$$
$$\nabla\times\nabla\varphi = \mathbf{0}$$

KRONECKER DELTA

Definition

$$\delta_{ij} = 1 \,, \text{ if } i = j$$
$$\delta_{ij} = 0 \,, \text{ if } i \neq j$$

Hence, (δ_{ij}) is the *identity matrix*.

Substitution property

If a Kronecker delta in a term has a free and a repeated (dummy) subscript, then the free subscript replaces the repeated subscript elsewhere in the term and the Kronecker delta is dropped, as, for example, in:

$$\delta_{ij} T_{jk} = T_{ik}$$

This does not apply to numerical subscripts; thus, for example:

$$\delta_{21} T_{13} = 0 \neq T_{23}$$

or:

$$\delta_{33} T_{13} = T_{13}$$

Orthogonality

Because cartesian coordinates are orthogonal:

$$\frac{\partial x_i}{\partial x_j} = \delta_{ij}$$

$\epsilon\delta$ Rule

The Kronecker delta and the alternating matrix are interrelated, as, for example, in:

$$\epsilon_{ijk} \epsilon_{pqk} = \delta_{ip} \delta_{jq} - \delta_{iq} \delta_{jp}$$

where one subscript is repeated in the product of two alternating matrices. Because of the great number of possible cyclic permutations, it is preferable to state the rule by identifying the four free subscripts by the order in which they occur on the left-hand side:

$$\epsilon_{\text{first second dummy}} \, \epsilon_{\text{third fourth dummy}}$$
$$= \delta_{\text{first third}} \, \delta_{\text{second fourth}}$$
$$- \delta_{\text{first fourth}} \, \delta_{\text{second third}}$$

Coordinate transformations

Transformation matrix, an array of direction cosines referring unit vectors along "new" coordinate axes (rows and first index) to the "old" axes (columns and second index).

"old"

		x_1	x_2	x_3
	x'_1	a_{11}	a_{12}	a_{13}
"new"	x'_2	a_{21}	a_{22}	a_{23}
	x'_3	a_{31}	a_{32}	a_{33}

Each row and each column of this matrix is a unit vector; hence:

$$a_{ij} a_{ij} = a_{ji} a_{ji} = 1 \;(\text{no sum on } i)$$

Also:

$$|a_{ij}| = \pm 1$$

Several corollaries follow from these conditions on the transformation matrix:

$$a_{1i} = \pm \epsilon_{ijk} a_{2j} a_{3k}$$
$$a_{2i} = \pm \epsilon_{ijk} a_{3j} a_{1k}$$
$$a_{3i} = \pm \epsilon_{ijk} a_{1j} a_{2k}$$
$$a_{i1} = \pm \epsilon_{ijk} a_{j2} a_{k3}$$
$$a_{i2} = \pm \epsilon_{ijk} a_{j3} a_{k1}$$
$$a_{i3} = \pm \epsilon_{ijk} a_{j1} a_{k2}$$

hence:

$$a_{11} = \pm \left(a_{22} a_{33} - a_{23} a_{32} \right)$$
$$a_{12} = \pm \left(a_{23} a_{31} - a_{21} a_{33} \right)$$
etc.

where the negative sign implies a change of handedness from right to left or vice versa. The components of the transformation matrix are not independent of each other; they are interrelated by the:

Orthogonality relations:

$$a_{ik}\, a_{jk} = \delta_{ij},$$

$$a_{ki}\, a_{kj} = \delta_{ij}$$

A consequence of orthogonality is:

$$a_{ij}^{-1} = a_{ij}^{\mathsf{T}} = a_{ji}$$

Transformation rule for vectors:

$$V'_i = a_{ij}\, V_j$$

Sequence of transformations

The combined transformation resulting from a first transformation A_{ij} followed by a second B_{ij} is:

$$C_{ij} = B_{ik}\, A_{kj}$$

The sequence in which the incremental transformations are written is arbitrary in subscripted notation; however, if the dummy subscript, here k, appears last in the first, first in the second matrix, then the equation may be replaced by one in subscript-free matrix notation:

$$\mathbb{C} = \mathbb{B}\,\mathbb{A}$$

where:

$$\mathbb{B}\,\mathbb{A} \neq \mathbb{A}\,\mathbb{B}$$

The cumulative transformation is calculated by premultiplying the transformation matrix representing the earlier increment with that representing the later increment.

Coordinate conversion

From cartesian to spherical on the unit sphere

The spherical coordinate ρ on the unit sphere is identically $\rho = 1$, and by convention cartesian axes are taken to intercept that sphere at the following θ, φ:

	θ	φ
x_1	0°	90°
x_2	90°	90°
x_3		0°

Direction cosines l_i in cartesian coordinates have the following spherical counterparts:

$$\rho = 1$$
$$\theta = \cos^{-1}\left(l_1/\sin\varphi\right)$$
$$= \sin^{-1}\left(l_2/\sin\varphi\right)$$
$$= \tan^{-1}\left(l_2/l_1\right)$$
$$\varphi = \cos^{-1} l_3$$

where the second expression for θ breaks the sign ambiguity of the first and the third yields greater accuracy for small φ. With reference to coordinates having axes N, E, and down, the trend of a line, in geological parlance, is $\mathsf{T} = \theta$; the plunge is $\mathsf{P} = 90° - \varphi$.

From spherical on the unit sphere to cartesian

Spherical coordinates at $\rho = 1$ are equivalent to the following direction cosines:

$$l_1 = \cos\theta \sin\varphi$$
$$l_2 = \sin\theta \sin\varphi$$
$$l_3 = \cos\varphi$$

Tensors of the second rank

Invariants of second-rank tensors:

$${}^1I = T_{ii}$$

$${}^2I = \begin{vmatrix} T_{22} & T_{23} \\ T_{32} & T_{33} \end{vmatrix} + \begin{vmatrix} T_{11} & T_{13} \\ T_{31} & T_{33} \end{vmatrix}$$
$$+ \begin{vmatrix} T_{11} & T_{12} \\ T_{21} & T_{22} \end{vmatrix}$$

$${}^3I = \left| T_{ij} \right|$$

Second-rank tensors are three-by-three matrices for which these invariants do not change with a coordinate transformation. The first invariant, the sum of the diagonal components, is also called the *trace* of the tensor. Invariants other than these three can be formed by algebraic manipulation or combination of the standard invariants.

Transformation rule for second-rank tensors:

$$T'_{ij} = a_{ik} a_{jl} T_{kl}$$

Second-rank tensor interrelating two vectors:

$$p_i = T_{ij} q_j$$

$$q_i = T^{-1}_{ij} p_j$$

Tensor (or matrix) inversion by Cramer's rule:

$$T^{-1}_{ij} = A_{ji} / D$$

where the *cofactors* are:

$$A_{ij} = \epsilon_{ikl} \epsilon_{jmn} T_{mk} T_{nl} / 2$$

and D is the determinant of the tensor:

$$D = |T_{ij}|$$

Each cofactor A_{ij} is the determinant, multiplied by a factor of $(-1)^{i+j}$, that remains after deleting the ith column and jth row of T_{ij}, as, for example:

$$A_{23} = \begin{vmatrix} T_{11} & T_{12} \\ T_{31} & T_{32} \end{vmatrix} (-1)^5$$

Secular equation (eigenvalue problem)

The principal values of a symmetric second-rank tensor are found by solving the equation:

$$|S_{ij} - \lambda \delta_{ij}| = 0$$

or:

$$\begin{vmatrix} S_{11} - \lambda & S_{12} & S_{13} \\ S_{12} & S_{22} - \lambda & S_{23} \\ S_{13} & S_{23} & S_{33} - \lambda \end{vmatrix} = 0$$

or explicitly:

$$\lambda^3 - (S_{11} + S_{22} + S_{33}) \lambda^2$$
$$+ \left(\begin{array}{c} S_{22} S_{33} + S_{11} S_{33} + S_{11} S_{22} \\ - S_{23} S_{23} - S_{13} S_{13} - S_{12} S_{12} \end{array} \right) \lambda$$
$$+ \left(\begin{array}{c} S_{11} S_{22} S_{33} + 2 S_{23} S_{13} S_{12} \\ - S_{11} S_{23} S_{23} - S_{22} S_{13} S_{13} \\ - S_{33} S_{12} S_{12} \end{array} \right) = 0$$

Note that the three coefficients of this cubic equation are the three standard invariants of the tensor. The second invariant can also be stated as:

$$^2I = A_{ii}$$

the trace of the tensor's cofactor matrix.

Eigendirection problem

Unit vectors l_i in the principal or eigendirections of a symmetric tensor $S_{ij} = S_{ji}$ are subject to the equation:

$$S_{ij} l_j = \lambda l_i$$

Hence, the equations for the three eigendirections are:

$$S_{ij} \,^1X_j = \,^1\lambda \,^1X_i$$

$$S_{ij} \,^2X_j = \,^2\lambda \,^2X_i$$

$$S_{ij} \,^3X_j = \,^3\lambda \,^3X_i$$

where the ^iX_j are vectors of arbitrary magnitude parallel to the principal directions of the tensor; because of their uncertain magnitude, it is impossible to solve for them directly. Instead, one solves for the ratios of two of the components of each. Define:

$$^i\widehat{X}_1 \equiv {}^iX_1 / {}^iX_3$$

$$^i\widehat{X}_2 \equiv {}^iX_2 / {}^iX_3$$

and arbitrarily set:

$$^i\widehat{X}_3 = \pm 1$$

Find the direction cosines (components of a unit vector) in the ith eigendirection by normalization:

$$^il_j = {}^i\widehat{X}_j \Big/ \left({}^i\widehat{X}_1^2 + {}^i\widehat{X}_2^2 + 1 \right)^{1/2}$$

Cartesian coordinates parallel to the principal directions are called *principal coordinates*. The off-diagonal components of a second-rank tensor in principal coordinates are zero, and the principal values of a tensor are occasionally shown as if they were components of a vector:

$$\left[{}^iS \right] = \left[{}^1S \quad {}^2S \quad {}^3S \right]$$

where:

$$^1S \equiv S_{11}, \quad {}^2S \equiv S_{22}, \quad {}^3S \equiv S_{33}$$

and the invariants are:

$$^1I = {}^1S + {}^2S + {}^3S$$
$$^2I = {}^2S\,{}^3S + {}^3S\,{}^1S + {}^1S\,{}^2S$$
$$^3I = {}^1S\,{}^2S\,{}^3S$$

Note that a number of arbitrarily accurate numerical methods exist for the determination of eigenvalues and eigendirections.

Magnitude

For a specific direction with the direction cosines l_i, the magnitude of a second-rank tensor is:

$$T = T_{ij}\, l_i\, l_j$$

Hence, for the direction of a vector V_i with arbitrary magnitude:

$$T = T_{ij} \frac{V_i}{V} \frac{V_j}{V}$$

In principal coordinates, a symmetric tensor has the magnitude:

$$S = {}^iS$$

Representation quadric

A symmetric second-rank tensor may be represented by the quadric centered on the coordinate origin:

$$S_{ij}\, x_i\, x_j = \pm 1$$

with semiaxes $S_i^{-1/2}$. The sign is chosen to avoid imaginary branches of the quadric. Let $[\,r_i\,]$ be the radius vector to x_i; then the magnitude of the tensor is:

$$S = 1/r^2$$

Axial vectors

In contradistinction to polar vectors, axial vectors are actually degenerate second-rank tensors. Whereas the transformation rule for a polar vector is $p'_i = a_{ij}\, p_j$, for an axial vector it is:

$$r_i = \big| a_{km} \big|\, a_{ij}\, r_j$$

An antisymmetric second-rank tensor, such as $V_{ij} = -V_{ji}$, may be written in terms of the axial vector r_i, which in turn is the cross product of two polar vectors p_i and q_i:

$$r_i = \epsilon_{ijk}\, p_j\, q_k$$

or:

$$r_1 = p_2\, q_3 - p_3\, q_2$$
$$r_2 = p_3\, q_1 - p_1\, q_3$$
$$r_3 = p_1\, q_2 - p_2\, q_1$$

then the second-rank tensor is:

$$[V_{ij}] = \begin{bmatrix} 0 & -r_3 & r_2 \\ r_3 & 0 & -r_1 \\ -r_2 & r_1 & 0 \end{bmatrix}$$

Generator of rotations

It is obtained by normalizing r_i to unit magnitude and expanding the resulting axial unit vector l_i to its antisymmetric second-rank tensor form:

$$R_{ij} = \frac{V_{ij}}{\left(p_k p_k q_l q_l - p_m q_m p_n q_n \right)^{1/2}}$$

or:

$$[R_{ij}] = \begin{bmatrix} 0 & -l_3 & l_2 \\ l_3 & 0 & -l_1 \\ -l_2 & l_1 & 0 \end{bmatrix}$$

and represents a 90° clockwise rotation *relative to a fixed reference frame* about a line with direction cosines l_i, looking in the positive direction of that vector. Because both the generator of rotations R_{ij} and the rotation matrix a_{ij} represent rotations, they are interrelated. A clockwise rotation through an angle θ about a line with direction cosines l_i is represented by the following transformation matrix representing the *corresponding counterclockwise rotation of the reference frame* itself:

$$a_{ij} = \delta_{ij} \cos\theta + \epsilon_{ijk} l_k \sin\theta + l_i l_j (1 - \cos\theta)$$

or, in terms of the generator of rotations:

$$a_{ij} = \delta_{ij} - R_{ij} \sin\theta + R_{ik} R_{kj} (1 - \cos\theta)$$

From a known transformation matrix a_{ij}, $\sin\theta$, $\cos\theta$, and l_i, and thus R_{ij}, can be calculated:

$$\sin\theta = \frac{1}{2} \left\langle \begin{array}{c} (a_{23} - a_{32})^2 \\ + (a_{31} - a_{13})^2 \\ + (a_{12} - a_{21})^2 \end{array} \right\rangle^{1/2}$$

$$\cos\theta = (a_{11} + a_{22} + a_{33} - 1)/2$$

$$l_1 = (a_{23} - a_{32})/(2\sin\theta)$$

$$l_2 = (a_{31} - a_{13})/(2\sin\theta)$$

$$l_3 = (a_{12} - a_{21})/(2\sin\theta)$$

Note that the generator of rotations describes a physical rotation with respect to a fixed reference frame, whereas the transformation matrix a_{ij} describes a change of reference frame for a fixed object.

Stress

Sign convention

Tensile normal stress is taken to be positive. (The opposite convention is commonly used in geology.)

Equilibrium conditions

In a domain under stress, but in the absence of a force field that generates moments, equilibrium of forces exists if the following conditions are met:

$$\sigma_{ji} = \sigma_{ij}$$

Thus the stress tensor must be symmetric, and:

$$\sigma_{ij,j} + F_i = \rho \left(\dot{v}_i + v_j v_{i,j} \right)$$

where F_i is the body force per unit volume, ρ the density, and v_i the velocity. More explicitly, the second condition is:

$$\frac{\partial \sigma_{ij}}{\partial x_j} + F_i$$

$$= \rho \left(\frac{d^2 u_i}{d t^2} + \frac{d u_j}{d t} \frac{\partial}{\partial x_j} \frac{d u_i}{d t} \right)$$

where u_i are displacement components and t is time.

Pressure

is the average normal stress, taken positive for compression; hence:

$$p = -\sigma_{ii}/3$$

Deviator of stress

is the part of the stress not due to pressure:

$$\Delta_{ij} = \sigma_{ij} - \delta_{ij}\left(\sigma_{kk}/3\right)$$
$$\equiv \sigma_{ij} + p\,\delta_{ij}$$

Its invariants are:

$$^1I = \Delta_{ii} = 0$$

$$^2I = \begin{vmatrix} \Delta_{22} & \Delta_{23} \\ \Delta_{23} & \Delta_{33} \end{vmatrix} + \begin{vmatrix} \Delta_{11} & \Delta_{13} \\ \Delta_{13} & \Delta_{33} \end{vmatrix}$$
$$+ \begin{vmatrix} \Delta_{11} & \Delta_{12} \\ \Delta_{12} & \Delta_{22} \end{vmatrix}$$

$$^3I = \left| \Delta_{ij} \right|$$

Traction and Cauchy's equation

The traction ^nT_i acts inward across a plane with the outward-normal unit vector n_i. In a domain subject to the stress σ_{ij}, it is:

$$^nT_i = \sigma_{ij}\, n_j$$

Infinitesimal strain and rotation

Displacement gradient

If u_i are continuous displacements in a domain, then their gradient is:

$$e_{ij} = u_{i,\,j} \equiv \frac{\partial u_i}{\partial x_j}$$

If both displacements and gradient are infinitesimally small, then the following holds for the strain and rotation tensors ε_{ij} and ϖ_{ij}:

$$e_{ij} = \varepsilon_{ij} + \varpi_{ij}$$

where:

$$\varepsilon_{ij} = \left(e_{ij} + e_{ji}\right)/2$$

and:

$$\varpi_{ij} = \left(e_{ij} - e_{ji}\right)/2$$

The spatial derivatives of these two tensors are interdependent:

$$\varpi_{ij,\,k} = \varepsilon_{ik,\,j} - \varepsilon_{jk,\,i}$$

Compatibility equations

For compatibility, continuity must exist throughout the deformed domain and originally distinct material points may not arrive at the same spatial point; these conditions are fulfilled if the following equations hold:

$$\varepsilon_{ij,\,kl} + \varepsilon_{kl,\,ij} = \varepsilon_{ik,\,jl} + \varepsilon_{jl,\,ik}$$

Many of these 81 equations, which assure that the strain components can be derived from the displacements, are identical or tautological. Only six of them are *essential*, and they are:

$$\varepsilon_{11,\,23} + \varepsilon_{23,\,11} = \varepsilon_{31,\,12} + \varepsilon_{12,\,31}$$

$$\varepsilon_{22,\,31} + \varepsilon_{31,\,22} = \varepsilon_{12,\,23} + \varepsilon_{23,\,12}$$

$$\varepsilon_{33,\,12} + \varepsilon_{12,\,33} = \varepsilon_{23,\,31} + \varepsilon_{31,\,23}$$

$$2\varepsilon_{12,\,12} = \varepsilon_{11,\,22} + \varepsilon_{22,\,11}$$

$$2\varepsilon_{23,\,23} = \varepsilon_{22,\,33} + \varepsilon_{33,\,22}$$

$$2\varepsilon_{31,\,31} = \varepsilon_{33,\,11} + \varepsilon_{11,\,33}$$

Infinitesimal linear dilatation

is the average normal infinitesimal strain:

$$D = \varepsilon_{ii}/3$$

Infinitesimal dilatation

is the change in volume caused by an infinitesimal strain:

$$\Delta = \varepsilon_{ii}$$

Infinitesimal distortion

or *deviator of strain* measures the change of shape and is the difference of infinitesimal strain and linear dilatation:

$$d_{ij} = \varepsilon_{ij} - D\delta_{ij} \equiv \varepsilon_{ij} - \delta_{ij}\,\varepsilon_{kk}/3$$

Rotation vector

A rotation *with respect to a fixed reference frame* may be represented by the axial vector ω_i:

$$\omega_i = \left(\epsilon_{ijk}\,\varpi_{kj}\right)/2$$

hence:

$$[\omega_i] = [\varpi_{32}\ \varpi_{13}\ \varpi_{21}]$$

and:

$$\omega_i = \omega l_i$$

where ω is the magnitude of $[\omega_i]$ and l_i are the direction cosines of a vector about which the rotation with respect to the reference frame through the angle ω (in radians) is clockwise.

Rigid-body rotation

In the absence of strain, the following is valid for the tensor and the axial vector representing an infinitesimal rotation *with respect to a fixed reference frame*:

$$d u_i = \varpi_{ji}\,dx_j = -\varpi_{ij}\,dx_j$$
$$= -\epsilon_{ijk}\,\omega_k\,dx_j$$

which implies that an element of displacement equals the cross product of the rotation vector ω with an element of the position vector **x**. Furthermore, if the displacements themselves are infinitesimal as well, the following also holds:

$$\omega_i = \epsilon_{ijk}\,u_{k,j}/2$$

which implies that the rotation vector equals half the curl of **u**.

Infinitesimal homogeneous displacement gradients

have components that are independent of position and thus are constant. This simplifies their relationship to the displacements to:

$$u_i = {}^\circ u_i + e_{ij}\,x_j$$
$$= {}^\circ u_i + \varpi_{ij}\,x_j + \varepsilon_{ij}\,x_j$$

where ${}^\circ u_i$ is the displacement at the coordinate origin.

Finite strain and rotation

Green's and Almansi's tensors

In the *Lagrangian* reference frame, positions of material points after deformation x_i are considered to be the function of original positions a_i of the same material points:

$$dx_i = \frac{\partial x_i}{\partial a_j}\,da_j$$

In the *Eulerian* reference frame, original positions are treated as functions of positions in the deformed state:

$$da_i = \frac{\partial a_i}{\partial x_j}\,dx_j$$

Coordinates a_i and x_i may coincide. Let ds be the original, dS the final length of the same material-line element. Then:

$$ds^2 = da_k\,da_k = \delta_{ij}\,da_i\,da_j$$
$$= \delta_{ij}\frac{\partial a_i}{\partial x_l}\frac{\partial a_j}{\partial x_m}\,dx_l\,dx_m$$

Similarly for the final length:

$$d\,S^2 = dx_k\,dx_k = \delta_{ij}\,dx_i\,dx_j$$

$$= \delta_{ij}\frac{\partial x_i}{\partial a_l}\frac{\partial x_j}{\partial a_m}\,da_l\,da_m$$

The difference between original and final squares of elemental length is:

$$dS^2 - ds^2$$

$$= \left(\delta_{kl}\frac{\partial x_k}{\partial a_i}\frac{\partial x_l}{\partial a_j} - \delta_{ij}\right)da_i\,da_j$$

$$= \left(\delta_{ij} - \delta_{kl}\frac{\partial a_k}{\partial x_i}\frac{\partial a_l}{\partial x_j}\right)dx_i\,dx_j$$

where the upper right-hand-side expression uses a Lagrangian and the lower an Eulerian reference frame. Green's and Almansi's tensors are based on these two measures of the change of length. Green's tensor is:

$$\mathfrak{E}_{ij} = \mathfrak{E}_{ji}$$

$$= \tfrac{1}{2}\left(\delta_{kl}\frac{\partial x_k}{\partial a_i}\frac{\partial x_l}{\partial a_j} - \delta_{ij}\right)$$

and Almansi's tensor is:

$$e_{ij} = e_{ji}$$

$$= \tfrac{1}{2}\left(\delta_{ij} - \delta_{kl}\frac{\partial a_k}{\partial x_i}\frac{\partial a_l}{\partial x_j}\right)$$

Thus the difference between the squares of lengths caused by a strain is:

$$dS^2 - ds^2 = 2\mathfrak{E}_{ij}\,da_i\,da_j$$

$$= 2e_{ij}\,dx_i\,dx_j$$

Displacements being $u_i = x_i - a_i$, Green's tensor also is:

$$\mathfrak{E}_{ij} = \tfrac{1}{2}\left(\frac{\partial u_i}{\partial a_j} + \frac{\partial u_j}{\partial a_i} + \frac{\partial u_k}{\partial a_i}\frac{\partial u_k}{\partial a_j}\right)$$

and Almansi's tensor is:

$$e_{ij} = \tfrac{1}{2}\left(\frac{\partial u_i}{\partial x_j} + \frac{\partial u_j}{\partial x_i} + \frac{\partial u_k}{\partial x_i}\frac{\partial u_k}{\partial x_j}\right)$$

where $\partial u_i/\partial x_j$ is a generally asymmetric tensor, the *displacement gradient*. The relative change in length of an original line element is:

$$\mathfrak{E} \equiv \frac{dS - ds}{ds} = (1 + 2\mathfrak{E}_{ii})^{1/2} - 1$$

$$(\text{no sum on } i)$$

and:

$$dS = (1 + \mathfrak{E})\,ds = (1 + 2\mathfrak{E}_{ii})^{1/2}ds$$

$$(\text{no sum on } i)$$

and the same length change relative to the final, deformed length is:

$$e \equiv \frac{dS - ds}{dS} = 1 - (1 - 2e_{ii})^{1/2}$$

$$(\text{no sum on } i)$$

and:

$$ds = (1 - e)\,dS$$

$$= 1 - (1 - 2e_{ii})^{1/2}dS$$

$$(\text{no sum on } i)$$

Transition from finite to infinitesimal strain

As the displacement gradient becomes infinitesimal, Almansi's tensor becomes identical with the infinitesimal strain tensor:

$$e_{ij} \to \varepsilon_{ij} \text{ as } u_{i,j} \to 0$$

and, as both the displacement gradient and the displacements themselves become infinitesimal, Green's tensor also becomes identical with the infinitesimal strain tensor:

$$\mathfrak{E}_{ij} \to \varepsilon_{ij}$$

$$\text{as } u_{i,j} \to 0 \text{ and } u_i \to 0$$

Volume of a parallelepiped

Let a parallelepiped be defined by three nonparallel edge vectors $^pA_i\,(p = 1, 2, 3)$, then its volume is:

$$V = \epsilon_{ijk} \, {}^1A_i \, {}^2A_j \, {}^3A_k \equiv \left| {}^pA_i \right|$$

Finite dilatation

is defined as:

$$\Delta \equiv \frac{dV - dv}{dv}$$

It is the *Jacobian* in a Lagrangian framework, a determinant that can be calculated in several ways, minus unity:

$$\Delta = \left| \frac{\partial x_i}{\partial a_j} \right| - 1$$

$$= \left| \delta_{ij} + 2\mathfrak{E}_{ij} \right|^{1/2} - 1$$

$$= \left| \frac{\partial x_k}{\partial a_i} \frac{\partial x_k}{\partial a_j} \right|^{1/2} - 1$$

$$= \left| \frac{\partial u_i}{\partial a_j} + \delta_{ij} \right| - 1$$

and the inverse Jacobian in an Eulerian framework minus unity:

$$\Delta = \left| \frac{\partial a_i}{\partial x_j} \right|^{-1} - 1$$

$$= \left| \delta_{ij} - 2e_{ij} \right|^{-1/2} - 1$$

$$= \left| \frac{\partial a_k}{\partial x_i} \frac{\partial a_k}{\partial x_j} \right|^{1/2} - 1$$

$$= \left| \frac{\partial u_i}{\partial x_j} + \delta_{ij} \right|^{-1} - 1$$

where square roots are taken positive.

Finite linear dilatation

is the change of length exclusively due to dilatation:

$$D = (1 + \Delta)^{1/3} - 1$$

Finite distortion

is the change in shape produced by a general strain, omitting the effects of a dilatation. Green's tensor for distortion is:

$$\mathfrak{D}_{ij} = \mathfrak{E}_{ij} \left| \delta_{pq} + 2\mathfrak{E}_{pq} \right|^{-1/3}$$

$$+ \frac{1}{2} \left(\left| \delta_{pq} + 2\mathfrak{E}_{pq} \right|^{-1/3} - 1 \right) \delta_{ij}$$

Almansi's tensor for distortion is:

$$\mathfrak{d}_{ij} = e_{ij} \left| \delta_{pq} - 2e_{pq} \right|^{-1/3}$$

$$+ \frac{1}{2} \left(1 - \left| \delta_{pq} + 2e_{pq} \right|^{-1/3} \right) \delta_{ij}$$

Elongations

$\lambda \equiv dS/ds$, also called *length ratios*, refer to material lines with direction cosines l_i in the original and L_i in the final state, such that:

$$da_i = l_i \, ds$$
$$dx_i = L_i \, dS$$

(Jaeger, 1969; Jaeger & Cook, 1979; Ramsay, 1967; and others use the symbol λ for the quadratic elongation, here λ^2.) If $\mathfrak{E}$ is the magnitude of Green's tensor $\mathfrak{E}_{ij}$ in the direction l_i and e that of Almansi's tensor e_{ij} in the direction L_i, then the elongation of the specified material line is:

$$\lambda = (1 + 2\mathfrak{E})^{1/2}$$

$$\equiv \left(1 + 2\mathfrak{E}_{ij} \, l_i \, l_j \right)^{1/2}$$

$$= \left(\frac{\partial x_k}{\partial a_i} \frac{\partial x_k}{\partial a_j} \, l_i \, l_j \right)^{1/2}$$

or:

$$\lambda = (1 - 2e)^{-1/2}$$

$$\equiv \left(1 - 2e_{ij} \, L_i \, L_j \right)^{-1/2}$$

$$= \left(\frac{\partial a_k}{\partial x_i} \frac{\partial a_k}{\partial x_j} \, L_i \, L_j \right)^{-1/2}$$

with all square roots taken positive. Hence:

$$\mathfrak{E} = \left(\lambda^2 - 1 \right) / 2$$

and:

$$e = \left(1 - \lambda^{-2}\right)/2$$

Principal elongations, in the principal directions of the tensors, are:

$$^i\lambda = \left(1 + 2\ ^i\mathfrak{E}\right)^{1/2} = \left(1 - 2\ ^ie\right)^{-1/2}$$

Hence, $^i\mathfrak{E} > (-1/2)$ and $^ie < 1/2$.

Change of angle with finite strain

If two vectors from an original point a_i to neighboring points $a_i + da_i$ and $a_i + d\bar{a}_i$, of lengths ds and $d\bar{s}$, enclose an angle θ and the corresponding vectors in the deformed material enclose the angle Θ, then:

$$dS\,d\bar{S}\cos\Theta - ds\,d\bar{s}\cos\theta$$
$$= 2\mathfrak{E}_{ij}\,da_i\,d\bar{a}_j$$

and:

$$dS\,d\bar{S}\cos\Theta - ds\,d\bar{s}\cos\theta$$
$$= 2e_{ij}\,dx_i\,d\bar{x}_j$$

As a consequence, a right angle $\theta = 90°$ enclosed by original lines with directions l_i and $\bar{l}_i$ generally changes after deformation into the different angle Θ, enclosed by the deformed lines with directions L_i and $\bar{L}_i$. Θ is:

$$\Theta = \cos^{-1}\left(2e_{ij}\,L_i\,\bar{L}_j\right)$$

A right angle $\Theta = 90°$ enclosed by lines in the deformed body with directions L_i and $\bar{L}_i$ generally corresponds to a different original angle θ, enclosed by lines with directions l_i and $\bar{l}_i$. θ is:

$$\theta = \cos^{-1}\left(-2\mathfrak{E}_{ij}\,l_i\,\bar{l}_j\right)$$

Homogeneous finite strain

Let α_{ij}, γ_{ij}, and A_{ij} be constants related to each other by:

$$\delta_{ij} + \alpha_{ij} = A_{ij}$$

and:

$$\delta_{ij} - \gamma_{ij} = A_{ij}^{-1}$$

where A_{ij} is a transformation matrix. Because of its constant components A_{ij} implies, along with a possible rotation, a homogeneous strain for which the following equations hold:

$$x_i = A_{ij}\,a_j + \beta_i$$
$$= \left(\delta_{ij} + \alpha_{ij}\right)a_j + \beta_i$$
$$a_i = A_{ij}^{-1}\,x_j - \beta_i$$
$$= \left(\delta_{ij} - \gamma_{ij}\right)x_j - \beta_i$$

where β_i is a constant translation. If and only if this translation is zero does the matrix A_{ij} alone transform a_i into x_i. Green's and Almansi's tensors for homogeneous strain are:

$$^H\mathfrak{E}_{ij} = \left(A_{ki}\,A_{kj} - \delta_{ij}\right)/2$$
$$= \left(\alpha_{ij} + \alpha_{ji} + \alpha_{ki}\,\alpha_{kj}\right)/2$$

$$^He_{ij} = \left(\delta_{ij} - A_{ki}^{-1}\,A_{kj}^{-1}\right)/2$$
$$= \left(\gamma_{ij} + \gamma_{ji} - \gamma_{ki}\,\gamma_{kj}\right)/2$$

The homogeneous dilatation is:

$$^H\Delta = \left|A_{ij}\right| - 1 = \left|\delta_{ij} + \alpha_{ij}\right| - 1$$

or:

$$^H\Delta = \left|A_{ij}^{-1}\right|^{-1} - 1 = \left|\delta_{ij} - \gamma_{ij}\right|^{-1} - 1$$

Elongations in homogeneous finite strain

are the positive square roots of the magnitudes of the symmetric tensor $A_{ki}A_{kj}$:

$$\lambda = \left(A_{ki}\,A_{kj}\,l_i\,l_j\right)^{1/2}$$
$$= \left(A_{ki}^{-1}\,A_{kj}^{-1}\,L_i\,L_j\right)^{-1/2}$$

Stretch tensor and elongations

If A_{ij} represents only strain and no rotation, it is symmetric and becomes the stretch tensor, the magnitudes of which are the elongations:

$$\lambda = {}^{S}A$$
$$\equiv {}^{S}A_{ij}\, l_i\, l_j = \left({}^{S}A_{ij}^{-1}\, L_i\, L_j\right)^{-1}$$

The three principal elongations are:

$${}^{i}\lambda \equiv {}^{iS}A = \left(1 + 2\,{}^{i}\mathfrak{E}\right)^{1/2}$$
$$= \left(1 - 2\,{}^{i}e\right)^{-1/2}$$

Homogeneous finite dilatation:

$${}^{H}\Delta = {}^{1S}A\; {}^{2S}A\; {}^{3S}A - 1$$
$$= {}^{1}\lambda\; {}^{2}\lambda\; {}^{3}\lambda - 1$$

Homogeneous finite distortion

The stretch tensor for a general homogeneous strain must be normalized to represent the corresponding distortion:

$${}^{D}A_{ij} = {}^{S}A_{ij}\left|\,{}^{S}A_{pq}\right|^{-1/3}$$

Principal elongations and March densities

In a homogeneously deformed domain the distribution of orientations of material lines or planes differs from that in the original state. Assuming that the distribution was originally uniform, the principal elongations, in terms of the principal (so-called *March*) angular densities ${}^{iL}\rho$ of deformed linear markers, are:

$${}^{i}\lambda = {}^{iL}\rho^{1/3}$$

Assuming an originally uniform distribution of tabular markers, the principal elongations in terms of the principal densities ${}^{iT}\rho$ of the poles of deformed tabular markers are:

$${}^{i}\lambda = {}^{iT}\rho^{-1/3}$$

Comparison of matrix and subscript notation

General rule

Expressions in subscript notation can be adapted to the subscript-free matrix notation by placing dummy subscripts in each term proximally to each other. In subscript notation multiplication is commutative, but not in matrix notation:

$$A_{kj}\, B_{ik} = B_{ik}\, A_{kj}$$

is equivalent to:

$$\mathbb{B}\,\mathbb{A} \neq \mathbb{A}\,\mathbb{B}$$

If $\mathbb{A}$ and $\mathbb{B}$ represent transformation matrices, then the matrix product $\mathbb{B}\,\mathbb{A}$ implies the compound transformation produced by transformation $\mathbb{A}$ followed by transformation $\mathbb{B}$.

Transpose

The columns of a square matrix form the rows of its transpose:

$$A^{\mathsf{T}}_{ij} = A_{ji}$$

If ordering the factors of a term in subscript notation cannot place dummy subscripts proximally, this can be achieved by use of the transpose:

$$B_{ik}\, A_{jk} = B_{ik}\, A^{\mathsf{T}}_{kj} \;\rightarrow\; \mathbb{B}\,\mathbb{A}^{\mathsf{T}}$$

Finite incremental strain

Compound of finite strain increments

Unless two consecutive finite strain increments are mutually co-axial, the compound effect is a transformation that implies both a strain and a rigid-body rotation. Let incremental strains be represented by stretch tensors (symmetric transformation matrices) $\mathbb{A}$ and $\mathbb{B}$, and the compound, asymmetric transformation by $\mathbb{F}$:

$$\mathbb{B}\,\mathbb{A} = \mathbb{F}$$

Polar decomposition

The effect of the transformation $\mathbb{F}$ may be represented, neglecting its actual strain history, as the compound of a single stretch and a rotation matrix $\mathbb{R}$. Depending on whether it is assumed that the rotation occurred last or first, the decomposition is called *right* or *left polar* (from the position of the symbol for the stretch tensor in matrix notation):

$$\mathbb{B}\mathbb{A} = \mathbb{F} = \mathbb{R}\mathbb{U} = \mathbb{V}\mathbb{R}$$

where $\mathbb{U}$ is the right, $\mathbb{V}$ the left decomposed, symmetric stretch tensor. $\mathbb{U}$ and $\mathbb{V}$ differ only by their orientation with respect to the reference coordinates; in their respective principal coordinates they are identical:

$$^{P}\mathbb{U} = {}^{P}\mathbb{V}$$

The decomposed stretch tensors are:

and:
$$\mathbb{U} = \sqrt{\mathbb{F}^{\mathsf{T}}\,\mathbb{F}}$$
$$\mathbb{V} = \sqrt{\mathbb{F}\,\mathbb{F}^{\mathsf{T}}}$$

where the square root of a matrix is understood to be that matrix that, multiplied with itself, yields the radicand matrix. (For a symmetric matrix like $\mathbb{F}^{\mathsf{T}}\mathbb{F}$ or $\mathbb{F}\mathbb{F}^{\mathsf{T}}$, this root is most easily extracted by rotating the radicand to principal coordinates, extracting the positive square roots of its principal values, and then restoring the resulting matrix to the reference coordinates.) The rotation matrix is:

$$\mathbb{R} = \mathbb{F}\mathbb{U}^{-1} = \mathbb{V}^{-1}\mathbb{F}$$

and, because it is a rotation matrix:

$$\mathbb{R}^{-1} = \mathbb{R}^{\mathsf{T}}$$

Also because it is a rotation matrix, $\mathbb{R}$ describes a change of reference frame, not a rotation with respect to such a fixed frame. In terms of the incremental stretch tensors, the right and left compound stretches are:

$$\mathbb{U} = \sqrt{\mathbb{A}^{\mathsf{T}}\mathbb{B}^{\mathsf{T}}\,\mathbb{B}\mathbb{A}}$$
$$\mathbb{V} = \sqrt{\mathbb{B}\mathbb{A}\mathbb{A}^{\mathsf{T}}\mathbb{B}^{\mathsf{T}}}$$

APPENDIX A
INTERRELATIONS OF ELASTIC PARAMETERS FOR ISOTROPIC BODIES

The stiffness matrix for an ideally elastic isotropic material is:

$$\left(c_{ij}\right) = \begin{pmatrix} \lambda+2\mu & \lambda & \lambda & 0 & 0 & 0 \\ \lambda & \lambda+2\mu & \lambda & 0 & 0 & 0 \\ \lambda & \lambda & \lambda+2\mu & 0 & 0 & 0 \\ 0 & 0 & 0 & \mu & 0 & 0 \\ 0 & 0 & 0 & 0 & \mu & 0 \\ 0 & 0 & 0 & 0 & 0 & \mu \end{pmatrix},$$

where λ and μ are Lamé's constants, μ also being called the shear modulus. Instead of Lamé's constants, other pairs of parameters, or moduli, are used, depending on the context. All of them are, necessarily, interrelated. They can be paired arbitrarily, but the following combinations are most common:

	(λ, μ)	(k, μ)	(μ, ν)	(E, ν)	(E, μ)
λ	λ	$k-\dfrac{2\mu}{3}$	$\dfrac{2\mu\nu}{1-2\nu}$	$\dfrac{E\nu}{(1+\nu)(1-2\nu)}$	$\dfrac{\mu(E+2\mu)}{3\mu-E}$
μ	μ	μ	μ	$\dfrac{E}{2+2\nu}$	μ
k	$\lambda+\dfrac{2\mu}{3}$	k	$\dfrac{2\mu(1+\nu)}{3(1-2\nu)}$	$\dfrac{E}{3(1-2\nu)}$	$\dfrac{E\mu}{3(3\mu-E)}$
E	$\dfrac{(3\lambda+2\mu)\mu}{\lambda+\mu}$	$\dfrac{9k\mu}{3k+\mu}$	$2+2\mu\nu$	E	E
ν	$\dfrac{\lambda}{2(\lambda+\mu)}$	$\dfrac{3k-2\mu}{6k+2\mu}$	ν	ν	$\dfrac{E}{2\mu}-1$

where k is the *bulk modulus*, E is *Young's modulus*, and ν is *Poisson's ratio*.

❧

APPENDIX B
UNITS OF STRESS AND CONVERSION FACTORS

The unit of stress in the *Système International* or (m k s) system, the pascal, is usually too small for convenient application in the context of geology. The preferred units in geology are therefore the megapascal and the gigapascal. Listed are conversion factors to and from three of the most commonly encountered different systems of units. They are the earlier metric or (c g s) system, a system based on the standard atmospheric pressure, and the English system based on the pound and the inch. Conversion factors are also given from the kilogram-weight units, based on the standard weight of a mass of one kilogram (for a standard gravitational acceleration at sea level).

$$
\begin{aligned}
1\,\text{MPa} &= 10\,\text{bar} \\
&= 10^7\,\text{dyn cm}^{-2} \\
&= 9.869\,\text{atm} \\
&= 1.02\times10^5\,\text{kgwt m}^{-2} \\
&= 1.45\times10^5\,\text{psi}
\end{aligned}
\qquad
\begin{aligned}
1\,\text{atm} &= 0.10133\,\text{MPa} \\
&= 1.0133\,\text{bar} \\
&= 1.0133\times10^6\,\text{dyn cm}^{-2} \\
&= 1.033\times10^4\,\text{kgwt m}^{-2} \\
&= 14.7\,\text{psi}
\end{aligned}
$$

$$
\begin{aligned}
1\,\text{bar} &= 10^5\,\text{Pa} \\
&= 10^{-1}\,\text{MPa} \\
&= 10^6\,\text{dyn cm}^{-2} \\
&= 0.9869\,\text{atm} \\
&= 1.02\times10^4\,\text{kgwt m}^{-2} \\
&= 14.5\,\text{psi}
\end{aligned}
\qquad
\begin{aligned}
1\,\text{psi} &= 0.6895\,\text{MPa} \\
&= 0.6895\times10^{-1}\,\text{bar} \\
&= 0.6895\times10^5\,\text{dyn cm}^{-2} \\
&= 0.6805\times10^{-1}\,\text{atm} \\
&= 7.03\times10^2\,\text{kgwt m}^{-2}
\end{aligned}
$$

ల

References

Christensen, J. N., Rosenfeld, J.L., & DePaolo, D. J. (1989) Rates of tectono-metamorphic processes from rubidium and strontium isotopes in garnet. *Science* **244**, 1465-69.

Coulomb, C.A. (1776) Essai sur une application des règles de maximis et minimis à quelques problèmes de statique, relatifs à l'architecture. *Mémoires de mathématique et physique, présentés à l'Académie Royale des Sciences par divers savans* **1773**, 343-84.

Fung Y. C. (1965) *Foundations of Solid Mechanics.* Prentice-Hall, Englewood Cliffs.

Fung, Y.C. (1977) *A First Course in Continuum Mechanics.* 2nd edition, Prentice-Hall, Englewood Cliffs.

Glen, J.W. (1958) The flow law of ice. Internat. Union of Geodesy and Geophysics. Publ. **44**. Symposium of Chamonix. 171-83.

Jaeger, J.C. (1969) *Elasticity, Fracture and Flow with Engineering and Geological Applications.* Methuen, London.

Jaeger, J. C., & Cook, N. G.W. (1979) *Fundamentals of Rock Mechanics.* 3rd edition, Methuen, London.

Lipshie, S.R. (1984) Development of Phyllosilicate Preferred Orientation in Naturally and Experimentally Metamorphosed and Deformed Rocks. Unpubl. dissertation. University of California, Los Angeles.

March, A. (1932) Mathematische Theorie der Regelung nach der Korngestalt bei affiner Deformation. *Zeitschr. f. Kristallographie* **81**, 285-297.

Mohr, O. (1882) Ueber die Darstellung des Spannungszustandes und des Deformationszustandes eines Körperelementes und über die Anwendung derselben in der Festigkeitslehre. *Der Civilingenieur* **28**, 113-56.

Mohr, O. (1914) *Abhandlungen aus dem Gebiete der technischen Mechanik.* 2nd edition, W. Ernst und Sohn, Berlin, pp. 192-235.

Nye, J. F. (1953) The flow law of ice from measurements in glacier tunnels, laboratory experiments and the Jungfraufirn borehole experiment. *Proc. Roy. Soc. A* **219**, 477-89.

Nye, J.F. (1964) *Physical Properties of Crystals.* Clarendon Press, Oxford, revised printing.

Owens, W. H. (1973) Strain modification of angular density distributions. *Tectonophysics* **16**, 249-61.

Ramsay, J. G. (1967) *Folding and Fracturing of Rocks.* McGraw-Hill, New York.

Rosenfeld, J.L. (1970) *Rotated Garnets in Metamorphic Rocks.* Geological Society of America, Special Paper **129**.

Sokolnikoff, I. S. & Redheffer, R.M. (1958) *Mathematics of Physics and Modern Engineering,* McGraw-Hill, New York.

∽

INDEX

ablation, **88**, 234

acceleration, x, 9, **112**

algorithms, computer, 96

Almansi's tensor, x, **74**, 218, 281

alternating matrix, x, **7**, 146, 271

analytic geometry, 91, 254

angular distribution, 91

angular velocity, xi, **12**, 116

anisotropy, **14**, 25

antiparallel, **4**

antisymmetric tensor, xii, **25**, 35, 132, 136, 157

asymmetric tensor, **25**, 135

axial vector, **35**, 273, 277

axis density, **91**, 264, 284

axis of shear, **49**

balance of forces, **109**, 176, 278

bed slip, 237

biaxial stress, strain **49**, 66, 174

binomial theorem, **159**

body couple, **46**

body force, x, **44**, 47, 176, 183

body torque, x, 46, **174**, 183

bulk modulus, 201, **286**

cartesian coordinate system, xii, **4**, 270, 275

Cauchy's equation, **48**, 279

Cauchy's tensors, x, **75**

change of angle, **77**, 283

change of length, **77**, 281

Christensen, 90, 288

coaxiality, **94**

coefficient of internal friction, **89**, 243

cofactors, x, **16**, 124, 219, 276

cohesion, **89**, 243

commutative, **3**

compaction strain, 96, 265

compatibility equations, 61, **63**, 69, 199, 203, 279

compliance matrix, xi, **86**

compliance tensor, xi, **85**

confining pressure, xi, **87**, 89, 232, 242

constraint equation, **22**

continuum, vii, **44**, 56

conversion factors, 287

Cook, 78, 282, 288

coplanarity, 4, **101**

Coulomb, 89, 244, 288

Cramer's rule, **15**, 124, 276

crevasse, 89, 240

cross product, xii, **4**, 272

crystallographic component, direction, **27**

cubic equation, 21, 129, **270**

curl, **10**, 116, 273

cyclic permutations, **7**, 271

del (nabla), xii, **10**, 273

DePaolo, 288

determinant, xii, **4**, 30, 50, 52, 147, 171, 177, 179, 271

determinant decomposition, 177

deviator of stress, x, **50**, 71, 173, 212, 279

diagonal tensor components, **129**

differential operators, **10**, 273

dilatation, x, **65**, 82, 198, 222, 279, 282

direction angles, x, **7**, 16

direction cosines, x, **7**, 17, 274

direction vector, xi, **7**, 98, 144

discontinuity, **44**

discriminant, **270**

displacement, xii, **56**, 65, 197

289

displacement field, 66, 199
displacement gradient, x, **57**, 192, 279, 281
displacement gradient field, 196
distortion, x, **65**, 280, 282
distortion rate, 71, **212**
divergence, **10**, 273
dot product, xii, **3**, **6**, 272
double angle, 270
dummy subscript, **6**, 270
dynamic stress, 51

$\epsilon\delta$ rule, **19**, 274
eigendirection (principal direction), **32**, 276
eigenvalue (principal value), **21**, 129, 158, 276
Einstein summation convention, **6**, 270
elasticity, ideal, **56**, 65, 85, 286
elongation, xi, **78**, 282, 283, 284
engineering shear strain, x, **87**, 233
English System (of units), 287
equation of a plane, **5**, 103
equilibrium conditions, **47**, 278
Euler's stress principle, **45**
Eulerian, x, **72**, 75, 217, 280
exact differential, **63**

fiducial surface, **90**, 246
field tensor, **44**
finite dilatation, **75**, 218, 282
finite distortion, **282**
finite linear dilatation, **282**
finite rotation, 82, 223, 280
finite simple shear, 253
finite strain, **74**, 218, 249, 280
flow law, **71**, **86**, 231
force density, **45**
free subscript, **6**, 270
Fung, v, vi, 45, 64, 288

generator of rotations, xi, **278**
geological coordinate system, 7

Glen, 87, 231, 288
glide plane, direction, 54, 185
gradient **10**, 273
Green's tensor, x, **74**, 218, 281

heat flow, xi, **14**, 29, 119, 141
heterotropy, **14**
homogeneous displacement gradient, **60**, 196, 280
homogeneous finite dilatation, **283**, **284**
homogeneous finite distortion, **284**
homogeneous finite strain, **79**, 83, 228, 283
homogeneous infinitesimal strain, 60, 65, 192, **197**
homogeneous linear transformation, 229
homogeneous stress, **46**
Hooke's law, **56**, 85
Hubble's law, 12, **116**
hydrostatic pressure, **49**

identity matrix, x, **18**, 34, 274
implied spatial differentiation, x, **10**, 270
incompressibility, **71**, 88, 213, 232
incremental strain, stretch, 95, 284
infinitesimal distortion, **65**
infinitesimal rotation, xi, **59**, 201, 279
infinitesimal strain, x, **59**, 201, 279
inhomogeneous displacement gradient, **61**
integration path, **67**, 203
invariant, xi, **22**, 130, 145, 151, 275, 277, 279
inverse, **15**, 124, 276
inversion, **20**
inversion of axes, 19, **126**
irrotational vector fields, **11**, 118, 273
isotropy, **14**, 26, 65, 71, 85, 141, 286

Jacobian, **73**, 218, 229, 282

Jaeger, 78, 282, 288

Kronecker delta, x, **18**, 21, 34, 130, 151, 274

Lagrange multiplier, xi, **22**, 271
Lagrangian, x, **72**, 75, 216, 280
Lamé's constants, xi, 86, **286**
length ratio, **78**, 283
linear dilatation, **65**, 279, 282
linear equations, system, 16, 32, **123**
Lipshie, 264, 288

magnitude (tensor), xi, **15**, 277
magnitude (vector), **3**, 98, 131, 272
manufactured materials, 29
March, 91, 263, 284, 288
matrix (tensor) inversion, **15**, 124, 276
matrix notation, vii, x, **80**, 86, 135, 220, 284
matter tensor, **14**
modulo n, xi, **7**
Mohr, x, xi, xii, 37, 159, 288
Mohr circle, **37**, 159
Mohr space, xii, **38**
moment of force, **45**

nabla (del), xii, **10**, 273
Neumann's principle, **26**, 28, 136, 141
Newton's law, **9**, 113
Newtonian viscosity, 70, 87, 209, 231
no-sum specification, **134**
nondiagonal tensor components, 47, **129**
normal force (traction), xi, 54, **180**
normal stress, 47, 54, **181**, 189
null vector, x, **97**
Nye, v, vi, 14, 22, 25, 47, 58, 86, 87, 139, 288

octahedral shear stress, xii, **55**, 190
one-dimensional strain, 56
orthogonality, **18**, 34, 130, 153, 154, 274, 275

Owens, 91, 288

parallelepiped, volume of, 215, **281**
particle, **61**
pascal (unit of stress), xi, **48**, 287
Poisson's ratio, 286
polar decomposition, xi, xii, **94**, 266, 285
polar vector, **20**, 35, 277
pole density, **91**, 264, 284
position vector, x, xii, **4**, 11, 101, 114, 116
pre- and postmultiplication, **80**
pressure, xi, 49, 71, **173**, 174, 201, 212, 232, 279, 287
principal (eigen-) directions, **32**, 276
principal (eigen-) values, 21, 32, 129, 158, 276
principal axes, 21, **32**, 37, 150, 158
principal coordinates, 21, **32**, 151, 277
pure shear stress, strain, **49**, **66**, 174
Pythagoras' theorem, three-dimensional, **109**

quadratic elongation, **78**, 282
quadratic equation, **270**

radius-normal property, **24**
Ramsay, 78, 282, 288
Redheffer, v, 3, 288
reference frame (system), **20**, 72, 278, 280
reflection, 5, **20**, 104, 127
representation quadric, **23**, 133, 277
rigid-body rotation, **12**, **59**, 82, 112, 116, 192, 280
Rosenfeld, 90, 288
rotated garnet, 90, 246, 249, 254
rotation matrix, xi, xii, **17**, 26, 82, 125, 136, 274
rotation rate tensor, xi, **70**, 247
rotation tensor, xi, **59**, 279
rotation vector, xi, **59**, 280
rotation, physical, **136**, 193

scalar, 3, 10, 13, 273
scalar field, **10**, 273
secular equation, **21**, 32, 130, 150, 158, 276
sequence of transformations, **80**, 275, 284
shear modulus, xi, 201, **286**
shear stress, xii, **181**, 189
shear traction, xii, **181**
Shreve, v
simple shear strain, **66**, 253
Sokolnikoff, v, 3, 288
solenoidal fields, **11**, 118, 273
space derivatives, **10**, 61
spherical coordinates, **8**, 111, 275
stiffness matrix, x, **86**, 286
stiffness tensor, x, **85**
strain, x, **56,** 72, 279, 280
strain field, 197
strain history, 94
strain rate, x, **70**, 246
stress, xi, **44**, 278
stress field, **51**, 176
stress vector, **46**
stretch tensor, xi, xii, **82**, 284
strike and dip, x, xi, 108
subscripted variables, vii, **5**, 270
substitution property, **34**, 152, 274
symmetric tensor, xi, xii, **22**, 132, 135
symmetry operation, **20**, 136, 138
Système International (of units), 287

tangent planes and normal lines, **23**, 271
tangential force (traction), xii, **54**, 180
tangential stress, xii, **47**
tectonic strain, 96, 266
temperature (thermal) gradient, xii, 14, 119

tensile strength, **88**, 242
tensor, vii, xii, 13, **22**, **34**
tensor equation, **13**, 123
tensor field, **12**
tensor (matrix) inversion, **15**, 276
tensor rank, **13**, 22, **34**
tensor transformation, **20**, 22, 276
tensor, fourth-rank, **34**, 85
thermal conductivity tensor, xi, **14**, 29, 119, 142
thermal (temperature) gradient, xii, 14, 119
time derivative, xi, 8, 112, 273
torque, **45**
trace of tensor, **65**, 131, 276
traction, xii, **46**, 279
transformation (rotation) matrix, x, **17**, 26, 82, 125, 136, 274
translation, 57
transpose, **22**, 81, 147, 220, 271, 284
trend and plunge, xi, **110**, 275
triaxial stress, strain, **49**, 66, 174
trigonometry, 270
trisection, **3**, 98

uniaxial stress, strain, **49**, 66, 174
units of stress, 287

valley glacier, 89, 238
vector, xii, **3**, 272
vector field, **10**, 273
vector sum, 3, **6**, **97**, 272
velocity, 9, **112**
velocity gradient tensor, **70**, 237

well-behaved fields, **11**, 273

yield stress, **85**
Young's modulus, 286